Suguru Arimoto

Control Theory of Multi-fingered Hands

A Modelling and Analytical–Mechanics Approach for Dexterity and Intelligence

Suguru Arimoto, Dr. Eng.
Ritsumeikan University
Faculty of Science and Engineering
Department of Robotics
1-1-1 Nojihigashi
Kusatsu
Shiga 525-8577
Japan

ISBN 978-1-84800-062-9 e-ISBN 978-1-84800-063-6

DOI 10.1007/978-1-84800-063-6

British Library Cataloguing in Publication Data
Arimoto, Suguru
Control theory of multi-fingered hands : a modelling and
analytical-mechanics approach for dexterity and intelligence
1. Robot hands 2. Robots - Control systems
I. Title
629.8'933
ISBN-13: 9781848000629

Library of Congress Control Number: 2007941071

Cover design: eStudio Calamar S.L., Girona, Spain

Printed on acid-free paper

9 8 7 6 5 4 3 2 1

springer.com

Preface

In my previous book entitled "Control Theory of Non-linear Mechanical Systems" published in 1996 through Oxford University Press, I mentioned at its preface the difficulties of understanding human motor control and realizing in mechanical robots everyday powers inherent to humans. Regrettably, I could not discuss in that book any control–theoretic problem of dexterity in human or mechanical hands from not only biological but also computational viewpoints. Directly after my move to Ritsumeikan University in 1997, I started a research project on control of multi-fingered hands with the intention of exploring what is the underlying functionality of the human hand in prehension (stable grasping). Indeed, there was a dearth of papers that discussed the derivation of any dynamic model of grasping under rolling constraints.

In this book I attempt to provide a study of robotic prehension (stable grasping and object manipulation) from computational perspectives based upon Newtonian mechanics. The principal approach is grounded on the derivation of a faithful mathematical model of grasping that is a physical interaction between the fingerends and the object through rolling contacts. In the sequel, Lagrange's equation of motion of the overall fingers/object system is formulated, together with holonomic or non-holonomic constraints of contact and rolling, on the basis of the variational principle developed in analytical mechanics. The most essential functionality of prehension that is referred to for designing a coordinated control signal is the fingers–thumb opposability that distinguishes the mankind from the chimpanzee and other primates, as claimed in anthropology. Stable grasping is regarded in a dynamic sense as a transient behaviour of a solution to the closed-loop equation of system dynamics that should converge to an equilibrium state or manifold satisfying the balance of forces and torques exerted on the object.

I hope that this book will facilitate further indepth research works that unveil the secrets of dexterity and versatility of the human hand and make a contribution to the technological development of dexterous robot hands.

Acknowledgements. There is never enough space to thank all the people involved in the research on multi-fingered hands conducted in Ritsumeikan University since 1997. However, I am especially grateful to Dr. Morio Yoshida of Biomimetics Center of Nagoya RIKEN, who contributed to the progress of this research project as a ph.d. student during 2004–2007 and provided invaluable results from simulations and experiments on control of multi-fingered hands. Other previous ph.d. and m.s. students at Ritsumeikan University contributed greatly through many discussions and publication of our joint works. Special thanks to go to Professor Nguyen Anh of Hanoi Institute of Techonology and Professor Kenji Tahara of Kyushu University. My secretary Ms Yuko Murata helped me to materialize this work by preparing typescripts in Latex and cleared patiently a lot of corrections found in the proof-reading.

Finally, I wish to thank my wife Noriko for her patience and support rending me to devote to this work over the last decade of my academic life.

Kusatsu and Ohtsu, Japan,
November 2007

Suguru Arimoto

Contents

1

Characterisations of Human Hands

It is said that the hand is an agency of the brain. It reflects activities of the brain and thereby it is a sort of mirror to the mind. It is the hand that is the most intriguing and most human of appendages.

This chapter firstly discusses why the human hand has attracted so many research workers from a variety of different scientific domains including developmental psychology, neuro-physiology, kinesiology, anthropology, biomechanics and robotics. Napier's book entitled "Hands" points out that mankind separated several millions of years ago from the apes, who are intrinsically brachiators. In the process of evolution, the human thumb has become fat and long and acquired opposability to other fingers (index or/and middle fingers). In the process of the perfection of fingers–thumb opposition, humans have acquired dexterity, versatility, and multi-purpose functionality of the hand, which has driven humans from tool-users to tool-makers. The second part of this chapter focuses on a study of the functionality of the human hand in grasping and prehension based upon such fingers–thumb opposability. One important question is addressed, that is what kind of physical (or mechanics) and neuro-physiological principles might be involved in the execution of precision prehension. Another question is also posed, that is, whether a complete mathematical model of grasping can be developed and used to validate control models of prehensile functions. In the third part, the problem of everyday physics is discussed in relation to Bernstein's degree-of-freedom (DOF) problem. The human hand has many joints and is therefore redundant in DOF, yet it is blessed with dexterity. In the last part, the least but necessary fundamentals of analytical mechanics based upon the variational principle are summarised, based upon Newton's laws of motion.

1.1 What Has Evolved the Human Hand?

What distinguishes humans from other primates? Anthropology differentiates humans by four hallmarks 1) biped walking, 2) tool-making, 3) use of fire, and 4) speech communication. The analysis of sequences of genes in human deoxyribonucleic acid which are coincident with those of the chimpanzee at the rate of 98% reveals that mankind separated around several millions of years ago from chimpanzees. According to the analysis of fossil remains of bones that belonged to *Australopithecus afarensis* (which appeared of a few millions years ago, the oldest mankind according to the present state of the art of fossilology based upon fossil bones to date), the ancesters of mankind already walked in a bipedal manner adduced by the shape of leg bones (crura). The South African *Australopitheceins* were already adapted to ground living by having become free from brachiation of locomotion style, swinging from branch to branch as among most apes. Differently from living in the trees in a forest, ground-living in a steppe of the Great Rift Valley of Africa did not allow our human ancesters to obtain a regular supply of food. They were forced to hunt small animals like hare and deer. They needed to develop a strong digit for gripping sticks and stones and vigorous pounding and throwing. According to John Napier's book "Hands" [1-1], a fossil thumb metacarpal of the South African *Australopithecs* found complete and undamaged has a good saddle joint at the base of the bone. The bone itself was exceptionally robust and the strong muscular markings indicate powerful action of the thumb. However, no stone tools we found in association with the fossil bones, and stone artifacts do not appear in the fossil record until 2.5 million years ago.

The transition from tool-users to tool-makers certainly ocurred between the period of the fossil *Australopithecus afarensis* (3 millions of years ago) and that of the so-called "handyman", the fossil *Homo habilis* (1.75 million years old), which was discovered in 1960 at Olduvai Gorge, Tanzania. After explaining that "the most striking human features are the breadth and potential power of the terminal phalanges, particularly of the thumb, which clearly carried a broad flat nail" and showing a photograph of the hand bones of *Homo habilis*, Napier says in the book "Hands":

> "In combination with a very well developed saddle joint, the thumb gave the appearance of a notably strong digit.
>
> Functionally, it is very probable that the power grip was well developed and effective but there is some doubt about precision grip that, while undoubtedly possible, may not have been as fully evolved as in present-day humans."

Handymen could already make simple stone artifacts called pebble choppers and handy stone axes.

The most important movement of the human hand is opposition as was claimed by Napier in the same book "Hands", which gives the following definition:

> "Opposition is a movement by which the pulp surface of the thumb is placed squarely in contact with – or diametrically opposite to – the terminal pads of one or all of the remaining digits"

Fig. 1.1. A statue of the human hand symbolising the power and greatness of the thumb. This status was installed in Tokyo by the Japanese Association of Finger-Pressing Therapists

During the transition from tool-using to tool-making, the human hand must have evolved steadily. Evolution of the saddle joint of the thumb promotes its movements of flexion and rotation toward making easy contact with another digit (index or middle finger). Fingers–thumb opposability enables the human hand to carry out prehensile movements of the fingers to hold an object securely. There are two main patterns in prehensile grasping, the precision grip (or grasp) and the power grip (or grasp). The two grips are defined as follows:

The precision grip occurs between the terminal digital pad of the opposed thumb and the pads of the fingertips. Large objects held in this way involve all the digits, and small ones require only the thumb, the index, and the middle digits. Smaller objects may be pinched specifically between the thumb and index (or middle) finger, which should be adapted to perform fine control.

The power grip is executed between the surface of the fingers and the palm of the thumb, which acts as a buttressing and reinforcing agent.

Today we live without being conscious of the power or skill of the thumb. Even though we notice the power of the thumb, we easily grasp and manipulate all kinds of objects in our everyday life without being aware of the functions of precision grip that underlie in execution of object grasping and manipulation. According to the dictionary of English, "he is all thumbs" means "he is quite clumsy" However, nowadays not only youngsters but also elderly people use primarily the thumb in pushing buttons of their cell phones or remote

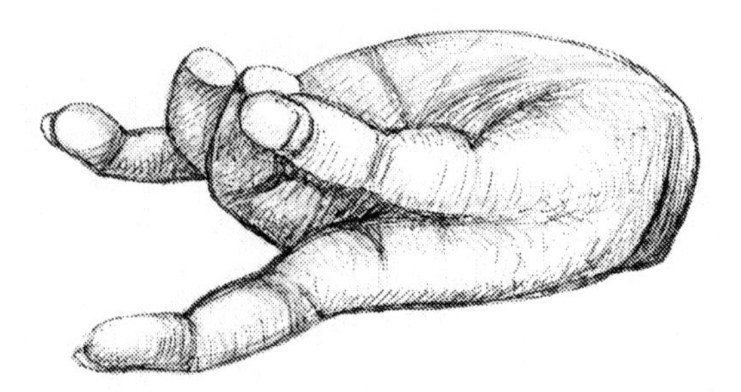

Fig. 1.2. A sketch of the hand of Amidabutsu which is considered to have been separated from an old wooden buddha made in the eighth century. Presently the hand is owned by the gallery of Harvard University

Fig. 1.3. A photograph of human execution of stable and beautiful precision grasp based upon fingers–thumb opposability

control for channel selection or volume control of their television sets. Figure 1.1 shows a photograph of a statue of the human thumb, symbolising the power and sophistication of the thumb in finger-pressing therapy. It has been installed in front of the Dentsuin temple in Tokyo, dedicated by the Japanese association of finger-pressure therapists. Figure 1.2 shows a sketch of the hand of Amidabutsu which was an appendage to an old wooden statue of Buddah presently belonged to the gallery of Harvard University, USA. A terminal pad of its middle finger was missing, unfortunately. Nevertheless, it is possible to suppose from the religious meaning of the shapes of Buddah's fingers that the fingertip of the middle finger might have been lightly contacting with the thumb's terminal pad in a beautiful example of fingers–thumb opposition.

Figure 1.3 shows how elegantly and beautifully a human hand can pinch an object by unconsciously using the thumb and index finger in everyday living.

1.2 Dexterity in Redundancy of Finger Joints

The hand of *Homo sapiens* represents millions of years of evolutionary pressures and changes. Again, let us quote Napier's book [1-1], in which he lists names of his three heroes, Charles Bell, John Hunter and Charles Darwin, and citing John Hunter's principle that structure was the intimate expression of function and function was conditioned by the environment, he writes:

"John Hunter turned our attention from the structure of the hand to its function; Bell related the function of the hand to the environment; and Darwin demonstrated that the environment, by process of natural selection, gave birth to structure."

Let us also quote a few phrases from the book entitled "The Grasping Hand" by Christine L. Mackenzie and Thea Iberall [1-2]:

"The human hand is a highly complex structure that in many ways defies understanding." "The hand consists of five digits made up of a collection of bones, muscles, ligaments, tendons, nails, and vascular structures encapsulated by skin." "Thousands of sensors in the skin, muscles, and joints let the brain know its current state." "Yet, what are the functions supported by that form?"

This last sentence is really our incentive to pursue research on multi-fingered hand from the robotics veiwpoint. It is the goal of this book to identify the underlying functionality of the human hand in prehension. It attempts to unveil which physical and/or mathematical principles might work in the derivation of computational models of prehensile functions and the construction of coordinated control signals (that must emanate from the central nervous system, CNS, in the case of human hands) towards the perfection of precision grasping and object manipulation. Our approach is based on the assumption that a complete model of grasping must be developed and validated both computationally and experimentally through creation of a robot hand. In other words, the book is devoted to the first trial of understanding and exploring what are the functions of the human hand from the standpoint of Turing (or AI, artificial intelligence) and robotics. A designed and constructed multi-fingered robot hand must eventually be controlled by a computer, an artificial central nervous system.

In fact, there is no need to resort to Alan Turing's computer and Alonzo Church's lamda calculus to argue that a line cannot be drawn between software and mathematical expression [1-3].

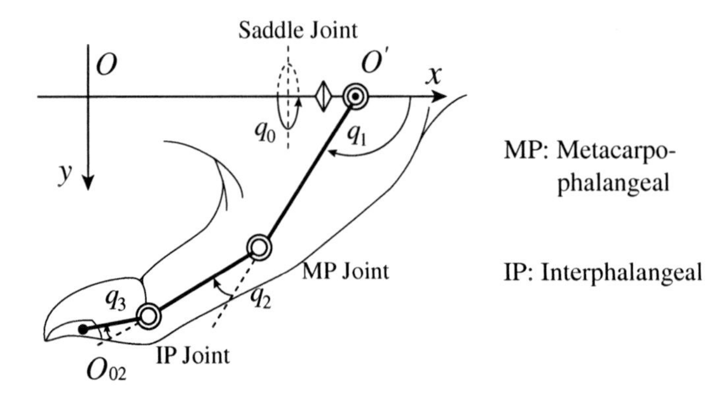

Fig. 1.4. Configuration of the joint variables of the thumb, where q_0 denotes the angle of rotation around the x-axis and q_1, q_2 and q_3 denote the angles of rotation around the z-axis perpendicular to the xy-plane

In this book, we focus on the functional aspects of the problem of precision prehension with the intention of implementing such functions in a robotic hand to promote its dexterity. In particular, our attention is focused on the problem of what kind of functionally effective forces from the fingertips should be applied to an object under numerous constraints for a given task with specified demands such as force/torque balance, stability, orientation control *etc.* We must bear in mind the fact that the forces to be applied to an object should be originally generated at finger joints as a rotational moment torque, that is, the object is not directly regulated from finger-joint torques but indirectly controlled by interactive constraint forces between fingertips and object surfaces under existence of external forces such as gravity that affect the object.

For the purpose of developing a mathematically faithful but physically simplified model of grasping, we devote our attention to the fact that the human thumb has three joints but it has four DOFs since its third joint, named the carpometacarpal joint, is of saddle type and is almost freely movable as a ball-and-socket joint (see Figure 1.4). The index finger can be flexed or straightened (extended) at the finger joints between the phalanges (see Figure 1.5). It has three joints and three DOFs that generate planar motions with rotational axes of the common direction, which is described in Figure 1.5 by the z-axis.

The ultimate goal of this book is to establish a mathematical model of the dynamics of object grasping by means of a pair of robotic fingers with similar mechanisms to the human thumb and index finger (see Figure 1.6). The dynamics of the setup of such a fingers–object system must be subject to contact constraints between the fingertips and object surfaces. The physical conditions of this contact must vary depending on rolling between the fingertip and object surfaces. The conditions also differ depending on whether the

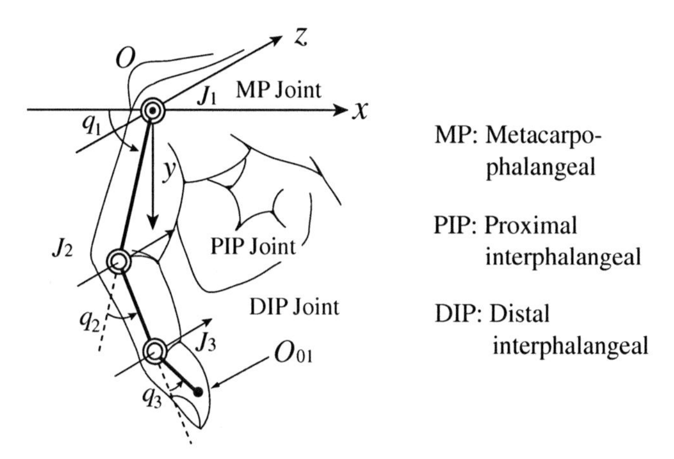

Fig. 1.5. Configuration of the joints of the index finger. All q_1, q_2 and q_3 stand for the joint angles of rotation around the z-axis

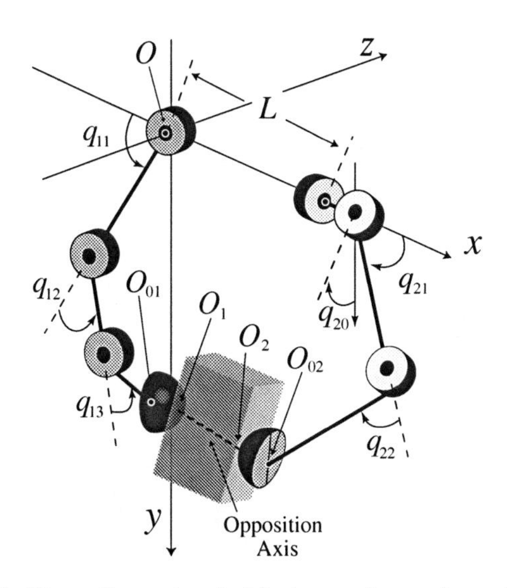

Fig. 1.6. Three-dimensional object grasping and manipulation

fingertips are rigid or soft and deformable. Behind the overall structure of the fingers–object system, there may arise a problem of redundancy of DOFs. In Figure 1.6 the right finger has only three DOFs, one joint with one DOF and another, saddle joint with two DOFs. Therefore, this right-hand finger is not a robotic thumb. However, is this physical setup of robotic fingers capable of grasping an object in a similar way to human pinching using the thumb and index finger? Numerous questions regarding the problem of prehensility may

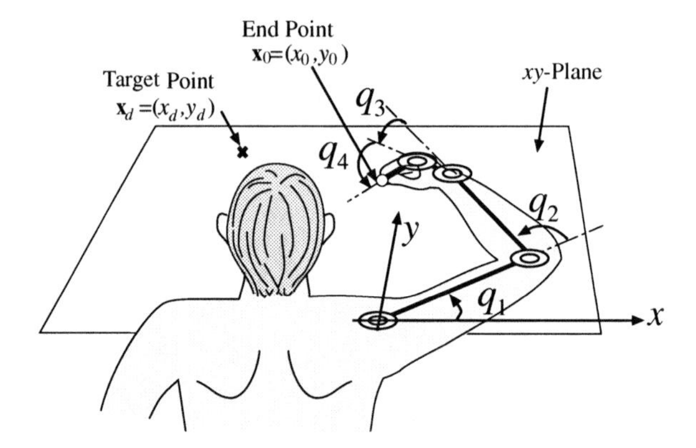

Fig. 1.7. Multi-joint reaching movements. The task is to move the endpoint of the whole arm with some still pose toward the target in the xy-plane

arise even if considerations are limited to grasping and object manipulation by means of robotic fingers.

1.3 Bernstein's DOF Problems

The source of the complexity in the structures of mechanisms and fingers–object system when it is involved in grasping and object manipulation is related to the famous Bernstein problem of overcoming excessive degrees of freedom (DOF) [1-4][1-5]. During virtually all voluntary movements of the limb, the number of kinematic degrees of freedom, which can be associated with the number of independent axes of joint rotation summed over all the joints of the limb concerned, is higher than the number of variables necessary to execute a motor task or to describe its execution. Specifically, in the problem of grasping an object by means of a pair of multi-jointed fingers, the number of DOFs should be discounted by the number of contact and rolling constraints. If other constraints that must be called non-holonomic constraints are involved in the motion of the overall system, it is uncertain how many DOFs are involved in the motion dynamics. Even in the simpleset case of single-arm movement of point-to-point reaching, in which motion is confined to a horizontal plane, the Bernstein problem has not yet been completely solved not only from the viewpoint of neuro-physiology but also from various viewpoints of developmental psychology, kinesiology, biomechanics and robotics. Indeed, consider the multi-joint reaching problem shown in Figure 1.7, where the task is to manoeuvre the planar arm with four joints (shoulder, elbow, wrist and finger-root joints, each of which has a single DOF) to let the arm endpoint reach a given target position $P = (x_d, y_d)$ in the xy-plane. Even if we assume that all joints have a common axis in z-direction perpendicular to the xy-plane, there exist an infinite number of possible solutions to inverse kinematics, that is, from the two-dimensional space (x, y) of the task description to the four-dimensional configuration space composed of all possible combinations of joint angles $q = (q_1, q_2, q_3, q_4)$ that satisfy $\boldsymbol{x}(q) = \boldsymbol{x}_d = (x_d, y_d)$, where $\boldsymbol{x} = (x, y)$.

There is a vast literature on this simple multi-joint point-to-point reaching problem not only from the areas of neuro-physiology, kinesiology and developmental psychology but also from robotics. In its history of almost a half century, a great number of papers have made various proposals for how the brain makes sensible choices among the myriad possibilities for movement that the limb offers and how we can figure out what signals the brain transmits to the many muscles involved in limb movement. Unfortunately, most joints of human limbs have more than one axis of rotation and are controlled by more than two muscles (agonist and antagonist muscles). In addition, a movement in one joint may lead to a change in the relation between the muscle force and joint torque. Thus, various active muscle forces may be required to balance a constant external force. In the following we summarise some important hypothetical proposals for understanding of human multi-joint movements:

1) Equilibrium-point hypothesis.
2) Minimum-jerk hypothesis.
3) Virtual-trajectory hypothesis.
4) Minimum-torque-change hypothesis.
5) Internal-model hypothesis.
6) Virtual spring/damper hypothesis.

The first hypothesis was originally proposed by Anatol G. Feldman in 1966 [1-6] and later in 1986 [1-7] called the λ-model. By introducing the controllable mechanical parameter λ of the zero length of the muscle, he hypothesised that shifts in the threshold λ of the stretch reflex result in active movements of joints and give rise to a shift in the equilibrium state. This hypothesis was reinterpreted by I.A. Bizzi in 1976 [1-8] and N. Hogan in 1984 [1-9] on the basis of observations of spring-like behaviours of muscles in such a way that net spring-like forces, each of which depends on the muscle length, determine the ultimate equilibrium state. From the standpoint of the interpretation of muscle activities evoked by neuro-motor signals from the CNS, the equilibrium point hypothesis has subtle differences from that of Feldman and Bizzi to that of Hogan. In 1981 [1-10], Morasso observed that skilled human movement of multi-joint point-to-point reaching had the following characteristics:

1) The profile of the endpoint trajectory in task space becomes a quasi-straight line,

2) the velocity profile becomes symmetric and bell-shaped,

3) the acceleration profile has double peaks,

4) each time history of joint angles $q_i(t)$ and angular velocities $\dot{q}_i(t)$ may differ for $i = 1, 2, \cdots, 4$.

Then, based on this observation, a formalism of using dynamic optimisation theory to determine the reaching movement was proposed by Hogan in 1884 [1-9] and the concept of a virtual trajectory was introduced, which could be applied by minimising the rate of change of acceleration (jerk) of the limb. This principle for a class of voluntary movements is called the virtual-trajectory or equilibrium-point trajectory hypothesis. Since then, a variety of criteria or performance indices has been proposed not only for the endpoint trajectory in task space but also at the level of joint trajectories in joint space such as minimum joint jerk, minimum torque change, and minimum driving force change. In parallel with this optimisation formalism for movements of multi-joint reaching, robotics was concerned with redundancy resolution to overcome the ill-posedness of inverse kinematics or dynamics in the case of excess DOFs. Thus, roboticists took advantages of full computations of the pseudo-inverse of a non-square $m \times n$ Jacobian matrix of task coordinates in m-dimensional task space with respect to joint coordinates in n-dimensional joint space when $n > m$. A variety of optimisation criteria for uniquely determining the arm endpoint trajectory were proposed such as the maximum manipulability index, minimum acceleration, minimum torque, *etc.* Nevertheless, no evidence has been found or observed to support the hypothesis that the brain executes complex computations of the pseudo-inverse of Jacobian matrix and mathematical optimisation in joint level, including computations of boundary-value problems.

On the other hand, the EP hypothesis based upon spring-like forces is superficially regarded as feedback-based control. However, experimental results and numerous observations obtained in neuro-physiology indicate that, in the case of fast voluntary movements of human limbs, open-loop (or feedforward) control based upon anticipation is predominant, because dynamic movements appear with a time delay relative to the electromyographic signal by a magnitude of about $70 \sim 100$ [ms]. Furthermore, the process of sensing through joint receptors, tendon receptors, and muscle spindles and transferring the sensed information to the spinal chord has latency times between 30 and 70 [ms]. More substantially, medium-scale muscle contractile actions last at least 100 [ms]. Decisively, M. Ito discovered the fact that the vestibulo-ocular reflex loop (see Figure 1.8) has no direct sensory feedoback path from the CNS and therefore the regulation of an ocular target against eyeball movements must be executed in a feedforward manner [1-11][1-12]. Thus, the EP hypothesis was caught in a dilemma of whether to avoid computational complexity or to counter-balance the lack of feedback loops from the CNS with some other formalisation [1-13].

Thus, the internal model hypothesis was proposed to support the predominancy of open-loop control on the basis of anticipation for fast voluntary movements. It postulates that an internal model that eventually sends neuromotor signals to muscles to faithfully reconstruct a joint trajectory can be organised in the cerebellum based on learning through a series of previous practices (see Figure 1.8). This hypothesis was extended by M. Kawato [1-14] to claim that "inverse dyamics through error-feedback learning" may be organised in the cerebellum. However, error-feedback learning does not match numerous observations of human learning provided by many developmental

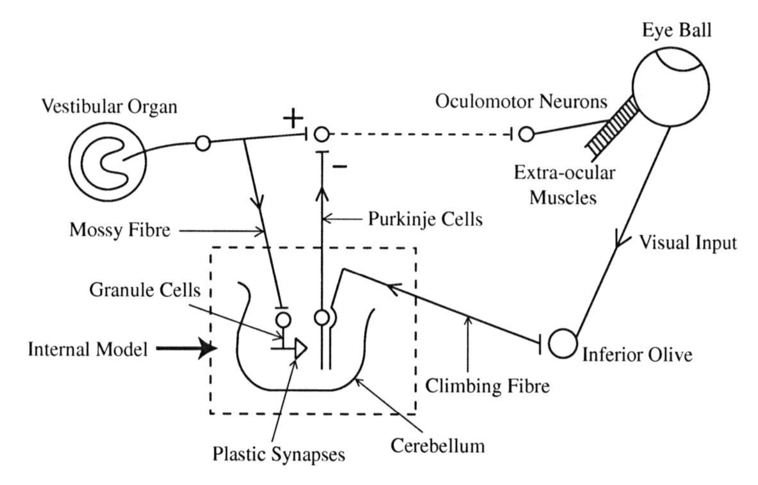

Fig. 1.8. A signal flow graph of the vestibulo-ocular reflex loop

psychologists [1-15]. Almost 60 years ago, N. Bernstein already wrote that:

"What develops is never a moving pattern, and the repetition of one movement pattern is by no means a guarantee that that pattern will be imprinted or more likely to occur in future. What one learns is how to solve a motor problem or how to act."

These statements are seen on p. 437 of the article written by E.S. Reed and B. Bril included in the book edited by M.L. Latash and M.T. Turvey and entitled "Dexterity and Its Development" published in 1996 [1-5]. It is surprising that the total of seven esseys included in the book had been written by Bernstein in Russian more than 60 years ago. In Bernstein's essay 6 in the book [1-5], he noted further:

"stable forms [of action] have all the prerequisites for being easily reproducible and, therefore, should be easily memorised. The result is that bad, unsuccessful movements are not fixed in memory, where successful solutions to motor problems tend to be firmly remembered."

Then, what are successful movements that can be easily memorised? Developmental phychologists led, by E. Thelen, observed in the 1980s [1-16] that,

"The infants modulated reaches in task-approporiate ways in the weeks following onset. Reaching emerges when infants can intentionally adjust the force and compliance of the arm, often using muscle coactivation. These results suggest that the infant central nervous system does not contain programs that detail hand trajectory, joint coordination, and muscle activation pattern.

Rather, these patterns are the consequences of the natural dynamics of the system and the active exploration of the match between those dynamics and the task."

Based upon these observations, Thelen proposes a new approach for "development", by claiming that the dynamic point of view postulates that new spatio-temporal orders emerge not from centrally prescribed programs but from the system dynamics [1-15]. It is quite interesting to compare this dynamics-based approach with the traditional approach based upon Piage's theory of schema that:

"Development arises from an increased control of the higher functions over the skeletomotor system. This is supposedly made possible either by maturation of the CNS allowing inhibition of the primitive responses and the development of voluntary cortical control, or by cognitive progress allowing increasing representations of schemes."

From the robotics viewpoint, there still remains a substantial uncertainty in what is the underlying dynamics that should match the task. Even in the simplest multi-joint reaching task, psychologists have not yet explored the details of the dynamics [1-15]. From the robotics viewpoint, some physical principles entailing computational details of successful movements that match the task must be found in order to acquire skills in robotic mechanisms with multi-joints like those of human arms and hands through an artificial CNS, a computer.

In view of all the arguments, the author and his group proposed very recently a surpringly simple physical principle called the virtual spring hypothesis [1-17] and then the virtual spring/damper hypothesis in 2005 [1-18][1-19], which is visualised as in Figure 1.9. More explicitly, in the former case, coordinated motor signals that should be generated by joint actuators take the form

$$u = -C\dot{q} - J^{\mathrm{T}}(q)k\Delta\boldsymbol{x}, \tag{1.1}$$

and in the latter case

$$u = -C_0\dot{q} - J^{\mathrm{T}}(q)\left\{c\dot{\boldsymbol{x}} + k\Delta\boldsymbol{x}\right\}, \tag{1.2}$$

where $\Delta\boldsymbol{x} = \boldsymbol{x} - \boldsymbol{x}_d$, $q = (q_1, \cdots, q_4)^{\mathrm{T}}$, $\dot{\boldsymbol{x}} = \mathrm{d}\boldsymbol{x}/\mathrm{d}t$, $\dot{q} = \mathrm{d}q/\mathrm{d}t$, t stands for the independent variable of time, k denotes a single stiffness parameter, c is a single damping factor, C and C_0 stand for diagonal matrices whose diagonal entries c_i and c_{0i} express positive damping factors and $J(q)$ denotes the Jacobian matrix of $\boldsymbol{x}(q)$ with respect to q, and $J^{\mathrm{T}}(q)$ the transpose of $J(q)$.

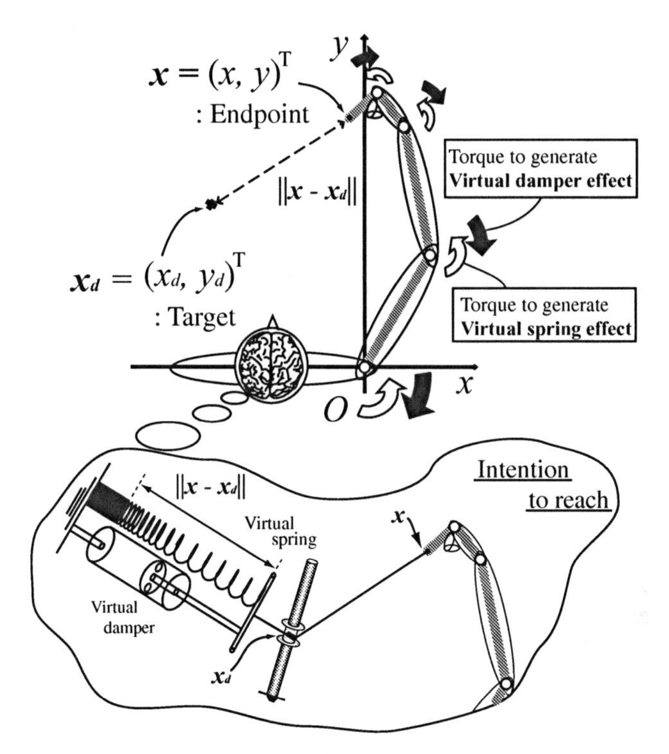

Fig. 1.9. A graphycal expression of the virtual spring/damper hypothesis

The physical principle of the formalism of Equation (1.1) or (1.2) is clear from Newtonian mechanics. The skilled reaching movement may likely arise as the endpoint of the whole arm is drawn toward the target position by the force virtually emanated from a spring/damper mechanism through a pulley (that is located at the target). More explicitly, each component of the term $-J^{\mathrm{T}}(q)\{c\dot{\boldsymbol{x}} + k\Delta \boldsymbol{x}\}$ in Equation (1.2) of the joint control signals is equivalent to the reaction torque at each corresponding joint that may arise owing to exertion of the virtual force drawing the arm endpoint to the target. Existence of damping effect $c\dot{\boldsymbol{x}}$ in Equation (1.2) is expected to decrease damping factors in C of Equation (1.1) to C_0 in (1.2) drastically. The most important finding of the formalism of Equation (1.2) is that, if only two parameters k and c are synergistically chosen, movements of reaching behave like the human skilled motion satisfying conditions 1–4 discussed on page 10 regardless of DOF redundancy. The formalism of Equation (1.1) or (1.2) is nothing but a simple development of Newton's third law of motion (the law of action and reaction) to this multi-joint reaching movement.

Table 1.1. Size effect

Physical Quantities / Robot	Length (link, radius)	Mass	Inertia Moment	World	Redundancy of DOF
Fingers & Hand	1~5 [cm]	0.5~50.0 $\times 10^{-2}$ [kg]	0.1~50.0 $\times 10^{-6}$ [kgm^2]	Centimetre World	Highly Redundant
Human Arm	10~30 [cm]	0.2~2.0 [kg]	0.5~5.0 $\times 10^{-2}$ [kgm^2]	Deca-Centimetre World	Universal Joints (Wrist & Shoulder)
Robot Manipulator	0.1~0.8 [m]	1.0~25.0 [kg]	0.5~50.0 $\times 10^{-2}$ [kgm^2]	Sub-Metre World	Non-Redundant

1.4 Physical Principles Underlying Functionality of the Human Arm and Hand

It was widely recognised among psychologists in the late 1980s that Bernstein's concept of dexterity is ecological [1-5]. In developing skills of action the CNS is learning, not to move the limb, but to solve motor problems by external circumstances. In other words, the development of dexterity takes place context-dependently. This psychological point of view suggests that action skills are hardly spelled out in the generic formulation of mathematics or control theory, even if their concrete dynamics as the limb moves through the environment are explicitly expressed in mathematical formulae. Indeed, in a specific task of multi-joint reaching under excess DOFs, the process of skill acquisition must be context-dependent as well as being confronted with redundancy resolution. In this case, differences among the physical scales of the upper arm, lower arm, hand palm, and index finger (see Figure 1.7 and Figure 1.9) are considerably noteworthy. According to Table 1.1 the discrepancy between the upper arm and index finger inertia moments is beyond the scale of 10^3–10^4 times. To gain a more physical insight into the acquisition of skills of reaching movements, it is necessary at this stage to introduce the dynamics of motion, expressed as

$$H(q)\ddot{q} + \left\{ \frac{1}{2}\dot{H}(q) + S(q,\dot{q}) \right\} \dot{q} = u, \tag{1.3}$$

where $q = (q_1, \cdots, q_4)^{\mathrm{T}}$, $u = (u_1, \cdots, u_4)^{\mathrm{T}}$, and $H(q)$ denotes the 4×4 inertia matrix of the whole arm shown in Figure 1.7. All the meanings of $\dot{H}(q)$ $(= \mathrm{d}H(q)/\mathrm{d}t)$ and $S(q,\dot{q})$ will be provided in the last section of this chapter. At this stage, we remark only the fact that $H(q)$ is symmetric and positive definite and that $S(q,\dot{q})$ is skew-symmetric. The most important meaning of the inertia matrix $H(q)$ is that the total kinetic energy of arm movement is expressed as the quadratic form of joint velocity vector $\dot{q}$ through $H(q)$ in the following way:

$$K = \frac{1}{2}\dot{q}^{\mathrm{T}} H(q)\dot{q}. \tag{1.4}$$

In this quantity, the (4,4)-entry h_{44} of $H(q)$ where $H(q) = (h_{ij}(q))$ stands for the inertia moment of the index finger, is considerably smaller than the other diagonal entries $h_{ii}(q)$ ($i = 1, 2$, and 3). Therefore, synergistic choice of damping factors C in Equation (1.1) and C_0 in Equation (1.2) must reflect different scales of inertia moments. To see this in detail more, let us substitute the control input u of Equation (1.2) into Equation (1.3), which results in the form:

$$H(q)\ddot{q} + \left\{\frac{1}{2}\dot{H}(q) + S(q, \dot{q})\right\}\dot{q} + C_0\dot{q} + J^{\mathrm{T}}(q)\left\{c\dot{\boldsymbol{x}} + k\Delta\boldsymbol{x}\right\} = 0. \tag{1.5}$$

Taking the inner product of this equation with the angular velocity vector $\dot{q}$ yields

$$\frac{\mathrm{d}}{\mathrm{d}t}E(q, \dot{q}) = -\dot{q}^{\mathrm{T}} C_0\dot{q} - c\|\dot{\boldsymbol{x}}\|^2, \tag{1.6}$$

where

$$\begin{aligned} E(q, \dot{q}) &= K + \frac{k}{2}\|\Delta\boldsymbol{x}\|^2 \\ &= \frac{1}{2}\dot{q}^{\mathrm{T}} H(q)\dot{q} + \frac{k}{2}\|\Delta\boldsymbol{x}\|^2. \end{aligned} \tag{1.7}$$

The quantity $E(q, \dot{q})$ is called the total energy. Equation (1.6) shows that the time rate of change of the total energy is equal to the instantaneous energy dissipation. Since the right-hand side of Equation (1.6) is negative definite in $\dot{q}$, Equation (1.6) itself may be expected to play a similar role of a Lyapunov relation as discussed in Section 1.9. However, $E(q, \dot{q})$ is not positive definite in q and $\dot{q}$, because E includes only a quadratic form of the two-dimensional position vector $\Delta\boldsymbol{x}$. Further, in the right-hand of Equation (1.6) there does not arise any quadratic form of $\Delta\boldsymbol{x}$. Therefore, it is possible to see that $\int_0^\infty \dot{q}^{\mathrm{T}} C_0\dot{q}\,\mathrm{d}t < +\infty$, $\int_0^\infty \|\dot{\boldsymbol{x}}\|^2\,\mathrm{d}t < +\infty$ but it is not possible to predict anything about the boundedness of the two metrics $\int_0^\infty \sqrt{\frac{1}{2}\sum_{i,j} h_{ij}(q)\dot{q}_i\dot{q}_j}\,\mathrm{d}t$, $\int_0^\infty \sqrt{\frac{k}{2}\|\Delta\boldsymbol{x}\|^2}\,\mathrm{d}t$. The former quantity is nothing but the Riemannian metric and the finiteness and the scale of the latter quantity are requisited to see whether self-motion pertinent to redundant systems remains for a long time. In order to check this, it is necessary to gain an indepth insight into the physical interactions among the inertia term $H(q)\ddot{q}$, the dissipation $C_0\dot{q}$, and the external torque $J^{\mathrm{T}}(q)(c\dot{\boldsymbol{x}} + k\Delta\boldsymbol{x})$. It is claimed at the present stage of research on redundant systems that the numerical orders of damping matrix

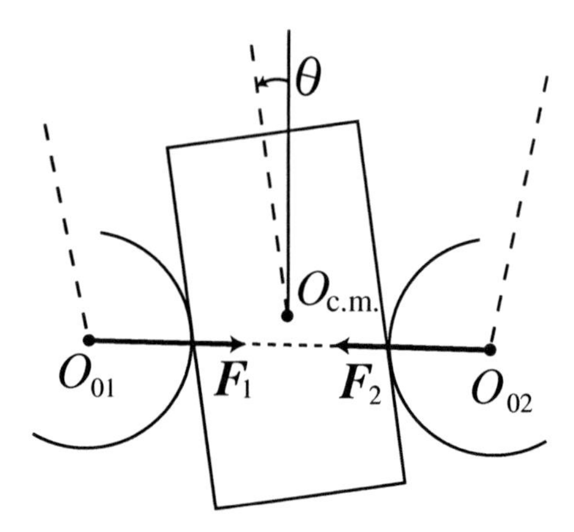

Fig. 1.10. Grasping of a 2-D object based on the coordination of the opposition forces $\boldsymbol{F}_1$ and $\boldsymbol{F}_2$ exerted from the centres O_{01} and O_{02} of finger-end spheres.

C or C_0 should be chosen synergistically in relation to the square root of the inertia matrix, $H^{1/2}(q)$, or roughly $c_i \approx \sqrt{h_{ii}}$ for $i = 1, 2, 3$ and 4.

In the case of grasp of a 2-D rigid object with parallel sides located on a flat table by a pair of multi-joint fingers, the principal term of control signals is composed on the basis of fingers–thumb opposability in the form

$$u_i = (-1)^i \frac{f_d}{2r} J_{0i}^{\mathrm{T}}(q) \begin{pmatrix} x_{01} - x_{02} \\ y_{01} - y_{02} \end{pmatrix}, \qquad i = 1, 2, \tag{1.8}$$

where r denotes the radius of the fingertip spheres, $J_{0i}(q)$ signifies the Jacobian matrix of the position vector $(x_{0i}, y_{0i})^{\mathrm{T}}$ of O_{0i} in finger joint vectors $q_i = (q_{i1}, q_{i2}, \cdots)^{\mathrm{T}}$, and $i = 1$ corresponds to the left-hand finger and $i = 2$ the right-hand finger. In order to press the object from the left the reaction torques at the finger joints ($i = 1$) should be generated to withstand the force $-\boldsymbol{F}_1$ whose direction is coincident with the straight line $\overline{O_{01}O_{02}}$. It should be noted that, in ordinary circumstances of grasping, it is difficult to know the exact locations of the contact points between the fingertips and object surfaces. Instead, the locations of O_{01} and O_{02} can be assumed to be known. In order to maintain coordination of pressing forces from the left and right, the direction of $\boldsymbol{F}_2$ must be opposite to that of $\boldsymbol{F}_1$ and the magnitudes of $\boldsymbol{F}_1$ and $\boldsymbol{F}_2$ must be equal also. The control signals of Equation (1.7) reflect these requirements. However, stabilisation of grasps should be analyzed context-dependently under various environmental conditions such as the existence or absence of rolling contact constraints, gravity effect, robustness against the arbitrariness of objects, viscoelasticity of fingertip material, *etc.*

The control signals given in Equation (1.2) and or (1.8) take the form of position feedback. The former corresponds to the potential function $(k/2)\|\Delta \boldsymbol{x}\|^2$ and the latter to $(f_d/2r)\|\boldsymbol{x}_{01} - \boldsymbol{x}_{02}\|^2$, where $\boldsymbol{x}_{0i} = (x_{0i}, y_{0i})^{\mathrm{T}}$. This may contradict the observation that fast voluntary movements of the human limb are executed in a feedforward manner based on anticipation, as discussed in the previous section. Nevertheless, if dynamic behaviours of coactivations of agonist and antagonist muscles are taken into account, both signals $-k\Delta \boldsymbol{x}$ in Equations (1.2) and (1.8) can be regarded to be exerted at joints in a feedforward manner. However, we do not discuss the details of muscle physiology further, because the signal of Equation (1.8) can easily be constructed in real time from the measured data on joint angles and the knowledge of finger parameters only, irrespective of how it is constructed in a feedback manner.

Before closing this section, we emphasise that in differential geometry Equation (1.3) is written in the following form:

$$\sum_{j=1}^{4} h_{ij}(q)\ddot{q}_j + \sum_{j,k=1}^{4} \Gamma_{kij}(q)\dot{q}_j\dot{q}_k = u_i, \tag{1.9}$$

where $\Gamma_{kij}(q)$ is a Christoffel symbol of the first kind. By multiplying this equation by $H^{-1}(q)$, we obtain another form of the differential equation:

$$\frac{\mathrm{d}^2}{\mathrm{d}t^2}q_i + \sum_{m,n} \Gamma^i_{nm}\frac{\mathrm{d}q_m}{\mathrm{d}t}\frac{\mathrm{d}q_n}{\mathrm{d}t} = \sum_j h^{ij}u_j, \tag{1.10}$$

where h^{ij} denotes the (i,j)-entry of $H^{-1}(q)$ and Γ^i_{nm} is called a Christoffel symbol of the second kind. When the external torque u_i is zero for all i, a solution trajectory $q(t)$ of Equation (1.9) or (1.10) starting from a given initial position $q(0) = q^0$ to a target position $q(1) = q^1$ is called the geodesic. Then, the Riemannian distance between two points q^0 and q^1 in the configuration manifold $q \in CM^4$ can be defined as

$$R(q^0, q^1) = \min_{q(t)} \int_0^t \sqrt{\frac{1}{2}\sum_{i,j} h_{ij}(q(t))\dot{q}_i(t)\dot{q}_j(t)}\,\mathrm{d}t, \tag{1.11}$$

where the minimisation is taken over all curves $q(t)$ parameterised by $t \in [0,1]$ in such a way that $q(0) = q^0$ and $q(1) = q^1$. This metric plays an important role in defining neighbourhoods of a given position in the configuration manifold that is composed of all the generalised position vector q, no matter how components of q are mixed with physical variables with different physical units such as length x [m] and angle θ with dimensionless units [radian].

1.5 Difficulty in the Development of Everyday Physics

We shall begin this section by quoting the abstract of the author's old article entitled "Robotics research toward explication of everyday physics", which was published in one of two special issues for the new millenium in the International Journal of Robotics Research [1-20].

Abstract. It is commonly recognized now at the end of the 20th century that a general 6- or 7-degree-of-freedom robot equipped with an endeffector with simple structure is clumsy in performing a variety of ordinary tasks that a human encounters in his or her everday life. In this paper, it is claimed that the clumsiness manifests the lack of our knowledge of everday physics. It is then shown that even dynamics of a set of dual fingers grasping and manipulating a rigid object are not yet formulated when the fingers' ends are covered by soft and deformable material. By illustrating this typical problem of everday physics, it is pointed out that explication of everyday physics in computational (or mathematical) languages is inevitable for consideration of how to endow a robot with dexterity and versatility. Once kinematics and dynamics involved in such everyday tasks are described, it is then possible to discover a simple but fine control structure without the need of much computation of kinematics and dynamics. Simplicity of the control structure implies robustness against parameter uncertainties, which eventually allows the control to perform tasks with dexterity and versatility by using visual or tactile sensing feedback. Thus, a key to uncover the hidden secret of dexterity is to characterize complicated dynamics of such a robotic task as seen when a set of multifingers with multijoints covered by deformable material interacts physically with objects or an enviounment. It is pointed out throughout the paper that some of the generic characteristics of dynamics that everday physics encounters must be "passivity", "approximate Jacobian matrix of coordinates transformation", "feedback loops from sensation to action", "impedance matching", and "static friction."

In the introductory section of that paper, it was reported that around the early 1980s there had been a dream and enthusiasm among robot engineers and roboticists to create "intelligent robots". This dream can be compared with optimistic views that appeared in the history of artificial intelligence spelled out as "within twenty years machines will be capable of doing any work a man can do" [1-21] and "within a generation the problem of creating 'artificial intelligence' will be substantially solved" [1-22]. Around the mid 1980s, American philosophers Huber Dreyhus and Stuart Dreyhus (Professors of University of California, Berkeley) asserted a heterogeneous opinion by claiming [1-23] that the commonsense knowledge problem, including everyday physics or commonsense physics, has dampened robot engineers' enthusiasm and also blocked progress in theoretical AI for the past decades. They spelled this out as "Can there be a theory of the everyday world as rationalist philosophers have always held? Or is the commonsense background rather a combination of skills, practices, discrimination, and so on, which are not intentional states and so, *a fortiori* do not have any representational content to be explicated in terms of elements and rules?" They added: "Commonsense physics has turned out to be extremely hard to spell out in a set of facts and rules."

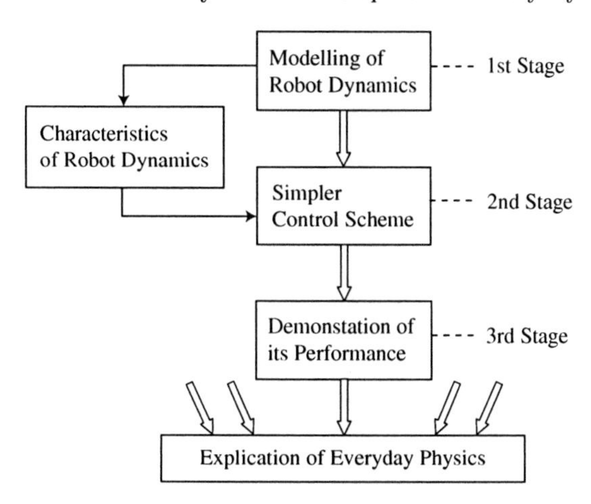

Fig. 1.11. Robotics research sould evolve from precise modelling of a robotic task to explication of everyday physics

The author's view was the following. If robotics is an integration of various domains involved in aiming to create an artificial reflection of the corresponding real world with which a human is concerned, robotics must naturally aim to account for the intelligibility of the real world or everyday physics. Here, the term "everyday physics" is used as a scientific domain related to the accountability of the dexterous accomplishment of ordinary tasks by manipulating things, with sensing and recognition, as seen in ordinary human beings. In this meaning of everyday physics, the first stage toward unveiling the secret of dexterity in a particular task assigned to a robotic system is to derive a detailed description of the dynamics of motion of the system that must accomplish the task. The second stage is to discover some of the key characteristics of such complicated dynamics to simplify the structure of control drastically (see Figure 1.11). Then, the third stage must be to demonstrate both theoretically and experimentally that such a simplified control scheme with the use of minimum knowledge and approximate values of physical parameters works well and accomplishes the task to some extent of satisfaction. The theoretical demonstration of how it works should be carried out on the basis of a complicated model of the full dynamics.

The present book intends to explore what is the core dynamics of human or robotic precision prehension and stable grasping of objects, believing that it must be the first step toward challenging everyday physics for the advancement of robotics. At the same time, the book aims to explore what the dexterity of robotic prehensility is and how such a dexterous way of grasping and object manipulation can function in multi-fingered robotic hands [1-24] [1-25]. In contrast to psychology and philosophy, robotics should challenge both the problems of the development of dexterity and everyday physics to make them explicable, visible, and feasible in computational languages [1-26].

1.6 Newton's Laws of Motion

If we turnned over the pages of Napier's book [1-1] we find at p. 51 the following two sentences:

"The thumb, the 'lesser hand' as Albinus called it, is the most specialized of the digits. Isaac Newton once remarked that, in the absence of any other proof, the thumb alone would convince him of God's existence."

The approach adopted in this book to explore of the dexterity and functionality of the multi-fingered hand is directly and solidly based on Newtonian mechanics.

We begin by introducing the concept of a point particle, which is regarded as a point endowed with mass m. In other words, a point particle is an idealisation in which the mass is conceived to be concentrated at a single point. As a matter of course, we must bear in mind that in many practical situations, as discussed later, a material body with a volume can be well approximated by such an idealised model of a point particle. Hence we will use the term particle or body conventionally, instead of point particle, on occasions when there is no possibility of misunderstanding. In classical mechanics a point particle preserves its identity and its mass does not vary with time or motion.

Next, we define the momentum $\boldsymbol{p}$ of a particle as the product of its mass and its velocity, *i.e.*,

$$\boldsymbol{p} = m\boldsymbol{v}. \tag{1.12}$$

Note that momentum is a vector since velocity is a vector.

Now we state Newton's laws of motion in their conventional forms.

Newton's first law: a body continues in a state of rest or constant velocity (zero acceleration) unless it is acted on by an external force.

Newton's second law: the rate of change of momentum of a body is proportional to the force acting on the body and is effective in the direction of the force.

Newton's third law: the mutual actions of two bodies are always equal in magnitude and opposite in direction.

The first law is called the law of inertia, and was originally deduced by Galileo. It introduces the concept of a force as the cause of non-uniform motion. In relation to this law, let us recall that the velocity depends on the choice of reference frame, as noted in the previous section. For Newton's laws to have a physically consistent meaning, it is important to seek a frame of reference which is non-accelerating. It is also important to know that, if the laws hold in a frame at rest, they will also hold in any frame moving with uniform velocity with respect to that frame. Such a non-accelerating frame of reference is called an inertial frame.

The second law derives a quantitative definition of force, which may be represented by

$$\boldsymbol{f} = K\frac{\mathrm{d}}{\mathrm{d}t}(m\boldsymbol{v}) = Km\frac{\mathrm{d}}{\mathrm{d}t}\boldsymbol{v} = Km\boldsymbol{a}, \tag{1.13}$$

where we have assumed in the second and third equalities that m is a constant. In this formula, K is a constant of proportionality. Since we are free to choose suitable units for this new quantity $|\boldsymbol{f}|$, we conveniently choose our units so that $K = 1$. In SI units, m is measured in kilograms (kg), a in metres per second squared (ms^{-2}), and f in newtons (N). One newton is the force which gives to a mass of one kilogram an acceleration of one metre per second squared:

$$1\mathrm{N} = 1\mathrm{kg} \times 1\mathrm{ms}^{-2} = 1\mathrm{kg\,ms}^{-2}. \tag{1.14}$$

With this choice of units, the mathematical formula of Newton's second law is

$$\frac{\mathrm{d}}{\mathrm{d}t}(m\boldsymbol{v}) = \boldsymbol{f}. \tag{1.15}$$

This is called the equation of motion of the body. For a particle or body of constant mass, the equation can be rewritten in the form

$$m\boldsymbol{a} = \boldsymbol{f}. \tag{1.16}$$

The third law is described mathematically by

$$\boldsymbol{f}_{12} = -\boldsymbol{f}_{21}, \tag{1.17}$$

where $\boldsymbol{f}_{12}$ denotes the force that body 1 exerts on body 2, and $\boldsymbol{f}_{21}$ denotes the force that body 2 exerts on body 1. This law gives a basis for the law of conservation of momentum, which states that:

> For an isolated system that is subject only to internal forces between members of the system, the total momentum of the system does not change in time.

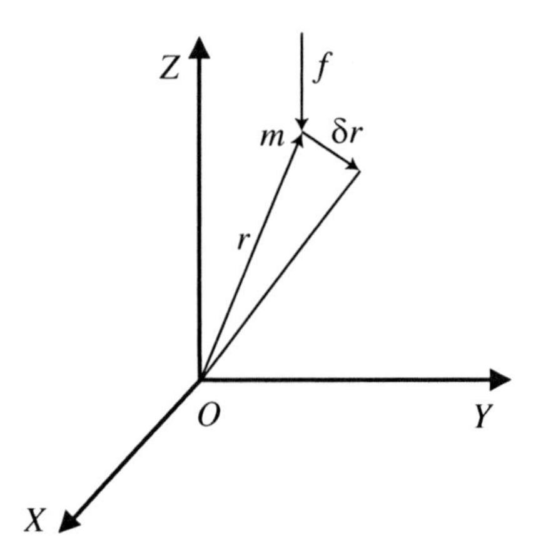

Fig. 1.12. A particle with mass m undergoes an infinitesimal small displacement $\delta \boldsymbol{r}$ due to a force $\boldsymbol{f}$ acting on it

When a force $\boldsymbol{f}$ acts on a particle with mass m and thereby the particle undergoes an infinitesimal displacement δr (see Figure 1.12), the inner product

$$\delta W = \boldsymbol{f}^{\mathrm{T}} \delta \boldsymbol{r} \tag{1.18}$$

is called the work increment. If the particle undergoes a displacement r in a constant direction under the constant force f, the quantity

$$W = \boldsymbol{f}^{\mathrm{T}} \boldsymbol{r} = |\boldsymbol{f}||\boldsymbol{r}| \cos \theta \tag{1.19}$$

is called the work, where θ stands for the angle between the vectors $\boldsymbol{f}$ and $\boldsymbol{r}$. However, the force exerted on the particle is dependent on the particle's position and hence it should be denoted by $\boldsymbol{f}(\boldsymbol{r})$. Then the work done by the force $\boldsymbol{f}(\boldsymbol{r})$ as the particle moves from position P to position Q is defined to be

$$W(P \rightarrow Q) = \int_P^Q \boldsymbol{f}^{\mathrm{T}}(\boldsymbol{r}) \, \mathrm{d}\boldsymbol{r}, \tag{1.20}$$

where the integration is taken over the path along which the particle moves. The SI unit for work is the newon-metre (Nm) or joule (J), where 1 [J] = 1 [Nm]. The time rate of change of work

$$P = \frac{\mathrm{d}W}{\mathrm{d}t} \tag{1.21}$$

is called the power. The SI unit of power is the joule per second, which is called the watt (1 [W] = 1 [Js^{-1}]).

The work can be expressed as the integral of the power

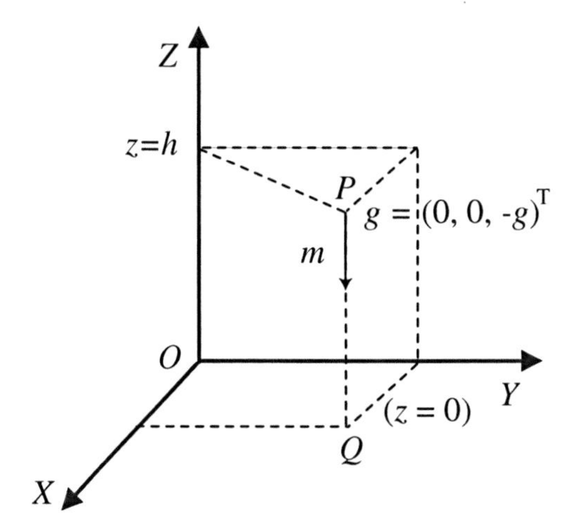

Fig. 1.13. The work done on a particle with mass m by the gravitational field is equal to $W = mgh$ when it falls from height $z = h$ to the ground $z = 0$

$$W(t_1 \to t_2) = \int_{t_1}^{t_2} P(t)\,\mathrm{d}t. \tag{1.22}$$

On the other hand, since the free motion of a particle m on which the force $\boldsymbol{f}$ acts implies the equation of motion $\boldsymbol{f} = m\,\mathrm{d}\boldsymbol{v}/\mathrm{d}t$, the work can be expressed as

$$W(P \to Q) = m \int_P^Q \frac{\mathrm{d}}{\mathrm{d}t}\boldsymbol{v}^{\mathrm{T}}\,\mathrm{d}\boldsymbol{r}. \tag{1.23}$$

Substitution of $\mathrm{d}\boldsymbol{r} = \dot{\boldsymbol{r}}\,\mathrm{d}t = \boldsymbol{v}\,\mathrm{d}t$ into this equation yields

$$\begin{aligned} m \int_P^Q \frac{\mathrm{d}}{\mathrm{d}t}\boldsymbol{v}^{\mathrm{T}}\,\mathrm{d}\boldsymbol{r} &= m \int_{t(P)}^{t(Q)} \left(\frac{\mathrm{d}}{\mathrm{d}t}\boldsymbol{v}^{\mathrm{T}}\right) \boldsymbol{v}\,\mathrm{d}t \\ &= \frac{m}{2} \int_{t(P)}^{t(Q)} \left(\frac{\mathrm{d}}{\mathrm{d}t}|\boldsymbol{v}|^2\right) \mathrm{d}t = \frac{m}{2}\left(|\boldsymbol{v}_Q|^2 - |\boldsymbol{v}_P|^2\right), \end{aligned} \tag{1.24}$$

where $t(P)$ and $t(Q)$ denote the instants of time when the particle is at positions P and Q, respectively. According to Equations (1.23) and (1.24), the work done by the particle's free motion can be expressed as

$$W(P \to Q) = \int_P^Q \boldsymbol{f}^{\mathrm{T}}\mathrm{d}\boldsymbol{r} = \frac{1}{2}m|\boldsymbol{v}_Q|^2 - \frac{1}{2}m|\boldsymbol{v}_P|^2 = K_Q - K_P. \tag{1.25}$$

This equation states that the work done on the free particle is equivalent to the change in the kinetic energy.

A region in three-dimensional space is called a force field whenever there is a force $\boldsymbol{f}(\boldsymbol{r}, \boldsymbol{v}, t)$ for each point $\boldsymbol{r}$ in the region. One good example is the electromagnetic force field. However, we restrict our consideration to a field such that the force exerted on a particle by the field depends only on the particle's position, *i.e.*, we consider only fields of the form $\boldsymbol{f}(\boldsymbol{r})$. A good example is the gravity field. In the neighbourhood of the surface of the Earth it is well approximated by a uniform field as

$$\boldsymbol{f} = m\boldsymbol{g}, \tag{1.26}$$

where $\boldsymbol{g} = (0, 0, -g)^{\mathrm{T}}$ as shown in Figure 1.13. In the case of a particle m falling freely from a height $z = h$ to the ground $z = 0$, the work done on the particle by this gravitational field is

$$W(h \to 0) = (m\boldsymbol{g})^{\mathrm{T}}\boldsymbol{r} = m(0, 0, -g)\begin{pmatrix} 0 \\ 0 \\ -h \end{pmatrix} = mgh, \tag{1.27}$$

where the work done between two points in space is

$$W(P \to Q) = \int_P^Q \boldsymbol{f}^{\mathrm{T}}(\boldsymbol{r})\,\mathrm{d}\boldsymbol{r}, \tag{1.28}$$

which is independent of the path along which the infinitesimal displacement is taken. This force field satisfying such a relation is said to be conservative. Fortunately the gravity field is conservative.

The work is a function of only the starting point P and terminal point Q for a conservative force field. Therefore, if a starting point P is fixed as a standard point (taken on the ground in the case of the gravity field), the work of integral (1.20) depends only on the present position Q. The negative of Equation (1.20), *i.e.*, the function

$$U(\boldsymbol{r}_Q) = -\int_P^Q \boldsymbol{f}^{\mathrm{T}}\,\mathrm{d}\boldsymbol{r} \tag{1.29}$$

is said to be the potential energy. From this definition it follows that $U(\boldsymbol{r}_P) = 0$. It should be noted that U is a scalar function of the position $\boldsymbol{r}$ once the starting point is fixed in space.

Directly from the definition of potential energy it follows that

$$\delta U = U(\boldsymbol{r} + \mathrm{d}\boldsymbol{r}) - U(\boldsymbol{r}) = \int_{\boldsymbol{r}}^{\boldsymbol{r}+\mathrm{d}\boldsymbol{r}} \boldsymbol{f}^{\mathrm{T}}\,\mathrm{d}\boldsymbol{r} = -\boldsymbol{f}^{\mathrm{T}}\,\mathrm{d}\boldsymbol{r}. \tag{1.30}$$

This implies that

$$\boldsymbol{f} = -\nabla U, \tag{1.31}$$

which means that the potential's negative gradient is the force exerted by the force field on the particle. The value of the potential energy depends on the selection of the starting point P, but the relations (1.30) and (1.31) are independent of the choice of P.

Finally we consider the relation of the potential to the kinetic energy. When we apply Newton's second law for a particle m that undergoes a free motion, Equation (1.31) can be written as

$$m\frac{\mathrm{d}^2}{\mathrm{d}t^2}\boldsymbol{r} = -\nabla U. \tag{1.32}$$

Taking the inner product of this with $\boldsymbol{v} = \mathrm{d}\boldsymbol{r}/\mathrm{d}t$, we have

$$\frac{\mathrm{d}}{\mathrm{d}t}\left(\frac{1}{2}m\left|\frac{\mathrm{d}}{\mathrm{d}t}\boldsymbol{r}\right|^2\right) = -\left(\frac{\mathrm{d}}{\mathrm{d}t}\boldsymbol{r}^{\mathrm{T}}\right)\nabla U. \tag{1.33}$$

Since the time rate of the potential function can be expressed as

$$\frac{\mathrm{d}}{\mathrm{d}t}U(\boldsymbol{r}) = \left(\frac{\mathrm{d}}{\mathrm{d}t}\boldsymbol{r}^{\mathrm{T}}\right)\nabla U \tag{1.34}$$

Equation (1.33) implies

$$\frac{\mathrm{d}}{\mathrm{d}t}\left(\frac{1}{2}m|\boldsymbol{v}|^2 + U\right) = 0. \tag{1.35}$$

Integration of this with respect to time yields

$$\frac{1}{2}m|\boldsymbol{v}|^2 + U(\boldsymbol{r}) = E = \mathrm{const}. \tag{1.36}$$

Thus it is concluded that the total energy E (which is the sum of the kinetic energy and the potential energy) remains constant throughout the free motion. This is called the law of conservation of mechanical energy.

1.7 Kinetic Energy of a System of Particles

First, we prove the law of conservation of momentum in the case of the dynamic behaviour of a system of particles, which is a collection of material of fixed identity. For this purpose, it is useful to distinguish between the forces of interaction with other particles within the system and forces due to the interaction with bodies external to the system or due to external factors, such as a gravitational or magnetic field in which the system may exist. We therefore denote by $\boldsymbol{f}_{ij}$ the force acting on the ith particle due to the jth particle, and denote by $\boldsymbol{f}_{ie}$ the force acting on the ith particle due to external factors. Further, we denote the mass of the ith particle by m_i, which for each i does not change with time, and set a suitable inertial frame. Then, with the observed acceleration denoted by $\boldsymbol{a}_i$ for the ith particle, Newton's second law is written as

$$\boldsymbol{f}_{ie} + \sum_{j(\neq i)} \boldsymbol{f}_{ij} = m_i \boldsymbol{a}_i, \qquad i = 1, 2, \cdots . \tag{1.37}$$

Summing all these equations results in

$$\sum_i \boldsymbol{f}_{ie} + \sum_i \sum_{j(\neq i)} \boldsymbol{f}_{ij} = \sum_i m_i \boldsymbol{a}_i. \tag{1.38}$$

Since $\boldsymbol{f}_{ij} = -\boldsymbol{f}_{ji}$ for $i \neq j$ owing to Newton's third law, the second term on the left-hand side of Equation (1.38) vanishes. Hence, denoting the resultant external force by $\boldsymbol{f}$ $(= \sum_i \boldsymbol{f}_{ie})$, we have

$$\sum_i m_i \boldsymbol{a}_i = \boldsymbol{f}. \tag{1.39}$$

At this stage, we must remark again upon the law of conservation of momentum for the system of particles. In the absence of external forces, Equation (1.39) is reduced to

$$0 = \sum_i m_i \boldsymbol{a}_i = \frac{\mathrm{d}}{\mathrm{d}t} \sum_i m_i \boldsymbol{v}_i = \frac{\mathrm{d}}{\mathrm{d}t} \left(\sum_i \boldsymbol{p}_i \right). \tag{1.40}$$

This demonstrates the conservation of momentum for the system of particles.

Now we introduce the concept of the centre of mass of the system, which is defined by the position vector

$$\boldsymbol{r}_c = \sum_i m_i \boldsymbol{r}_i \Big/ \sum_i m_i, \tag{1.41}$$

where $\boldsymbol{r}_i$ denotes the position vector of the ith particle. In other words, the centre of mass of the system is an average position vector for all the particles within the system, weighted according to particle masses. Denoting the total mass $\sum_i m_i$ of the system by m, we observe from Equation (1.41) that

$$\sum_i m_i \boldsymbol{r}_i = m \boldsymbol{r}_c. \tag{1.42}$$

Differentiation of this with respect to time yields

$$\sum_i m_i \boldsymbol{v}_i = m \boldsymbol{v}_c, \tag{1.43}$$

which on repeated differentiation yields

$$\sum_i m_i \boldsymbol{a}_i = m \boldsymbol{a}_c. \tag{1.44}$$

Taking into account Equations (1.39) and (1.44), we finally obtain the formula

$$\boldsymbol{f} = m \boldsymbol{a}_c. \tag{1.45}$$

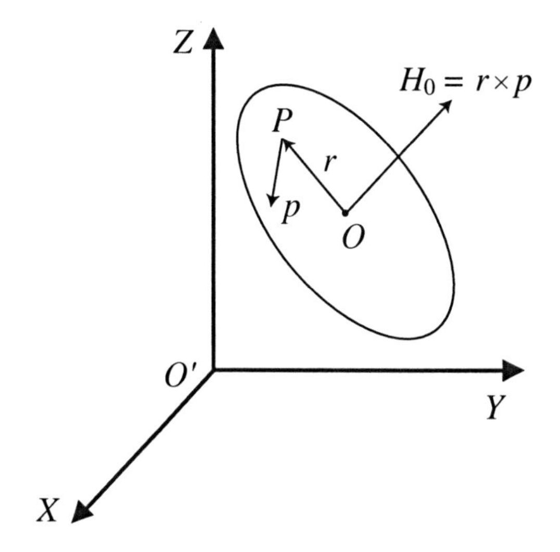

Fig. 1.14. Angular momentum H_0 about point O

This is a remarkable result, which can be interpreted as follows:

> The vector sum of external forces influences the motion of the centre of mass as if the total mass of the system were concentrated there and the resultant external force acted on it.

This principle plays an important role particularly in the anaylsis of rigid-body dynamics to be treated in the next section.

In the absence of external forces, Equation (1.45) is reduced to

$$\boldsymbol{v}_c = \text{const.} \tag{1.46}$$

This is another remarkable property of the centre of mass. It states that:

> The velocity of the centre of mass does not vary with time in the absence of external forces.

Next we introduce the important concept of angular momentum. The dimension of angular momentum is different from that of momentum, which is termed the linear momentum when it is necessary to distinguish between the two concepts.

Let O be an arbitrary point fixed in an inertial reference frame. For a single particle, the angular momentum H_0 about point O is defined as

$$H_0 = \boldsymbol{r} \times \boldsymbol{p} = \boldsymbol{r} \times m\boldsymbol{v}, \tag{1.47}$$

where $\boldsymbol{r}$ is the position vector of the particle relative to point O, and $\boldsymbol{p}$ is the linear momentum (see Figure 1.14). If a force $\boldsymbol{f}$ acts on the particle, the momentum of force, or torque, is defined as

$$T_0 = \boldsymbol{r} \times \boldsymbol{f}. \tag{1.48}$$

As predicted from Newton's second law, there is a relationship between these two quantities. In fact, differentiation of Equation (1.47) in time yields

$$\frac{\mathrm{d}}{\mathrm{d}t} H_0 = \dot{\boldsymbol{r}} \times \boldsymbol{p} + \boldsymbol{r} \times \dot{\boldsymbol{p}} = \boldsymbol{v} \times m\boldsymbol{v} + \boldsymbol{r} \times \boldsymbol{f}. \tag{1.49}$$

Since $\boldsymbol{v} \times \boldsymbol{v} = 0$ by the definition of outer product, this equation leads to the important result

$$\frac{\mathrm{d}}{\mathrm{d}t} H_0 = T_0. \tag{1.50}$$

This states that:

The rate of change of angular momentum is equal to the torque.

For a system of particles, the angular momentum about O is defined as

$$H_0 = \sum_i \boldsymbol{r}_i \times \boldsymbol{p}_i \tag{1.51}$$

and the total torque is written in the form

$$T_{\text{total}} = \sum_i \boldsymbol{r}_i \times \boldsymbol{f}_i. \tag{1.52}$$

Similar to Equation (1.50), we obtain the relationship

$$\frac{\mathrm{d}}{\mathrm{d}t} H_0 = T_{\text{total}}. \tag{1.53}$$

However, it is possible to show that the sum of internal torques that may arise from interaction forces among particles in the system does not contribute to the total torque T_{total}. To see this, we resolve the force acting on the ith particle into the form

$$\boldsymbol{f}_i = \boldsymbol{f}_{ie} + \sum_{j(\neq i)} \boldsymbol{f}_{ij} \tag{1.54}$$

as discussed in the previous section. Corresponding to this expression, the total torque is resolved into the form

$$T_{\text{total}} = T_0 + T_{\text{int}}, \tag{1.55}$$

where T_0, defined by

$$T_0 = \sum_i \boldsymbol{r}_i \times \boldsymbol{f}_{ie}, \tag{1.56}$$

is the torque due to external forces, and T_{int}, defined by

$$T_{\text{int}} = \sum_i \sum_{j(\neq i)} \boldsymbol{r}_i \times \boldsymbol{f}_{ij}, \tag{1.57}$$

is the torque due to internal forces among particles in the system. Now we note that the latter expression can be rewritten in the form

$$\begin{aligned} T_{\text{int}} &= \sum_i \sum_{j(\neq i)} \boldsymbol{r}_i \times \boldsymbol{f}_{ij} = \frac{1}{2} \sum_{i \neq j} (\boldsymbol{r}_i \times \boldsymbol{f}_{ij} + \boldsymbol{r}_j \times \boldsymbol{f}_{ji}) \\ &= \frac{1}{2} \sum_{i \neq j} (\boldsymbol{r}_i - \boldsymbol{r}_j) \times \boldsymbol{f}_{ij} \end{aligned} \tag{1.58}$$

in which the last equality follows from Newton's third law. Since $\boldsymbol{f}_{ij}$ is parallel to $\boldsymbol{r}_i - \boldsymbol{r}_j$ (this is true if the forces between the particles are central forces), this implies

$$T_{\text{int}} = 0. \tag{1.59}$$

In view of Equations (1.53), (1.55), and (1.59), we conclude that

$$\frac{\mathrm{d}}{\mathrm{d}t} H_0 = T_0. \tag{1.60}$$

The total angular momentum of a system of particles can also be split into two components as

$$H_0 = \sum_i m_i (\boldsymbol{r}_i - \boldsymbol{r}_c) \times \boldsymbol{v}_i + \sum_i m_i \boldsymbol{r}_c \times \boldsymbol{v}_i = H_{\text{c.m.}} + \boldsymbol{r}_c \times \boldsymbol{p}. \tag{1.61}$$

where $\boldsymbol{p} = \sum_i m_i \boldsymbol{v}_i$ denotes the total linear momentum. The term $\boldsymbol{r}_c \times \boldsymbol{p}$ is the angular momentum due to the motion of the centre of mass about the fixed point O. This term depends on the choice of point O, while the angular momentum about the centre of mass, denoted by $H_{\text{c.m.}}$ in Equation (1.61), does not. If the point O is chosen at the centre of mass, then Equation (1.60) may be written as

$$\frac{\mathrm{d}}{\mathrm{d}t} H_{\text{c.m.}} = T_0. \tag{1.62}$$

Since only external forces contribute to T_0, Equation (1.62) implies that the rotation about the centre of mass is determined by the total external torque.

Finally we introduce the concept of kinetic energy for a particle or a system of particles. For a particle P_i with mass m_i, the quantity

$$K_i = \frac{1}{2} m_i |\boldsymbol{v}_i|^2 \tag{1.63}$$

is called the kinetic energy of particle P_i. The kinetic energy for a system of particles is defined as

$$K = \sum_i \frac{1}{2} m_i |\boldsymbol{v}_i|^2. \tag{1.64}$$

Notice that this form can be rewritten as

$$\begin{aligned} K &= \sum_i \frac{1}{2} m_i \boldsymbol{v}_i^{\mathrm{T}} \boldsymbol{v}_i = \sum_i \frac{1}{2} m_i \left\{ (\boldsymbol{v}_i - \boldsymbol{v}_c) + \boldsymbol{v}_c \right\}^{\mathrm{T}} \left\{ (\boldsymbol{v}_i - \boldsymbol{v}_c) + \boldsymbol{v}_c \right\} \\ &= \sum_i \frac{1}{2} m_i (\boldsymbol{v}_i - \boldsymbol{v}_c)^{\mathrm{T}} (\boldsymbol{v}_i - \boldsymbol{v}_c) + \frac{1}{2} m \boldsymbol{v}_c^{\mathrm{T}} \boldsymbol{v}_c + \sum_i m_i (\boldsymbol{v}_i - \boldsymbol{v}_c)^{\mathrm{T}} \boldsymbol{v}_c. \end{aligned} \tag{1.65}$$

According to Equation (1.43), we see that

$$\sum_i m_i (\boldsymbol{v}_i - \boldsymbol{v}_c)^{\mathrm{T}} \boldsymbol{v}_c = \left\{ \left(\sum_i m_i \boldsymbol{v}_i \right) - m \boldsymbol{v}_c \right\}^{\mathrm{T}} \boldsymbol{v}_c = 0. \tag{1.66}$$

Substituting this into Equation (1.65), we obtain another important result:

$$K = \sum_i \frac{1}{2} m_i \left| \boldsymbol{v}_i - \boldsymbol{v}_c \right|^2 + \frac{1}{2} m |\boldsymbol{v}_c|^2 = K_{\text{c.m.}} + \frac{1}{2} m |\boldsymbol{v}_c|^2. \tag{1.67}$$

This states that:

> The total kinetic energy of a system of particles is expressed as the sum of two components, one of which is the total kinetic energy of particles relative to the centre of mass and the other is the kinetic energy of the centre of mass with total mass m.

1.8 Kinematics and Dynamics of a Rigid Body

A rigid body may be treated as a special case of a system consisting of a very large number of discrete particles. Hence, it is possible to apply Equations (1.60), (1.61), (1.62), and (1.67) to rigid-body dynamics. Motion of a rigid body can be completely characterised by the knowledge of translational velocity $\boldsymbol{v}_{\text{c.m.}}$ of the centre $O_{\text{c.m.}}$ of mass of the body and the angular velocity vector $\boldsymbol{\omega}$ whose axis goes through $O_{\text{c.m.}}$ (see Figure 1.15). In parallel with Equation (1.62), first we derive angular momentum about the centre of mass of the rigid body. This is defined as

$$\begin{aligned} H_{\text{c.m.}} &= \sum_i m_i (\boldsymbol{r}_i \times \boldsymbol{v}_i) = \sum_i m_i \boldsymbol{r}_i \times (\boldsymbol{\omega} \times \boldsymbol{r}_i) \\ &= \int_V \left\{ (\boldsymbol{r}_P^{\mathrm{T}} \boldsymbol{r}_P) \boldsymbol{\omega} - (\boldsymbol{\omega}^{\mathrm{T}} \boldsymbol{r}_P) \boldsymbol{r}_P \right\} \mathrm{d}m_P, \end{aligned} \tag{1.68}$$

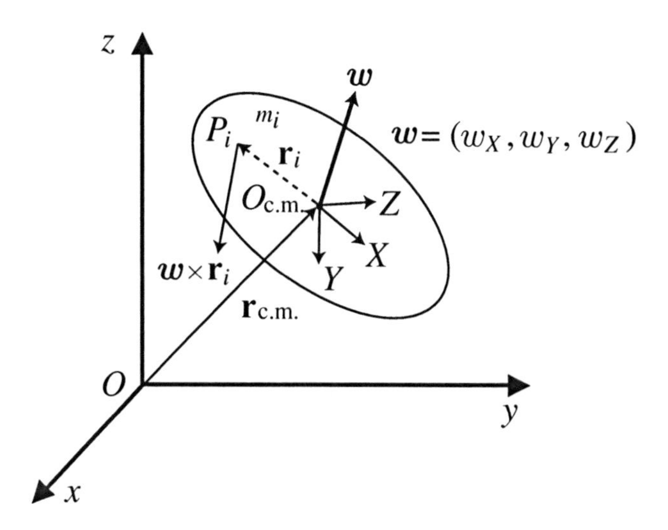

Fig. 1.15. Motion of a rigid body is characterised by the velocity $\boldsymbol{v}_{\text{c.m.}}$ of the body mass centre relative to the frame coordinates $O-xyz$ and the angular velocity vector $\boldsymbol{\omega}$ originates from the body mass centre $O_{\text{c.m.}}$

where we used the well-known vector formula

$$\boldsymbol{a} \times (\boldsymbol{b} \times \boldsymbol{c}) = (\boldsymbol{a}^{\mathrm{T}}\boldsymbol{c})\boldsymbol{b} - (\boldsymbol{b}^{\mathrm{T}}\boldsymbol{a})\boldsymbol{c}. \tag{1.69}$$

Here, $\mathrm{d}m_P$ denotes the mass element of a particle at position P in the body and the integration is taken over all volume elements that constitute the body. Second, let us fix the coordinates of X, Y, Z axes at $O_{\text{c.m.}}$, denote the unit vectors along the X-, Y-, Z-axes by $\boldsymbol{r}_X$, $\boldsymbol{r}_Y$, $\boldsymbol{r}_Z$, and define

$$\begin{cases} H_{\text{c.m.}} = H_X\boldsymbol{r}_X + H_Y\boldsymbol{r}_Y + H_Z\boldsymbol{r}_Z, \\ \boldsymbol{\omega} = \omega_X\boldsymbol{r}_X + \omega_Y\boldsymbol{r}_Y + \omega_Z\boldsymbol{r}_Z, \\ \boldsymbol{r}_P = x_P\boldsymbol{r}_X + y_P\boldsymbol{r}_Y + z_P\boldsymbol{r}_X. \end{cases} \tag{1.70}$$

Substituting these into Equation (1.68) and noting that

$$\boldsymbol{r}_P^{\mathrm{T}}\boldsymbol{r}_P = x_P^2 + y_P^2 + z_P^2 \tag{1.71}$$

we obtain that

$$\begin{cases} H_X = \omega_X \displaystyle\int (y_P^2 + z_P^2)\mathrm{d}m_P - \omega_Y \int x_P y_P \mathrm{d}m_P - \omega_Z \int x_P z_P \mathrm{d}m_P, \\ H_Y = -\omega_X \displaystyle\int y_P x_P \mathrm{d}m_P + \omega_Y \int (z_P^2 + x_P^2)\mathrm{d}m_P - \omega_Z \int y_P z_P \mathrm{d}m_P, \\ H_Z = -\omega_X \displaystyle\int z_P x_P \mathrm{d}m_P - \omega_Y \int z_P y_P \mathrm{d}m_P + \omega_Z \int (x_P^2 + y_P^2)\mathrm{d}m_P. \end{cases} \tag{1.72}$$

Then, by defining

$$\begin{cases} I_{xx} = \int (y_P^2 + z_P^2) \mathrm{d}m_P, & I_{yy} = \int (z_P^2 + x_P^2) \mathrm{d}m_P, \\ I_{zz} = \int (x_P^2 + y_P^2) \mathrm{d}m_P, & I_{xy} = -\int x_P y_P \mathrm{d}m_P, \quad etc. \end{cases} \tag{1.73}$$

we obtain

$$\boldsymbol{H}_{c.m.} = \begin{bmatrix} I_{xx} & I_{xy} & I_{xz} \\ I_{yx} & I_{yy} & I_{yz} \\ I_{zx} & I_{zy} & I_{zz} \end{bmatrix} \boldsymbol{\omega} = H\boldsymbol{\omega}. \tag{1.74}$$

Here, the 3×3 matrix H is called the inertia tensor or inertia matrix. Note that H is symmetric and positive definite by definition.

Next, we derive total kinetic energy of the rigid body in such a way that

$$\begin{aligned} K &= \frac{1}{2}\int \boldsymbol{v}_P^{\mathrm{T}} \boldsymbol{v}_P \, \mathrm{d}m_P = \frac{1}{2}\int \left(\boldsymbol{v}_{\text{c.m.}} + \boldsymbol{\omega} \times \boldsymbol{r}_P\right)^{\mathrm{T}} \left(\boldsymbol{v}_{\text{c.m.}} + \boldsymbol{w} \times \boldsymbol{r}_P\right) \mathrm{d}m_P \\ &= \frac{1}{2}\int \left\{\boldsymbol{v}_{\text{c.m.}}^{\mathrm{T}} \boldsymbol{v}_{\text{c.m.}} + 2\boldsymbol{v}_{\text{c.m.}}^{\mathrm{T}} (\boldsymbol{\omega} \times \boldsymbol{r}_P) + (\boldsymbol{\omega} \times \boldsymbol{r}_P)^{\mathrm{T}} (\boldsymbol{\omega} \times \boldsymbol{r}_P)\right\} \mathrm{d}m_P \\ &= \frac{1}{2} M |\boldsymbol{v}_{\text{c.m.}}|^2 + \frac{1}{2}\boldsymbol{\omega}^{\mathrm{T}} \int \boldsymbol{r}_P \times (\boldsymbol{\omega} \times \boldsymbol{r}_P) \, \mathrm{d}m_P \\ &= \frac{1}{2} M |\boldsymbol{v}_{\text{c.m.}}|^2 + \frac{1}{2}\boldsymbol{\omega}^{\mathrm{T}} H \boldsymbol{\omega} \end{aligned} \tag{1.75}$$

where the well-known formula $\boldsymbol{a}^{\mathrm{T}}(\boldsymbol{b} \times \boldsymbol{x}) = \boldsymbol{b}^{\mathrm{T}}(\boldsymbol{c} \times \boldsymbol{a})$ is used in the derivation of the third equation, and M denotes the total mass of the body. Equation (1.75) is an extended version of Equation (1.67) for the case of a rigid body with distributed mass.

1.9 Variational Principle and Lagrange's Equation

Given a system of particles or a system of rigid bodies like a robot arm or dual robot fingers grasping a rigid object, a set of physical variables that can conveniently express the position of the system are called generalised coordinates. A given set of generalised coordinates of the system is said to be complete, if, for any configuration of the system, the coordinates can specify the position corresponding to the configuration by appointing a set of corresponding values for the coordinates, and moreover for any different configuration the coordinates take different values. If any subset of a given set of generalised coordinates is fixed while the remaining coordinates can vary continuously with the configuration of the system, then the generalised coordinates are said to be independent. In general, the number of degrees of freedom of the system is defined as the number of physical variables of generalised coordinates that are complete and independent.

Given a system of particles or rigid bodies with complete and independent generalised coordinates $\boldsymbol{q} = (q_1, \cdots, q_n)^{\mathrm{T}}$, its equation of motion can be derived by the variational principle described as

$$\int_{t_1}^{t_2} \left[\delta(K - P) + \sum_{i=1}^{n} F_i \delta q_i \right] \mathrm{d}t = 0, \tag{1.76}$$

where K denotes the total kinetic energy of the system, P the potential energy, δ means to take a variation of $(K - U)$ in terms of $\boldsymbol{q}$, $\delta\boldsymbol{q}$ denotes an arbitrary vector-valued function of infinitesimally small increments satsifying $\delta\boldsymbol{q}(t_0) = 0$ and $\delta\boldsymbol{q}(t_1) = 0$. In general, F_i comes from the assumption that a force $\boldsymbol{f}_j$ acts at the corresponding point P_j expressed as the position vector $\boldsymbol{r}_j(\boldsymbol{q})$ for $j = 1, \cdots, m$ and the increment of the total work done by the force can be evaluated by

$$F_i = \sum_{j=1}^{n} \left(\frac{\partial \boldsymbol{r}_j}{\partial q_i} \right)^{\mathrm{T}} \boldsymbol{f}_j, \quad i = 1, \cdots, n, \tag{1.77}$$

$$\sum_{j=1}^{m} \boldsymbol{f}_i^{\mathrm{T}} \delta \boldsymbol{r}_j = \sum_{j=1}^{m} \sum_{i=1}^{n} \boldsymbol{f}_j^{\mathrm{T}} \frac{\partial \boldsymbol{r}_j}{\partial q_j} \delta q_i = \sum_{i=1}^{n} \left\{ \sum_{j=1}^{m} \left(\frac{\partial \boldsymbol{r}_j^{\mathrm{T}}}{\partial q_i} \right) \boldsymbol{f}_j \right\}^{\mathrm{T}} \delta q_i. \tag{1.78}$$

Lagrange's equation of motion for the system follows from the variational principle, *i.e.*, it follows from Equation (1.76) that

$$\frac{\mathrm{d}}{\mathrm{d}t} \left(\frac{\partial L}{\partial \dot{q}_i} \right) - \frac{\partial L}{\partial q_i} = F_i, \quad i = 1, \cdots, n, \tag{1.79}$$

where

$$L = K - P. \tag{1.80}$$

The scalar quantity L is called the Lagrangian. Equation (1.79) can be expressed in the vector form

$$\frac{\partial}{\mathrm{d}t} \left(\frac{\partial L}{\partial \dot{\boldsymbol{q}}} \right) - \frac{\partial L}{\partial \boldsymbol{q}} = \sum_{j=1}^{m} J_j^{\mathrm{T}}(\boldsymbol{q}) \boldsymbol{f}_j, \tag{1.81}$$

where $J_j^{\mathrm{T}}(\boldsymbol{q})$ signifies the transpose of the $3 \times n$ Jacobian matrix of $\boldsymbol{r}_j(\boldsymbol{q})$ with respect to the generalised position coordinates vector $\boldsymbol{q}$, *i.e.*,

$$J_j(\boldsymbol{q}) = \left(\frac{\partial \boldsymbol{r}_j}{\partial q_1}, \cdots, \frac{\partial \boldsymbol{r}_j}{\partial q_n} \right) = \frac{\partial \boldsymbol{r}_j}{\partial \boldsymbol{q}^{\mathrm{T}}}. \tag{1.82}$$

Throughout the book, vectors are expressed by columns and hence the gradient vector of the scalar L with respect to the column vector $\dot{q}$ or q is expressed

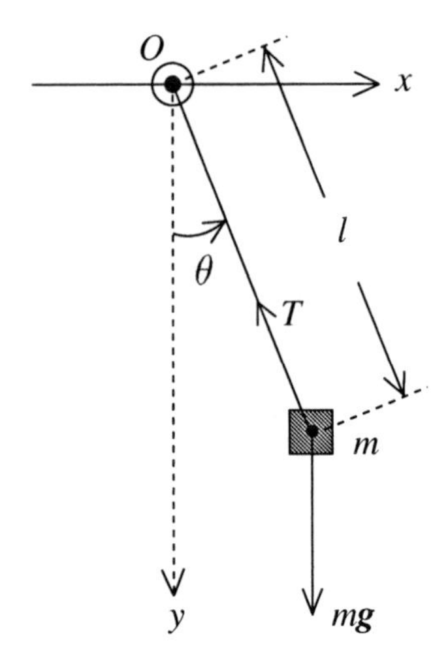

Fig. 1.16. A single-degree-of-freedom pendulum

by $\partial L/\partial \dot{\boldsymbol{q}}$ or $\partial L/\partial \boldsymbol{q}$, which is also a column vector. This rule is also applied for expressing Jacobian matrices of position vectors with respect to the generalised coordinates q of column vector as shown in Equation (1.82).

When the generalised position coordinate $\boldsymbol{q} = (q_1, \cdots, q_n)$ is complete but not independent, there may be a number of physical constraints, each of which can be described by an algebraic equation:

$$h_j(q_1, \cdots, q_n) = h_j(\boldsymbol{q}) = 0, \quad j = 1, \cdots, k. \tag{1.83}$$

In this case, Lagrange's equation of motion can be derived in a similar way to that of Equation (1.81) from Equation (1.76) by replacing $L = K - P$ with

$$L = K - P + \sum_{j=1}^{k} \lambda_j h_j(\boldsymbol{q}). \tag{1.84}$$

In fact, it follows that

$$\frac{\mathrm{d}}{\mathrm{d}t}\left(\frac{\partial L}{\partial \dot{\boldsymbol{q}}}\right) - \frac{\partial L}{\partial \boldsymbol{q}} = \sum_{j=1}^{k} \lambda_j \nabla_{hj}(\boldsymbol{q}) + \sum_{j=1}^{m} J_j^{\mathrm{T}}(\boldsymbol{q}) \boldsymbol{f}_j, \tag{1.85}$$

where ∇_{hj} means the gradient vector of h_j with respect to the column vector $\boldsymbol{q}$. The total number of DOFs of the system becomes $n - m$ in this case if all gradient vectors ∇_{hj} for $j = 1, \cdots, k$ are independent and it is possible to find $(n - m)$ physical variables from the original n variables $\boldsymbol{q} = (q_1, \cdots, q_n)^{\mathrm{T}}$ so that the $(n - m)$ variables are independent.

In order to understand the important meaning of algebraic constraint in relation to Lagrange's equation of motion, let us take a simple example. Consider a simple system of the pendulum shown in Figure 1.16. In this case, the kinetic energy and the potential energy are easily evaluated as follows:

$$K = \frac{m}{2}(\dot{x}^2 + \dot{y}^2), \quad P = mg(l - y), \tag{1.86}$$

where we have ignored the mass of a string with length l [m] from which the concentrated weight with mass m is hanging. The constraint equation can be described as

$$h(x, y) = \sqrt{x^2 + y^2} - l = 0, \tag{1.87}$$

which means that the centre of the weight mass is constrained on a circle with centre O and radius l (see Figure 1.16). Thus, the Lagrangian can be given in the following way:

$$\begin{aligned} L &= K - P + \lambda h(\boldsymbol{q}) \\ &= \frac{m}{2}(\dot{x}^2 + \dot{y}^2) - mg(l - y) + \lambda\left(\sqrt{x^2 + y^2} - l\right). \end{aligned} \tag{1.88}$$

If we use the Cartesian coordinates $\boldsymbol{q} = (x, y)^{\mathrm{T}}$ to denote the position of the mass centre of the weight as the generalised coordinates, then Lagrange's equation can easily be obtained from Equation (1.85), as

$$\begin{cases} m\ddot{x} = \lambda \dfrac{x}{\sqrt{x^2 + y^2}} = \lambda l^{-1} x, \\ m\ddot{y} - mg = \lambda \dfrac{y}{\sqrt{x^2 + y^2}} = \lambda l^{-1} y, \end{cases} \tag{1.89}$$

where the quantity $\sqrt{x^2 + y^2}$ is replaced with l according to the constraint equation (1.87). The pendulum system of Figure 1.17 superficially has two degrees of freedom if the generalised coordinates $\boldsymbol{q} = (x, y)$ is adopted. However, since the motion of the weight is subject to the constraint equation (1.87), the number of degrees of freedom of the system must be one [= 2 (variables) – 1 (constraint)].

If we express the position of the centre of mass of the weight by

$$\boldsymbol{r} = (x, y)^{\mathrm{T}} = (l \sin\theta, l \cos\theta)^{\mathrm{T}} \tag{1.90}$$

then the velocity and acceleration can be given in the following form:

$$\begin{cases} \boldsymbol{v} = \dot{\boldsymbol{r}} = l\dot{\theta}(\cos\theta, -\sin\theta)^{\mathrm{T}}, \\ \boldsymbol{a} = \dot{\boldsymbol{v}} = \ddot{\boldsymbol{r}} = l\ddot{\theta}(\cos\theta, -\sin\theta)^{\mathrm{T}} - l\dot{\theta}^2(\sin\theta, \cos\theta)^{\mathrm{T}}. \end{cases} \tag{1.91}$$

Hence, the tensile force of the string (denoting its magnitude by T) and the gravity force affecting the weight are expressed as (see Figure 1.16)

$$\boldsymbol{f} = m(0, g)^{\mathrm{T}} = m\boldsymbol{g}, \quad \boldsymbol{T} = -T(\sin\theta, \cos\theta)^{\mathrm{T}}. \tag{1.92}$$

Thus, Newton's second law of motion $m\boldsymbol{a} = \boldsymbol{f}$ can be described as

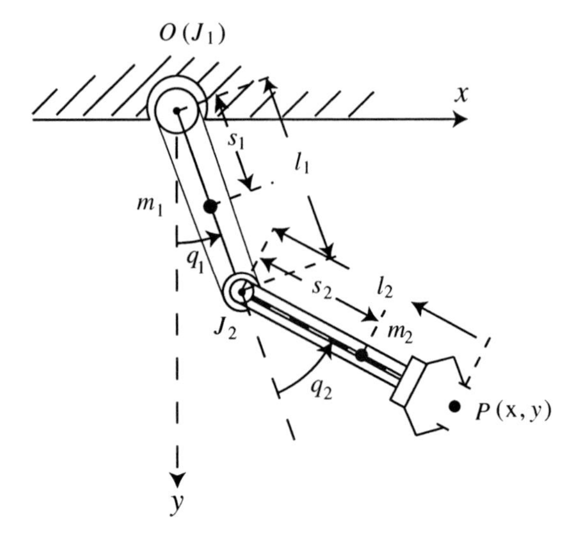

Fig. 1.17. A pendulum-type robot arm with two degrees of freedom

$$ml\ddot{\theta}\begin{pmatrix}\cos\theta\\ -\sin\theta\end{pmatrix} - ml\dot{\theta}^2\begin{pmatrix}\sin\theta\\ \cos\theta\end{pmatrix} = m\begin{pmatrix}0\\ g\end{pmatrix} - T\begin{pmatrix}\sin\theta\\ \cos\theta\end{pmatrix}. \tag{1.93}$$

Taking the inner product of this equation by the vector $(\cos\theta, -\sin\theta)^{\mathrm{T}}$ yields

$$ml\ddot{\theta} = -mg\sin\theta \tag{1.94}$$

from which we obtain the equation of motion as follows:

$$\ddot{\theta} + (g/l)\sin\theta = 0. \tag{1.95}$$

On the other hand, if we substitute $x = l\sin\theta$, $y = l\cos\theta$ into Equation (1.89), we have

$$ml\ddot{\theta}\begin{pmatrix}\cos\theta\\ -\sin\theta\end{pmatrix} - ml\dot{\theta}^2\begin{pmatrix}\sin\theta\\ \cos\theta\end{pmatrix} - m\begin{pmatrix}0\\ g\end{pmatrix} = \lambda\begin{pmatrix}\sin\theta\\ \cos\theta\end{pmatrix}. \tag{1.96}$$

Taking the inner product of this equation with the vector $(\cos\theta, -\sin\theta)^{\mathrm{T}}$ yields

$$ml\ddot{\theta} + mg\sin\theta = 0, \tag{1.97}$$

which is equivalent to Equation (1.94). If we take the inner product of Equation (1.96) with the vector $(\sin\theta, \cos\theta)^{\mathrm{T}}$, then we obtain

$$\lambda = -ml\dot{\theta}^2 - mg\cos\theta. \tag{1.98}$$

This shows that λ as a Lagrange multiplier can be regarded as $-T$, *i.e.*, the reaction force to the tensile force of the string.

Next, consider the two-DOF planar robot arm shown in Figure 1.17, which consists of two rigid bodies connected at the hinge joint J_2. The first joint J_1 is fixed at the origin O of the frame coordinates and pivots around the z-axis, which is perpendicular to the xy-plane. All the physical parameters of the rigid bodies are specified in Figure 1.17. In order to derive Lagrange's equation of motion for the system, we first evaluate the total kinetic energy of the system. The kinetic energy K_1 of the first link is easily obtained from Equation (1.75) as

$$K_1 = \frac{1}{2}\left(m_1 s_1^2 + I_{1z}\right)\dot{q}_1^2, \tag{1.99}$$

where s_1 denotes the distance from J_1 to the centre of mass of the first rigid body and I_{1z} the inertia moment around the z-axis through its centre of mass. For the sake of evaluating the kinetic energy of the second rigid link connected with the first link at joint J_2, let us denote the position of its mass centre by $\boldsymbol{r}_{2c}$, which is given by

$$\boldsymbol{r}_{2c} = \left(l_1 \sin q_1 + s_2 \sin(q_1+q_2),\ l_1 \cos q_1 + s_2 \cos(q_1+q_2)\right)^{\mathrm{T}}. \tag{1.100}$$

Hence, its velocity becomes

$$\begin{aligned}\boldsymbol{v}_{2c} = \dot{\boldsymbol{r}}_{2c} = (&l_1\dot{q}_1 \cos q_1 + s_2(\dot{q}_1+\dot{q}_2)\cos(q_1+q_2),\\ &-l_1\dot{q}_1 \sin q_1 - s_2(\dot{q}_1+q_2)\sin(q_1+q_2))^{\mathrm{T}}.\end{aligned} \tag{1.101}$$

Then, it is easy to see that

$$\|\boldsymbol{v}_{2c}\|^2 = \boldsymbol{v}_{2c}^{\mathrm{T}}\boldsymbol{v}_{2c} = l_1\dot{q}_1^2 + s_2^2(\dot{q}_1+\dot{q}_2)^2 + 2l_1 s_2\dot{q}_1(\dot{q}_1+\dot{q}_2)\cos q_2. \tag{1.102}$$

Thus, by referring to Equation (1.75), we obtain the kinetic energy of the second link as follows:

$$\begin{aligned}K_2 = &\frac{1}{2}m_2\left\{l_1^2 + \dot{q}_1^2 + s_2^2(\dot{q}_1+\dot{q}_2)^2 + 2l_1 s_2\dot{q}_1(\dot{q}_1+\dot{q}_2)\cos q_2\right\}\\ &+\frac{1}{2}I_{2z}(\dot{q}_1+\dot{q}_2)^2,\end{aligned} \tag{1.103}$$

where I_{2z} signifies the moment of inertia of the second link about the z-axis through the centre of mass. On the other hand, the total potential energy can be evaluated in the following way:

$$P = g\left\{(m_1 s_1 + m_2 l_1)(1-\cos q_1) + m_2 s_2(1-\cos(q_1+q_2))\right\}. \tag{1.104}$$

If we denote by u_i the driving torque at joint i generated by its corresponding joint actuator, the variational principle can be expressed as

$$\int_{t_0}^{t_1}(\delta L + u^{\mathrm{T}}\delta q)\,\mathrm{d}t = 0, \tag{1.105}$$

where $L = K_1 + K_2 - P$, $u = (u_1, u_2)^{\mathrm{T}}$ and $q = (q_1, q_2)^{\mathrm{T}}$. Then, the Lagrange equation follows from Equation (1.105) in the following way:

$$\begin{aligned} H(q)\ddot{q} - m_2 l_1 s_2 \begin{pmatrix} 2\dot{q}_1\dot{q}_2 + \dot{q}_2^2 \\ -\dot{q}_1^2 \end{pmatrix} \sin q_2 \\ + g \begin{pmatrix} (m_1 s_1 + m_2 l_1)\sin q_1 + m_2 s_2 \sin(q_1 + q_2) \\ m_2 s_2 \sin(q_1 + q_2) \end{pmatrix} = \begin{pmatrix} u_1 \\ u_2 \end{pmatrix}, \end{aligned} \tag{1.106}$$

where

$$H(q) = \begin{pmatrix} I_{J1} + m_2 l_1^2 + I_{J2} + 2m_2 l_1 s_2 \cos q_2 & I_{J2} + m_2 l_1 s_2 \cos q_2 \\ I_{J2} + m_2 l_1 s_2 \cos q_2 & I_{J2} \end{pmatrix} \tag{1.107}$$

and, for simplicity, we put

$$I_{J1} = \frac{m_1}{2} s_1^2 + I_{c1}, \quad I_{J2} = \frac{m_2}{2} s_2^2 + I_{c2}. \tag{1.108}$$

Here, it is interesting to note that I_{Ji} signifies the moment of inertia of the ith rigid link about the joint i, and $K = K_1 + K_2 = (1/2)\dot{q}^{\mathrm{T}} H(q)\dot{q}$. More interestingly, the second term of the left-hand side of Equation (1.106) can be expressed as the sum of $(1/2)\dot{H}(q)\dot{q}$ and $S(q, \dot{q})\dot{q}$ with a skew-symmetric matrix $S(q, \dot{q})$. This can be verified as follows (in detail, see [1-27]):

$$\begin{aligned} &-\frac{1}{2}\dot{H}(q)\dot{q} - m_2 l_1 s_2 \begin{pmatrix} 2\dot{q}_1\dot{q}_2 + \dot{q}_2^2 \\ -\dot{q}_1^2 \end{pmatrix} \sin q_2 \\ &= m_2 l_1 s_2 \begin{pmatrix} \dot{q}_2 & \frac{1}{2}\dot{q}_2 \\ \frac{1}{2}\dot{q}_2 & 0 \end{pmatrix} \begin{pmatrix} \dot{q}_1 \\ \dot{q}_2 \end{pmatrix} \sin q_2 - m_2 l_1 s_2 \begin{pmatrix} 2\dot{q}_1\dot{q}_2 + \dot{q}_2^2 \\ -\dot{q}_1^2 \end{pmatrix} \sin q_2 \\ &= m_2 l_1 s_2 \sin q_2 \begin{pmatrix} -\dot{q}_1\dot{q}_2 - \frac{1}{2}\dot{q}_2^2 \\ \frac{1}{2}\dot{q}_1\dot{q}_2 + \dot{q}_1^2 \end{pmatrix} \\ &= \frac{m_2 l_1 s_2 (2\dot{q}_1 + \dot{q}_2)}{2} \sin q_2 \begin{pmatrix} 0 & -1 \\ 1 & 0 \end{pmatrix} \begin{pmatrix} \dot{q}_1 \\ \dot{q}_2 \end{pmatrix} = S(q, \dot{q})\dot{q}. \end{aligned} \tag{1.109}$$

Thus, from this expression, Equation (1.106) can be rewritten as

$$H(q)\ddot{q} + \left\{ \frac{1}{2}\dot{H}(q) + S(q, \dot{q}) \right\} \dot{q} + g(q) = u, \tag{1.110}$$

where $g(q) = \partial P / \partial q$, *i.e.*, $g(q)$ expresses the gradient vector of the potential function with respect to the joint angle vector q. Note that $S(q, \dot{q})$ is linear and homogeneous in $\dot{q}$ and therefore the second term on the left-hand side of Equation (1.110) is quadratic in components of $\dot{q}$ with coefficients of sinusoidal functions of $\dot{q}$.

One of the most important advantages of such an expression of the Lagrange equation as in Equation (1.110) is that each term on the left-hand side retains its own physical meaning. The first term is called the inertia term and the third the potential term. Furthermore, the term $S(q, \dot{q})$ can be regarded as a gyroscopic term because it is irrelevant to energy consumption and conservation. Since the inner product of $H(q)\ddot{q}+(1/2)\dot{H}(q)\dot{q}$ with $\dot{q}$ yields

$$\dot{q}^{\mathrm{T}}\left\{H(q)\ddot{q}+(1/2)\dot{H}(q)\dot{q}\right\}=\frac{\mathrm{d}}{\mathrm{d}t}\left\{\frac{1}{2}\dot{q}^{\mathrm{T}}H(q)\dot{q}\right\} \tag{1.111}$$

it is easy to see, taking inner product of Equation (1.110) with $\dot{q}$, that

$$\frac{\mathrm{d}}{\mathrm{d}t}\left\{\frac{1}{2}\dot{q}^{\mathrm{T}}H(q)\dot{q}+P(q)\right\}=\dot{q}^{\mathrm{T}}u \tag{1.112}$$

from which it follows that

$$\int_0^t \dot{q}^{\mathrm{T}}(\tau)u(\tau)\,\mathrm{d}\tau = E(t)-E(0), \tag{1.113}$$

where

$$E(t)=\frac{1}{2}\dot{q}^{\mathrm{T}}(t)H(q(t))\dot{q}(t)+P(q(t)). \tag{1.114}$$

Since in this example we take the potential function $P(q)$ to be positive definite in q and zero if and only if $q_1 = q_2 = 0$, it follows from Equation (1.113) that

$$\int_0^t \dot{q}^{\mathrm{T}}(\tau)u(\tau)\,\mathrm{d}\tau \geq -E(0) \tag{1.115}$$

for any $t \geq 0$. This property is called passivity of the system with input u and output $\dot{q}$. If there is no external torque, *i.e.*, $u = 0$, then Equation (1.113) implies

$$E(t)=\frac{1}{2}\dot{q}^{\mathrm{T}}H(q)\dot{q}+P(q)=\text{const.} \tag{1.116}$$

This relation is nothing but the law of conservation of the mechanical energy as discussed in Equation (1.36).

2

Stability of Grasping in a Static or Dynamic Sense

The first part of this chapter studies the geometry of grasping or immobilisation of a solid object by a number of frictionless fingers or fixtures. It shows that at least four frictionless contact points or four fixtures are required to immobilise planar objects. In particular, we show that three contact points are necessary and sufficient for immobilising a two-dimensional (2-D) triangular object but that four frictionless contacts or four fixtures are necessary and sufficient to immobilise a parallelepiped or to establish force/torque closure grasp. In the case of 3-D polyhedra, seven frictionless contact points are sufficient to establish a force/torque closure grasp.

The latter half of this chapter addresses another type of problem of grasping or immobilisation of a 2-D rigid object, in a dynamic sense: the simplest but most fundamental problem for stopping or immobilising rotational motion of a 2-D object with a flat side surface by a multi-joint robot finger where the object can only pivot around a single axis. It is assumed that rotational motion of the object pivoted around the fixed axis is frictionless and the finger-end is hemispherical and therefore rolling between the finger-end and object surfaces is induced without incurring any slip. Lagrange's equation of motion of such a testbed finger/object system is derived together with two constraints, the point contact constraint and the rolling contact constraint. It is shown that there arises a rolling constraint force tangential to both the finger-end sphere and the object surface originates at the contact point. Another simple testbed problem of dynamic grasping of a 2-D object by two one-DOF fingers with spherical ends is proposed, where motion of the overall system is confined to the horizontal plane. Rolling contacts play an essential role in stabilisation of dynamic grasping through force/torque balance. In the final section, a class of coordinated motor-control signals based on fingers–thumb opposition is shown to establish force/torque balance in a blind manner without knowing the object kinematics or using visual or tactile sensing.

2.1 Immobilisation of 2-D Objects

The problem of achieving a firm grip on an object is one of the most fundamental issues underlying the design and control of multi-fingered hands. There are two approaches for defining the motion of a firm grip; form closure and immobilisation. In this section, we deal with the problem of immobilising a planar polygonal object. In the case of three-dimensional polyhedral objects, we only show theoretical results without giving the proof.

Given a planar shape P in the horizontal plane, a set of points S is said to immobilise P if any rigid motion of P in the plane forces at least one point of S to penetrate the interior of P. By shape we mean a two-dimensional set of points bounded by a Jordan curve. Evidently, any minimal S contains only points belonging to the boundary of P. For the sake of gaining physical intuition into the problem, we treat only polygonal objects (therefore, the disc is excluded from this consideration).

For example, consider the parallelepiped shown in Figure 2.1 and choose four points on the boundary as $S = \{P_1, P_2, P_3, P_4\}$. Apparently the set S immobilises this parallelepiped. However, if any one of points in S is excluded, then the set $S' = \{P_{i1}, P_{i2}, P_{i3}\}$ consisting of the remaining points does not immobilise the object. Indeed, for example, the set $S' = \{P_1, P_2, P_3\}$, any movement of P to the right does not make any point of S' penetrate the interior of P (that is, any fixed point P_i cannot get inside P by a certain infinitesimally small translational movement of P). Physically, a boundary point P_i of the set S for a shape P can be regarded as a contact with the boundary of P externally made by a rigid body, which is called a fixture (mainly, in the case of immobilisation) and a frictionless finger (in the case of form closure).

Next consider the problem of how many fixtures are necessary and sufficient to immobilise triangles. In general, four fixtures are enough to immobilise polygonal objects. We show the solution to the problem below.

Theorem 2.1. Three fixtures are necessary and sufficient to immobilise any triangle.

To find a set of such contact points $S = \{P_1, P_2, P_3\}$ for the triangle P, let us consider the maximally inscribed circle of P [see Figure 2.1(b)] and choose $S = \{P_1, P_2, P_3\}$ by picking every point at which the inscribed circle touches the boundary of P. Intuitively this set $S = \{P_1, P_2, P_3\}$ seems to immobilise the triangle P. Nevertheless, the proof is not trivial.

To confirm this, first observe that each contact point should lie on a different side of the triangle. Next, the three orthogonal lines to the boundary at the points P_1, P_2, and P_3 should meet at a common point. To show this, suppose that the three othogonal lines do not meet at a single point. Then, these three orthogonals constitute a triangle as shown in Figure 2.2, where we denote the original triangle by its vertices $\{A, B, C\}$. Let O be a point in the interior of the triangle constituted by the three orthogonals. Then, the three angles $\angle OP_1B$, $\angle OP_2C$ and $\angle OP_3A$ are all acute, as shown in Figure 2.2.

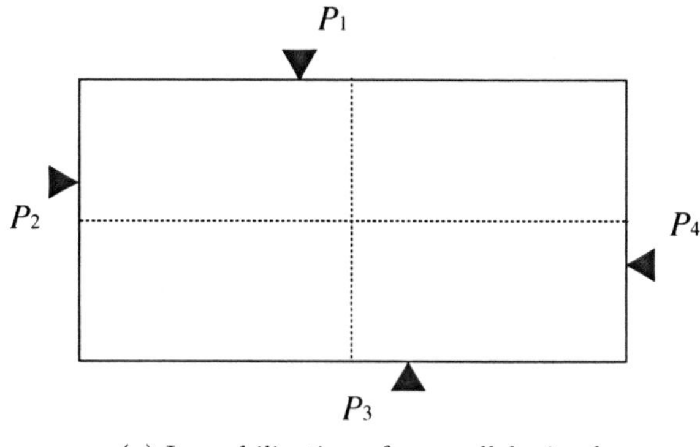

(a) Immobilisation of a parallelepiped

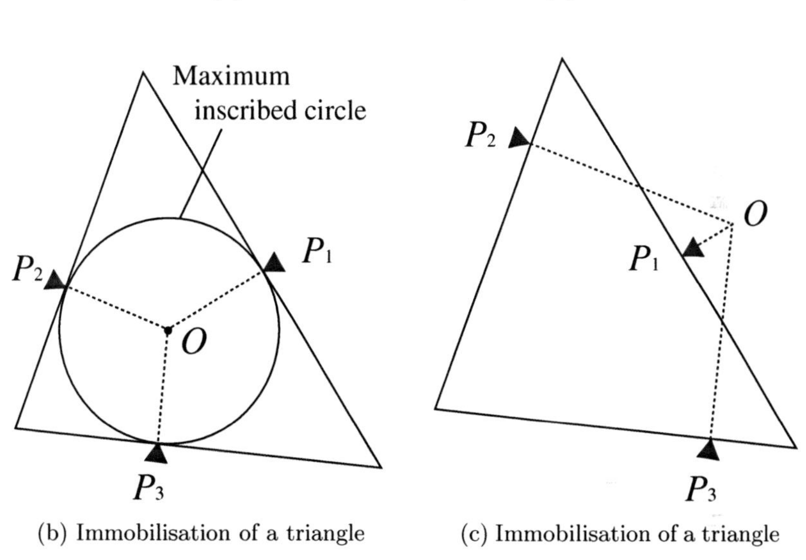

(b) Immobilisation of a triangle

(c) Immobilisation of a triangle

Fig. 2.1. Three fixtures are necessary and sufficient to immobilise triangles. In general, four fixtures are enough to immobilise polygonal objects.

Or, for some choices for $S = \{P_1, P_2, P_3\}$, all three of these angles become obtuse. Therefore, in the case shown in Figure 2.2, the triangle $\{A, B, C\}$ can be rotated counterclockwise by a small angle around O and the points $\{P_1, P_2, P_3\}$ remain outside the interior of $\{A, B, C\}$. In the latter case, the triangle $\{A, B, C\}$ can be rotated clockwise by a small angle. In either case, the existence of such a rotation around O contradicts immobilisation of the triangle. It is also possible to prove that the concurrency of the three orthogonals at the contact points $\{P_1, P_2, P_3\}$ is also a sufficient condition for immobilisation of the triangle. However, we omit the proof of this sufficiency but state this main result in the following theorem.

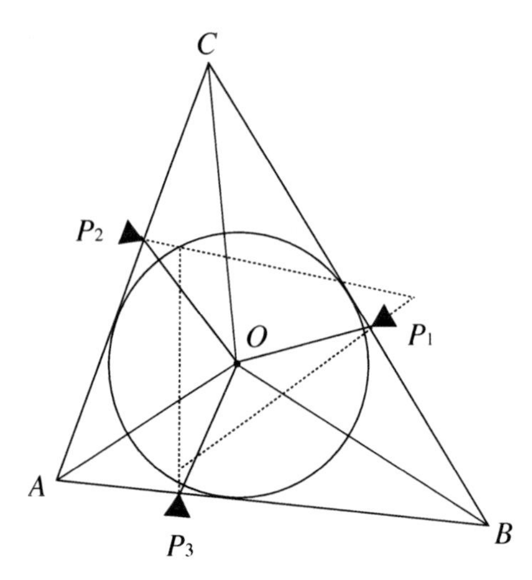

Fig. 2.2. The set $S = \{P_1, P_2, P_3\}$ cannot immobilise this triangle

Theorem 2.2. A necessary and sufficient condition for immobilising a triangle $\{A, B, C, \}$ by the contact points $S = \{P_1, P_2, P_3\}$ is that the three orthogonals at P_i $(i = 1, 2, 3)$ to the corresponding sides meet at a common single point.

Finally, we mention the following two results without proof.

Theorem 2.3. Any polygonal object in the plane can always be immobilised by using four fixtures (contact points).

Theorem 2.4. Any polygonal object containing no parallel sides can be immobilised by finding three fixtures.

Further, we summarize the theoretical results about immobilisation of polyhedra obtained and proved in the literature in the following two theorems.

Theorem 2.5. Any three-dimensional polyhedron can be immobilised by using six fixtures.

Theorem 2.6. Any n-dimensional polytope can be immobilised by using $2n$ fixtures.

2.2 Force/Torque Closure

Another concept of a firm grip on a rigid object is form closure (equivalently called force/torque closure), which is a finite set of wrench vectors (force–moment combinations) applied on the object with the property that any other wrench vector acting on the object can be balanced by a positive combination

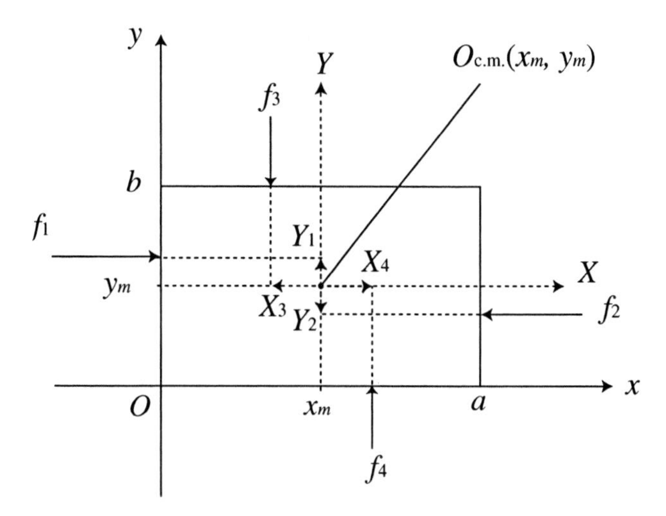

Fig. 2.3. The set of four forces $\{f_1, f_2, f_3, f_4\}$ directed normally to the four sides of the parallelepiped, respectively, achieves force/torque closure

of the original ones. It has already been pointed out that the form closure of a two-dimensional object requires at least four wrench vectors and that of a three-dimensional object requires at least seven wrench vectors. It is also known that these numbers can be achieved by wrench vectors realisable as forces normal to the surface of the object. Such wrench vectors are equivalent to the supposition of frictionless fingers contacting pointwise to the object surface. To gain physical insight into the concept of form closure, we discuss first a problem of firm grip of a parallelepiped as an illustrative example of two-dimensional objects.

Consider now a rigid parallelepiped lying on a horizontal xy-plane with four sides as shown in Figure 2.3, and suppose that four forces $\boldsymbol{f}_i$ $(i = 1, 2, 3, 4)$ are acting on different sides as in the figure. If we denote the magnitude of the force vector $\boldsymbol{f}_i$ by the positive number f_i and set the coordinates (x, y) as shown in Figure 2.3, the forces $\boldsymbol{f}_i$ $(i = 1, \cdots, 4)$ can be expressed by the vectors $\boldsymbol{f}_1 = (f_1, 0)^{\mathrm{T}}$, $\boldsymbol{f}_2 = (-f_2, 0)^{\mathrm{T}}$, $\boldsymbol{f} = (0, -f_3)^{\mathrm{T}}$ and $\boldsymbol{f}_4 = (0, f_4)^{\mathrm{T}}$. When the force $\boldsymbol{f}_1$ is exerted on the object, the rotational moment with a magnitude $|f_1 Y_1|$ arises clockwise around the object mass centre $O_{\mathrm{c.m.}}$ as shown in Figure 2.3. In the figure, all variables Y_i $(i = 1, 2)$ and X_i $(i = 3, 4)$ are defined. If we append such a rotational moment $f_i Y_i$ $(i = 1, 2)$ or $f_i X_i$ $(i = 3, 4)$ to $\boldsymbol{f}_i$ by letting the sign of a counter-clockwise moment be positive, then it is possible to consider the following four three-dimensional vectors:

$$\boldsymbol{w}_1 = \begin{pmatrix} f_1 \\ 0 \\ -f_1 Y_1 \end{pmatrix}, \quad \boldsymbol{w}_2 = \begin{pmatrix} -f_2 \\ 0 \\ f_2 Y_2 \end{pmatrix}, \quad \boldsymbol{w}_3 = \begin{pmatrix} 0 \\ -f_3 \\ -f_3 X_3 \end{pmatrix}, \quad \boldsymbol{w}_4 = \begin{pmatrix} 0 \\ f_4 \\ f_4 X_4 \end{pmatrix}. \tag{2.1}$$

Such a vector $\boldsymbol{w}_i$ is called a two-dimensional wrench vector. In the three-dimensional case, such a wrench vector is expressed by the six-dimensional vector $\boldsymbol{w} = (\boldsymbol{f}, \boldsymbol{r} \times \boldsymbol{f})^{\mathrm{T}}$, where $\boldsymbol{f}$ stands for the three-dimensional force vector acting at a contact point normally to the object boundary surface and $\boldsymbol{r}$ is the position vector originating from the object mass centre and terminating at the contact point. The symbol $\times$ means the vector outer product.

Definition 2.1. Suppose that n frictionless fingers are applied to a rigid object at different points with n wrench vectors $W = \{\boldsymbol{w}_1, \cdots, \boldsymbol{w}_n\}$. If any external wrench $\boldsymbol{w}_{\mathrm{ex}}$ applied to the object can be balanced by pressing fingertips against the object at the selected contact points, the grasp with the set W of wrench vectors is said to be form-closure (or force/torque closure).

In more detail, the grasp W is form-closure if and only if for any external wrench $\boldsymbol{w}_{\mathrm{ex}}$ it is possible to find a set of non-negative parameters $\alpha_i \geq 0$ $(i = 1, \cdots, n)$ that satisfy

$$\sum_{i=1}^{n} \alpha_i \boldsymbol{w}_i + \boldsymbol{w}_{ex} = 0. \tag{2.2}$$

This formula expresses the fact that external wrench $\boldsymbol{w}_{\mathrm{ex}}$ can be balanced by the set of original fingers through modifying the magnitudes of the forces f_i to $\alpha_i f_i$ $(i = 1, \cdots, n)$.

For example, in the case of the 2-D object shown in Figure 2.3, the set W of four wrench vectors given in Equation (2.1) achieves force/torque closure if f_i $(i = 1, \cdots, 4)$ satisfies

$$f_1 = f_2 > 0, \quad f_3 = f_4 > 0, \quad f_1 Y_1 = f_3 X_3 \tag{2.3}$$

provided that

$$Y_1 + Y_2 = 0, \quad X_3 + X_4 = 0. \tag{2.4}$$

This can be explicitly confirmed by the following argument.

For a given set of wrench vectors $W = \{\boldsymbol{w}_1, \cdots, \boldsymbol{w}_n\}$, let us consider the normalised set of wrenches $\bar{W} = \{\bar{\boldsymbol{w}}_1, \cdots, \bar{\boldsymbol{w}}_n\}$, where $\bar{\boldsymbol{w}}_i = \boldsymbol{w}_i / \|\boldsymbol{w}_i\|$, $i = 1, \cdots, n$, and $\|\boldsymbol{w}\|$ denotes the Euclidean norm of vectors $\boldsymbol{w}$. Then, consider a set of all points P such that

$$P = \sum_{i=1}^{n} \gamma_i \boldsymbol{w}_i, \quad \sum_{i=1}^{n} \gamma_i = 1, \quad \gamma_i \geq 0 \quad (i = 1, \cdots, n) \tag{2.5}$$

and denote the set by $H(W)$, and call it the convex hull of the set of original wrench vectors W.

Theorem 2.7. A necessary and sufficient condition for a grasp with a set of wrench vectors W to reach form-closure is that the origin of the wrench space lies exactly inside the convex hull $H(W)$ of the original set W of wrench vectors.

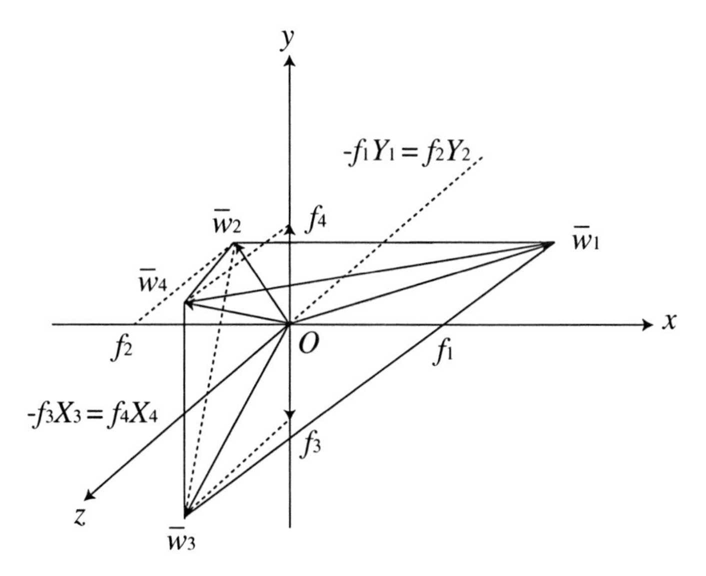

Fig. 2.4. The convex hull $H(\bar{W})$ composed of four normalised wrenches $\bar{W} = \{\bar{\boldsymbol{w}}_1, \bar{\boldsymbol{w}}_2, \bar{\boldsymbol{w}}_3, \bar{\boldsymbol{w}}_4\}$ includes the origin O as an interior point

In the case of the parallelepiped shown in Figure 2.3 with wrench vectors $W = \{\boldsymbol{w}_1, \boldsymbol{w}_2, \boldsymbol{w}_3, \boldsymbol{w}_4\}$ satisfying Equations (2.3) and (2.4), the convex hull of W can easily be constructed as shown in Figure 2.4, from which it is possible to see that the origin O is exactly inside $H(W)$. This shows that the grasp with this W achieves force/torque closure.

Straightforwardly from Theorem 2.2, it follows that

Theorem 2.8. The form closure of a two-dimensional object requires at least four wrenches and that of a three-dimensional object requires at least seven wrenches.

Theorem 2.9. Form closure of any two-dimensional bounded object (except a circle) can be achieved by four frictionless fingers. For three dimensions, form closure of three-dimensional objects with rotational symmetries can be achieved with seven frictionless fingers.

It is also shown that

Theorem 2.10. For any convex polygon, four contact points $S = \{P_1, \cdots, P_4\}$ at which the set of wrenches achieves form closure can immobilise a polygonal object.

It is interesting to note that any three wrench vectors constructed at three contact points $S = \{P_1, P_2, P_3\}$ of a triangle shown in Figure 2.1(b) cannot achieve form closure, even if the three orthogonals meet at a common point (that is, S immobilises the triangle).

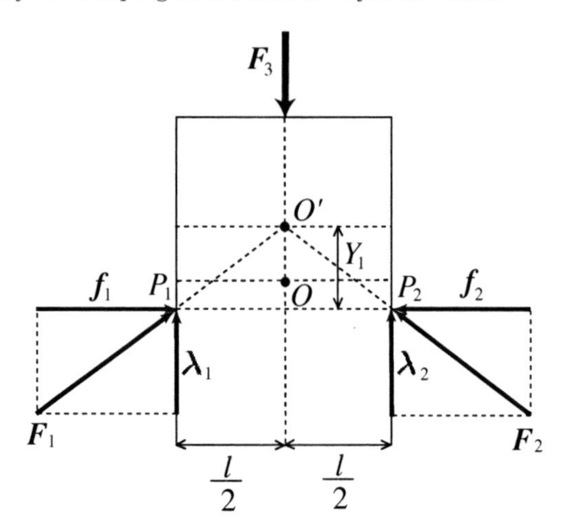

Fig. 2.5. Frictional fingers press the object against the sides in the directions of $\boldsymbol{F}_1$ and $\boldsymbol{F}_2$ due to the existence of the frictional forces $\boldsymbol{\lambda}_1$ and $\boldsymbol{\lambda}_2$ tangent to the object surface

2.3 Frictional Grasp of 2-D Objects

In our everyday life, we pick up small objects on a desk quite easily by using a thumb and index finger without dropping them. As a matter of course, our thumb and fingers are not rigid but soft and flexible. In fact, our finger-end has a curved surface and makes contact with a rigid object with some contacted area caused by deformation of the finger-end soft material. Then, there may arise a frictional force in the direction tangent to the contact area due to both static and viscous (dynamic) friction. Even if the finger-end is rigid but has a curved surface, it may cause rolling on the object surface without sliding or slipping. In the next section, we will show that such rolling constraint induces a constraint force in the direction tangential to both the finger-end and object surfaces. This tangential force is a constraint force and hence is irrelevant to energy consumption during motion, unlike the viscous friction.

In this section, we treat a force/torque closure problem for a 2-D rectangular object by using a pair of frictional fingers. Two frictional forces $\boldsymbol{F}_1$ and $\boldsymbol{F}_2$ acting at points P_1 and P_2, respectively, against the object can be regarded as a sum of components $\boldsymbol{f}_i$ normal and $\boldsymbol{\lambda}_i$ tangent to the object side $(i = 1, 2)$, as shown in Figure 2.5. Suppose that two straight lines drawn from P_1 and P_2 in the directions of $\boldsymbol{F}_1$ and $\boldsymbol{F}_2$, respectively, meet at a point O'. In other words, assume that the direction of $\boldsymbol{f}_2$ is just opposite to that of $\boldsymbol{f}_1$ and the straight line including the line segment $\overline{OO'}$ (where O denotes the centre of the rectangular) splits the rectangular into two parts with the same shape and size. If we regard the point O' as the origin of (x, y)-coordinates as in Figure 2.3, then the vectors $\boldsymbol{f}_i$ and $\boldsymbol{\lambda}_i$ $(i = 1, 2)$ can be expressed in the following way:

$$\boldsymbol{f}_1 = f_1 \begin{pmatrix} 1 \\ 0 \end{pmatrix}, \quad \boldsymbol{f}_2 = f_2 \begin{pmatrix} -1 \\ 0 \end{pmatrix}, \quad \boldsymbol{\lambda}_1 = \lambda_1 \begin{pmatrix} 0 \\ -1 \end{pmatrix}, \quad \boldsymbol{\lambda}_2 = \lambda_2 \begin{pmatrix} 0 \\ -1 \end{pmatrix}, \tag{2.6}$$

where all the f_i and λ_i $(i = 1, 2)$ are positive constants. Correspondingly to these four force vectors, the wrench vectors are given as follows:

$$\begin{aligned} \boldsymbol{w}_{f1} &= f_1 \begin{pmatrix} 1 \\ 0 \\ Y_1 \end{pmatrix}, \quad \boldsymbol{w}_{f2} = f_2 \begin{pmatrix} -1 \\ 0 \\ -Y_1 \end{pmatrix}, \\ \boldsymbol{w}_{\lambda 1} &= \lambda_1 \begin{pmatrix} 0 \\ -1 \\ -l/2 \end{pmatrix}, \quad \boldsymbol{w}_{\lambda 2} = \lambda_2 \begin{pmatrix} 0 \\ -1 \\ l/2 \end{pmatrix}. \end{aligned} \tag{2.7}$$

Evidently, the set of wrenches $W = \{\boldsymbol{w}_{f1}, \boldsymbol{w}_{f2}, \boldsymbol{w}_{\lambda 1}, \boldsymbol{w}_{\lambda 2}\}$ does not realise force/torque closure, because the sum of these four wrenches cannot be balanced except for the case $f_1 = f_2 = 0$ and $\lambda_1 = \lambda_2 = 0$. That is, the origin cannot be included inside $H(W)$, the convex hull of W. Then, let us consider the situation that the external force $\boldsymbol{F}_3$ is exerted on the rectangular object in the direction of the y-axis, down from the top, as shown in Figure 2.5. If this 2-D object with mass M is placed in a vertical plane and subjected to gravity, this external force $\boldsymbol{F}_3$ can be regarded as the gravity force expressed by $\boldsymbol{F}_3 = Mg(0, 1)^{\mathrm{T}}$. Then it is easy to see that the set of wrenches W and this additional wrench $\boldsymbol{w}_{F3} = Mg(0, 1, 0)^{\mathrm{T}}$ can be balanced by setting

$$f_1 = f_2 > 0, \quad \lambda_1 = \lambda_2 = Mg/2, \tag{2.8}$$

which yields

$$\sum_{i=1,2} (\boldsymbol{w}_{fi} + \boldsymbol{w}_{\lambda i}) + \boldsymbol{w}_{F3} = 0. \tag{2.9}$$

It is interesting to note that the set of five wrenches $W' = \{\boldsymbol{w}_{f1}, \boldsymbol{w}_{f2}, \boldsymbol{w}_{\lambda 1}, \boldsymbol{w}_{\lambda 2}, \boldsymbol{w}_{F3}\}$ achieves force/torque closure, because the convex hull $H(W')$ apparently includes the origin as an interior point. In fact, suppose that a small external force with wrench $\boldsymbol{w}_d = (\varepsilon_x, \varepsilon_y, \varepsilon_m)^{\mathrm{T}}$ is exerted on the rectangular object. In order to balance $\boldsymbol{w}_d$ by the set W', it is necessary and sufficient to choose $f_i > 0$ and $\lambda_i > 0$ while fixing the value Mg so that

$$\sum_{i=1,2} (\boldsymbol{w}_{fi} + \boldsymbol{w}_{\lambda i}) + \boldsymbol{w}_{F3} + \boldsymbol{w}_d = 0. \tag{2.10}$$

This equality can be satisfied by choosing $f_i > 0$ and $\lambda_i > 0$ so that they satisfy

$$\begin{cases} f_1 - f_2 = -\varepsilon_x, \quad \lambda_1 + \lambda_2 = Mg + \varepsilon_y \\ -\varepsilon_x Y_1 - l\lambda_1 + \dfrac{l}{2}(Mg + \varepsilon_y) + \varepsilon_m = 0 \end{cases} \tag{2.11}$$

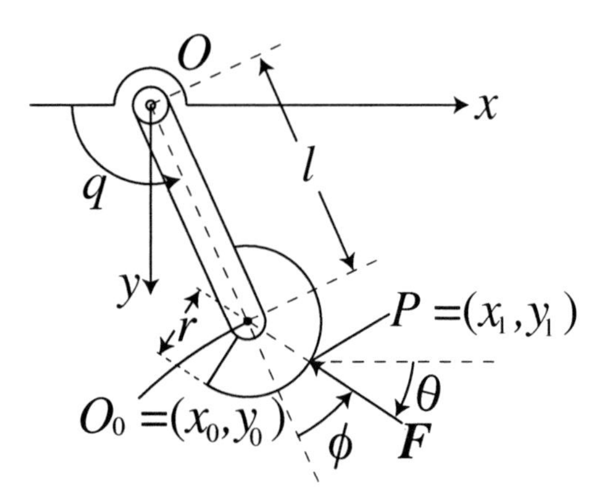

Fig. 2.6. A single-joint finger robot receives an external force $\boldsymbol{F}$ in the direction from P to O_0

from which it follows that

$$\begin{cases} f_i = f_d + (-1)^i \dfrac{\varepsilon_x}{2} \\ \lambda_i = \dfrac{1}{2}(Mg + \varepsilon_y) + (-1)^i \dfrac{\varepsilon_x Y_1 - \varepsilon_m}{l} \end{cases} \quad i = 1, 2. \tag{2.12}$$

where f_d is a certain appropriate positive constant. Conversely, it is easy to check that if f_i and λ_i are chosen by Equation (2.12) then the force/torque balance expressed by Equation (2.10) is realized. The arguments imply a potential way of controlling the grasp of a 2-D polygonal object by using a pair of fingers that exert not only pressing forces on the object normal to the object surface but also frictional (or rolling constraint) forces tangent to it.

2.4 Rolling Contact Constraint

Before discussing how rolling constraint force emerges from the physical interaction of rolling between a robot finger-end and a 2-D object surface and how effectively it is used to balance the force/torque, we show how an external force acting on a robot finger enters Lagrange's equation of motion for the finger. First, consider the single DOF finger depicted in Figure 2.6 on which the external force $\boldsymbol{F}$ is exerted at a fixed point P at a distance from the centre of curvature O_0 of the spherical finger-end r. The finger is composed of a rigid link wiht length l whose end is rigidly connected to the finger-end and therefore the link together with the finger-end is regarded as a single rigid body. Denote the pivoted origin of the robot finger by O and set the (x, y) coordinates as shown in Figure 2.6. Then, the positions O_0 and P can be expressed as

$$\begin{cases} O_0 = \begin{pmatrix} x_0 \\ y_0 \end{pmatrix} = \begin{pmatrix} -l\cos q \\ l\sin q \end{pmatrix} \\ P = \begin{pmatrix} x_1 \\ y_1 \end{pmatrix} = \begin{pmatrix} -l\cos q - r\cos(q+\phi) \\ l\sin q + r\sin(q+\phi) \end{pmatrix} \end{cases} \tag{2.13}$$

where the angles q and ϕ are defined in Figure 2.6. Since in this case the xy-plane is regarded as horizontal, the effect of gravity can be ignored. At the same time, we assume that rotational motion of the robot finger is frictionless. Evidently, the kinetic energy of the robot finger is denoted by $K = (1/2)I\dot{q}^2$ and the external force is expressed as

$$\boldsymbol{F} = F\left(\cos(q+\phi),\ -\sin(q+\phi)\right)^{\mathrm{T}}, \tag{2.14}$$

where I denotes the moment of inertia of the finger around O and F denotes the magnitude of $\boldsymbol{F}$. Then, applying the variational principle to the Lagrangian $L = K$ with the external force $\boldsymbol{F}$ yields

$$\int_{t_0}^{t_1} \left\{ \delta L + \frac{\partial(x_1, y_1)}{\partial q} \boldsymbol{F} \delta q \right\} \mathrm{d}t = 0 \tag{2.15}$$

from which it follows that

$$I\ddot{q} = J^{\mathrm{T}}(q)\boldsymbol{F}, \tag{2.16}$$

where

$$J(q) = \begin{pmatrix} \partial x_1/\partial q \\ \partial y_1/\partial q \end{pmatrix} = \begin{pmatrix} l\sin q + r\sin(q+\phi) \\ l\cos q + r\cos(q+\phi) \end{pmatrix}. \tag{2.17}$$

Hence, the right-hand side of Equation (2.15) can be calculated by using the expressions for $\boldsymbol{F}$ and $J(q)$ [Equations (2.13) and (2.16)] in such a way that

$$\begin{aligned} J^{\mathrm{T}}(q)\boldsymbol{F} &= -lF\left\{\sin q\cos(q+\phi) - \cos q\sin(q+\phi)\right\} \\ &\quad -rF\left\{\sin(q+\phi)\cos(q+\phi) - \cos(q+\phi)\sin(q+\phi)\right\} \\ &= -LF\sin\phi. \end{aligned} \tag{2.18}$$

Thus, Equation (2.15) can be written in the form

$$I\ddot{q} = -lF\sin\phi. \tag{2.19}$$

This shows that the external force $\boldsymbol{F}$ acting at the finger-end surface in the direction normal to it can be regarded as being acting at the centre of curvature of the finger-end sphere in the same direction. Furthermore, suppose that an appropriate servo-motor actuator is installed at the pivotal joint O to generate control torque signals, and denote the control input torque by u. Then, this control torque can be regarded as an external torque at joint O

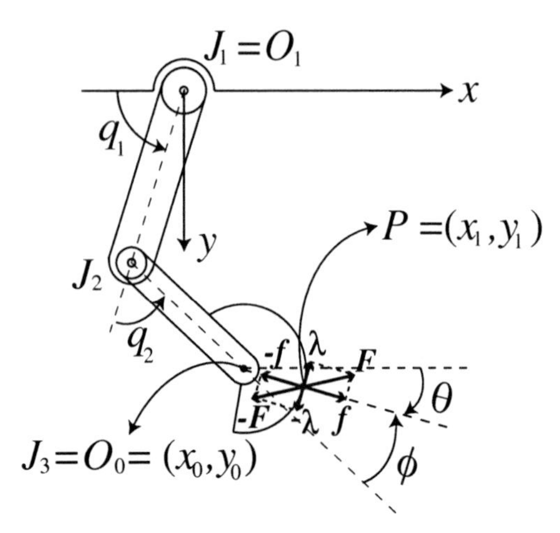

Fig. 2.7. A two-DOF finger robot receives an external force $-\boldsymbol{F}$ that can be regarded as the sum of component forces $-\boldsymbol{f}$ and $-\boldsymbol{\lambda}$ orthogonal to each other

and hence it is possible to express the equation of motion of the robot finger in the following form:

$$I\ddot{q} = -lF \sin\phi + u. \tag{2.20}$$

If all physical parameters l, F and ϕ are known, it is possible to design a control input u to stop the rotational motion of the finger by setting $u = lF\sin\phi$ so as to withstand and balance the external force $\boldsymbol{F}$.

Next consider the case of a two-DOF finger robot with a spherical end with an external force $-\boldsymbol{F}$ exerted at the point P $(= (x_1, y_1))$ as shown in Figure 2.7. For the convenience of discussions in subsequent sections, in this case we use a minus sign to express the external force. This external force can be regarded as the sum of two components $-\boldsymbol{f}$ and $-\boldsymbol{\lambda}$ as shown in Figure 2.7. Here the direction of $-\boldsymbol{f}$ is normal to the finger-end surface and that of $-\boldsymbol{\lambda}$ is tangential to it. In order to derive Lagrange's equation of motion, denote the kinetic energy of the finger robot with two joints by

$$K = K(q, \dot{q}) = \frac{1}{2}\dot{q}^{\mathrm{T}}(t)H(q(t))\dot{q}(t) \tag{2.21}$$

where $q = (q_1, q_2)^{\mathrm{T}}$ and $H(q)$ denotes the inertia matrix of the planar robot finger with two DOFs as expressed in Equation (1.107). Then, the variational principle in this case can be described by the following form:

$$\int_{t_0}^{t_1} \left[\delta K + \left\{ \frac{\partial(x_0, y_0)}{\partial q}\boldsymbol{f} + \frac{\partial(x_1, y_1)}{\partial q}\boldsymbol{\lambda} + u^{\mathrm{T}}\delta q \right\} \right] \mathrm{d}t = 0. \tag{2.22}$$

where $u = (u_1, u_2)^{\mathrm{T}}$ denotes the control torques generated at joint actuators. Note that $-\boldsymbol{f}$ and $-\boldsymbol{\lambda}$ can be described as

$$\begin{cases} -\boldsymbol{f} = f\begin{pmatrix} \cos(q_1+q_2+\phi) \\ -\sin(q_1+q_2+\phi) \end{pmatrix} = -f\begin{pmatrix} \cos\theta \\ -\sin\theta \end{pmatrix}, \\ -\boldsymbol{\lambda} = -\lambda\begin{pmatrix} \sin(q_1+q_2+\phi) \\ \cos(q_1+q_2+\phi) \end{pmatrix} = \lambda\begin{pmatrix} \sin\theta \\ \cos\theta \end{pmatrix}, \end{cases} \tag{2.23}$$

where θ denotes the angle from the x-axis to the direction of the external force $\boldsymbol{f}$. Since the sign of the angle in a counter-clockwise direction is taken to be positive, the angle θ in the case of Figure 2.7 is negative. Furthermore, it is possible to write the position P relative to the position O_0 in the following way:

$$\begin{pmatrix} x_1 \\ y_1 \end{pmatrix} = \begin{pmatrix} x_0 + r\cos(q_1+q_2+\phi) \\ y_0 - r\sin(q_1+q_2+\phi) \end{pmatrix} = \begin{pmatrix} x_0 \\ y_0 \end{pmatrix} + r\begin{pmatrix} \cos\theta \\ -\sin\theta \end{pmatrix}. \tag{2.24}$$

Futher, it follows that

$$\frac{\partial(x_1, y_1)}{\partial q} = J_0^{\mathrm{T}}(q) - r\begin{pmatrix} \sin\theta & \cos\theta \\ \sin\theta & \cos\theta \end{pmatrix}, \tag{2.25}$$

where

$$J_0^{\mathrm{T}}(q) = \frac{\partial(x_0, y_0)}{\partial q}. \tag{2.26}$$

Hence, from Equations (2.23) and (2.25) it follows that

$$-\frac{\partial(x_1, y_1)}{\partial q}\boldsymbol{f} = -fJ_0^{\mathrm{T}}(q)\begin{pmatrix} \cos\theta \\ -\sin\theta \end{pmatrix}, \tag{2.27}$$

$$-\frac{\partial(x_1, y_1)}{\partial q}\boldsymbol{\lambda} = \lambda\left\{J_0^{\mathrm{T}}(q)\begin{pmatrix} \sin\theta \\ \cos\theta \end{pmatrix} - r\begin{pmatrix} 1 \\ 1 \end{pmatrix}\right\}. \tag{2.28}$$

Thus, it follows from the variational principle expressed as Equation (1.85) that

$$\begin{aligned} H(q)\ddot{q} + \left\{\frac{1}{2}\dot{H}(q) + S(q,\dot{q})\right\} + fJ_0^{\mathrm{T}}(q)\begin{pmatrix} \cos\theta \\ -\sin\theta \end{pmatrix} \\ -\lambda\left\{J_0^{\mathrm{T}}(q)\begin{pmatrix} \sin\theta \\ \cos\theta \end{pmatrix} - r\begin{pmatrix} 1 \\ 1 \end{pmatrix}\right\} = u. \end{aligned} \tag{2.29}$$

It is important to note that, for the generation of two torque components to counter-balance the torques exerted by the external force $-\boldsymbol{F}$ whose direction is changing instantaneously (ϕ is not constant), two independent actuators installed separately at two joints are necessary. In other words, if u is a scalar control variable and hence $J_0^{\mathrm{T}}(q)$ is a 1×2 matrix, the total sum of the third and fourth terms of the left-hand side of Equation (2.29) cannot be restored into such two original terms.

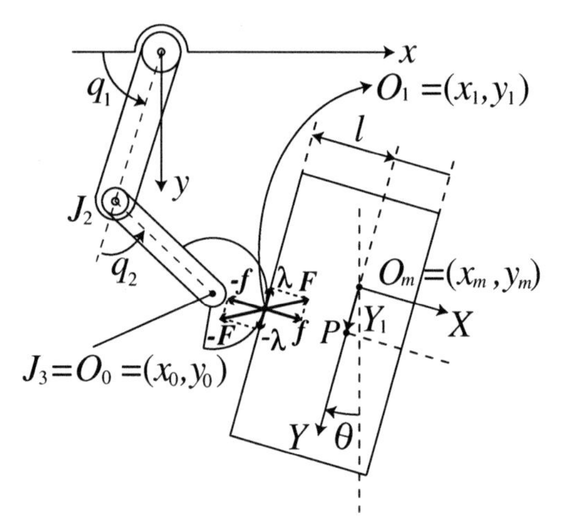

Fig. 2.8. A two-DOF finger robot immobilising a 2-D object pivoted at a fixed point

2.5 Testbed Problems for Dynamically Stable Grasp

For the study of stable grasping and dexterious manipulation by a human-like multi-fingered hand, a simple but canonical testbed problem underlying control of motion of the simplest robot mechanism is very useful for gaining both physical and mathematical insights into the problem. Such a prototype problem may play a similar role to the well-known problem of stability and control of the inverted pendulum that has contributed to advances in linear and non-linear control theory of mechanical systems. However, there is a noteworthy difference between the two testbed problems. Even in the case of the simpler testbed problems of stable grasp, Lagrange's equation of motion for a robot mechanism is subject to geometric constraints and therefore control inputs cannot enter explicitly into the equation of motion of the object to be controlled. Grasping an object should be controlled indirectly through constraint forces.

For the sake of mechanical simplicity but still to show the key role of the rolling constraint, let us consider the mechanical setup shown in Figure 2.8, which is composed of a two-DOF robot finger with a hemispherical finger-end and a rigid rectangular 2-D object pinned at a point O_m but pivoting around it. It is assumed that the xy-plane is horizontal and the whole motion of the finger and object is confined to this plane and therefore the effect of gravity can be ignored. For convenience and simplicity, we assume further that rotational motion of the object around the O_m is frictionless. Comparing Figure 2.8 with Figure 2.7, we can easily see that the equation of rotational motion of the object follows directly from Newton's law of action and reaction in the following form:

$$I\ddot{\theta} - fY_1 + \lambda l = 0, \tag{2.30}$$

where I denotes the inertia moment of the object around O_m, l the distance from O_m to the left rectilinear side of the object as shown in Figure 2.8 and Y_1 the Y-component of the (X, Y)-coordinates of position O_1 attached at the object. The equation of motion of the finger is the same as that described by Equation (2.29) in the previous section. These two equations are derived, however, by assuming that there arises a force $\boldsymbol{F}$ pressing the object in the direction shown in Figure 2.8 and a reactive force $-\boldsymbol{F}$ affecting the finger at the common point O_1 in the opposite direction.

We now show that these active and reactive forces $\boldsymbol{F}$ and $-\boldsymbol{F}$, which have tangential components $\boldsymbol{\lambda}$ and $-\boldsymbol{\lambda}$, actually arise at the contact point O_1 between the finger-end and the object surface. In reality, the physical situation of contacting of the finger-end with the object should be expressed firstly by the algebraic equation

$$r + l = (x_m - x_0)\cos\theta + (y_m - y_0)\sin\theta. \tag{2.31}$$

Since the length of the line from O_0 to P (see Figure 2.8) is $r + l$ it must also be equal to the right-hand side of Equation (2.31). Here, P is a point at which the extended straight line from O_0 to O_1 crosses the Y-axis originates from the origin O_m. Secondly, if the spherical finger-end is rolling on the object surface without slipping then the contact point velocity expressed on the finger-end must be equal to that expressed on the object surface. This can be written as

$$\frac{\mathrm{d}}{\mathrm{d}t}(r\phi) = -\frac{\mathrm{d}}{\mathrm{d}t}Y_1, \tag{2.32}$$

where ϕ denotes the angle specified in Figure 2.7. It is easy to verify that

$$\phi = \pi + \theta - q_1 - q_2 = \pi + \theta - q^{\mathrm{T}}\boldsymbol{e}, \tag{2.33}$$

$$Y_1 = (x_0 - x_m)\sin\theta + (y_0 - y_m)\cos\theta, \tag{2.34}$$

where $\boldsymbol{e} = (1, 1)^{\mathrm{T}}$ and $q = (q_1, q_2)^{\mathrm{T}}$. Note that Equation (2.32) can be integrated as follows:

$$r\phi(t) = -Y_1(t) - c_0, \tag{2.35}$$

where c_0 denotes a constant of integration. Thus, it follows from Equation (2.31) and substituting Equations (2.33) and (2.34) into Equation (2.35) that

$$\begin{aligned} Q(q, \theta) &= -(r + l) + (x_m - x_0)\cos\theta - (y_m - y_0)\sin\theta \\ &= 0, \end{aligned} \tag{2.36}$$

$$\begin{aligned} R(q, \theta) &= c_0 + Y + r\phi \\ &= c_0 + (x_0 - x_m)\sin\theta + (y_0 - y_m)\cos\theta + r(\pi + \theta - q^{\mathrm{T}}\boldsymbol{e}) \\ &= 0. \end{aligned} \tag{2.37}$$

We call Equation (2.36) the contact constraint and Equation (2.37) the rolling constraint. Both constraints can be regarded as holonomic. Then, by introducing Lagrange's multipliers f and λ for Equations (2.36) and (2.37), respectively, we can define the Lagrangian

$$L = K + fQ + \lambda R, \tag{2.38}$$

where K denotes the kinetic energy expressed by Equation (2.21). Applying the variational principle described by Equation (1.76) to this Lagrangian, we obtain the following Lagrange's equation of motion:

$$H(q)\ddot{q} + \left\{\frac{1}{2}\dot{H}(q) + S(q,\dot{q})\right\}\dot{q} - f\frac{\partial Q}{\partial q} - \lambda\frac{\partial R}{\partial q} = u, \tag{2.39}$$

$$I\ddot{\theta} - f\frac{\partial Q}{\partial \theta} - \lambda\frac{\partial R}{\partial \theta} = 0. \tag{2.40}$$

Obviously, Equation (2.39) expresses the motion of the robot finger and Equation (2.40) the rotational motion of the object. The gradient vectors $\partial Q/\partial q$ and $\partial R/\partial q$ and partial differentials of Q and R with respect to θ can be calculated as follows:

$$\left\{\begin{aligned} &\frac{\partial Q}{\partial q} = -\frac{\partial(x_0, y_0)}{\partial q}\begin{pmatrix}\cos\theta \\ -\sin\theta\end{pmatrix} = -J_0^{\mathrm{T}}(q)\begin{pmatrix}\cos\theta \\ -\sin\theta\end{pmatrix}, \\ &\frac{\partial R}{\partial q} = \frac{\partial(x_0, y_0)}{\partial q}\begin{pmatrix}\sin\theta \\ \cos\theta\end{pmatrix} - r\boldsymbol{e} = J_0^{\mathrm{T}}(q)\begin{pmatrix}\sin\theta \\ \cos\theta\end{pmatrix} - r\boldsymbol{e}, \\ &\frac{\partial Q}{\partial \theta} = Y_1, \qquad \frac{\partial R}{\partial \theta} = -l. \end{aligned}\right. \tag{2.41}$$

Thus, it is possible to confirm that substituting these partial differentials from Equation (2.41) into Equations (2.39) and (2.40) yields Equations (2.29) and (2.30). Evidently the Lagrange multiplier f acts at the contact point as a pressing force against the object and induces the rotational moment $-fY_1$ for the object around O_m. Similarly, another multiplier λ induces the rotational moment λl for the object around Q_m. Then, their two-dimensional wrenches acting on the object are written as follows:

$$\boldsymbol{w}_f = f\begin{pmatrix}\cos\theta \\ -\sin\theta \\ Y_1\end{pmatrix}, \qquad \boldsymbol{w}_\lambda = \lambda\begin{pmatrix}-\sin\theta \\ -\cos\theta \\ -l\end{pmatrix}. \tag{2.42}$$

Apparently, $\boldsymbol{w}_f$ acts at the contact point as a pressing force for the object in the direction normal to the object side and $\boldsymbol{w}_\lambda$ acts at the same contact point as a shear force along the object side. The typical testbed control problem for the set of motion Equations (2.29) and (2.30) that are subject to the constraint

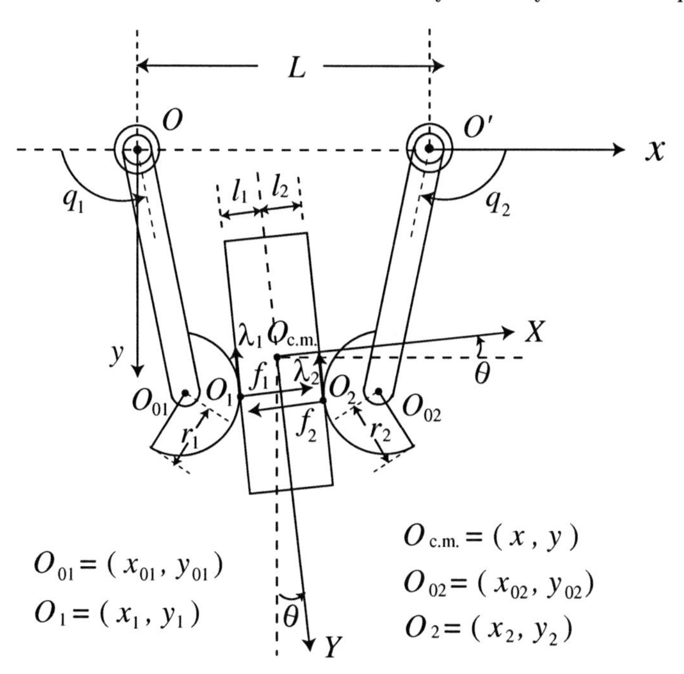

Fig. 2.9. A pair of single-DOF robot fingers grasping a 2-D rigid object with parallel surfaces

Equations (2.36) and (2.37) is to design a controller for stabilising rotational motion of the object by balancing induced moments of the object around O_m and maintaining some still state with a constant angle of θ and zero angular velocity $\dot{\theta} = 0$. Note that the control variables $u = (u_1, u_2)^{\mathrm{T}}$ do not enter into the dynamics of the object, *i.e.*, Equation (2.30). The motion of the object should be indirectly controlled and stabilised through the constraint forces f and λ. This problem will be investigated in detail in subsequent sections, which will help us in understanding physical meanings of control of precision prehension by a pair of human-like robot fingers.

Next we shall propose another testbed problem of stable grasp with a simple mechanical structure. Let us consider a pair of single-DOF robot fingers whose ends are spherical and a 2-D object which has parallel sides, as shown in Figure 2.9. The motion of the overall fingers-object is confined to the horizontal xy-plane. In this case, it is implicitly assumed that an object with a flat bottom is placed on a desk and both the translational and rotational motions of the object are frictionless. In contrast to the previous example of Figure 2.8, the centre of mass $O_{\text{c.m.}}$ of the object is free to move. Denote the mass and the inertia moment of the object around $O_{\text{c.m.}}$ by M and I, respectively. All other physical variables are specified in Figure 2.9, as in Figures 2.7 and 2.8 except that in Figure 2.9 the position of the object centre of mass is expressed by the vector $\boldsymbol{x} = (x, y)^{\mathrm{T}}$ in terms of the frame coordinates with origin O. Hence, the total kinetic energy of this overall fingers-object system is expressed as

$$K = \sum_{i=1,2} \frac{1}{2} I_i \dot{q}_i^2 + \frac{M}{2}(\dot{x}^2 + \dot{y}^2) + \frac{I}{2}\dot{\theta}^2, \tag{2.43}$$

where I_1 denotes the moment of inertia of the left finger around the origin O and I_2 that of the right finger around O'. In this case, the sign of angles q_1 and θ is taken to be positive in the counter-clockwise direction but that of q_2 is taken to be positive in the clockwise direction.

Now, in light of the arguments developed in the previous two sections, it is rather evident that Lagrange's equation of motion of this fingers-object system is governed by the following set of equations:

$$I_i \ddot{q}_i - f_i \frac{\partial Q_i}{\partial q_i} - \lambda_i \frac{\partial R_i}{\partial q_i} = u_i, \qquad i = 1, 2, \tag{2.44}$$

$$M \begin{pmatrix} \ddot{x} \\ \ddot{y} \end{pmatrix} - \sum_{i=1,2} \left\{ f_i \begin{pmatrix} \partial Q_i/\partial x \\ \partial Q_i/\partial y \end{pmatrix} + \lambda_i \begin{pmatrix} \partial R_i/\partial x \\ \partial R_i/\partial y \end{pmatrix} \right\} = 0, \tag{2.45}$$

$$I\ddot{\theta} - \sum_{i=1,2} \{ f_i(\partial Q_i/\partial \theta) + \lambda_i(\partial R_i/\partial \theta)\} = 0, \tag{2.46}$$

where

$$Q_i = -(l_i + r_i) - (-1)^i \{(x - x_{0i})\cos\theta - (y - y_{0i})\sin\theta\} = 0, \qquad i = 1, 2 \tag{2.47}$$

$$\frac{\mathrm{d}}{\mathrm{d}t} r_i \phi_i = -\frac{\mathrm{d}Y_i}{\mathrm{d}t}, \qquad i = 1, 2 \tag{2.48}$$

$$\begin{cases} Y_i = (x_{0i} - x)\sin\theta + (y_{0i} - y)\cos\theta \\ \phi_i = \pi - (-1)^i\theta - q_i \end{cases} \qquad i = 1, 2 \tag{2.49}$$

$$\begin{aligned} R_i &= Y_i + r_i(\phi_i - \pi) - c_{0i} \\ &= (x_{0i} - x)\sin\theta + (y_{0i} - y)\cos\theta - c_{0i} - r_i\left\{(-1)^i\theta + q_i\right\} \\ &= 0, \qquad i = 1, 2 \end{aligned} \tag{2.50}$$

and c_{0i} denotes an appropriate constant. Equation (2.47) for i signifies the contact constraint at the contact point O_{0i} and Equation (2.48) for i captures the rolling constraint at the same point O_{0i}. Since Equation (2.48) can be

Table 2.1. Partial differentials of Q_i and R_i

$$\begin{cases} \dfrac{\partial Q_i}{\partial q_i} = (-1)^i J_{0i}^{\mathrm{T}}(q_i) \begin{pmatrix} \cos\theta \\ -\sin\theta \end{pmatrix} \\ \dfrac{\partial R_i}{\partial q_i} = J_{0i}^{\mathrm{T}}(q_i) \begin{pmatrix} \sin\theta \\ \cos\theta \end{pmatrix} - r_i \end{cases} \quad i = 1, 2$$

$$\begin{cases} \dfrac{\partial Q_i}{\partial x} = (-1)^i \cos\theta, & \dfrac{\partial Q_i}{\partial y} = (-1)^i \sin\theta \\ \dfrac{\partial R_i}{\partial x} = -\sin\theta, & \dfrac{\partial R_i}{\partial y} = -\cos\theta \end{cases} \quad i = 1, 2$$

$$\begin{cases} \dfrac{\partial Q_i}{\partial \theta} = -(-1)^i Y_i \\ \dfrac{\partial R_i}{\partial \theta} = (-1)^i l_i \end{cases} \quad i = 1, 2$$

$$J_{0i}^{\mathrm{T}}(q_i) = \frac{\partial(x_{0i}, y_{0i})}{\partial q_i} = l_i \left((-1)^i \sin q_i, \cos q_i \right), \quad i = 1, 2$$

where l_1 = the length of $\overline{OO_{01}}$
and l_2 = that of $\overline{O'O_{02}}$

integrated and Y_i and ϕ_i can be expressed as in Equation (2.49), the rolling constraint can be rewritten in the form of the holonomic constraint shown in Equation (2.50). Therefore, the Lagrangian L of the overall system can be expressed as

$$L = K + \sum_{i=1,2} \{ f_i Q_i + \lambda_i R_i \} \tag{2.51}$$

and thereby Equations (2.44) to (2.46) follow from applying the variational principle for the Lagrangian L. For the sake of convenience, we have calculated all partial differentials of Q_i and R_i in q_i, x, y and θ, which are given in Table 2.1. Substituting all these partial differentials into Equations (2.44) to (2.46) leads to

$$I_i \ddot{q}_i - f_i (-1)^i J_{0i}^{\mathrm{T}}(q_i) \begin{pmatrix} \cos\theta \\ -\sin\theta \end{pmatrix} - \lambda_i \left\{ J_{0i}^{\mathrm{T}}(q_i) \begin{pmatrix} \sin\theta \\ \cos\theta \end{pmatrix} - r_i \right\} = u_i, \quad i = 1, 2 \tag{2.52}$$

$$M \begin{pmatrix} \ddot{x} \\ \ddot{y} \end{pmatrix} - (f_1 - f_2) \begin{pmatrix} \cos\theta \\ -\sin\theta \end{pmatrix} + (\lambda_1 + \lambda_2) \begin{pmatrix} \sin\theta \\ \cos\theta \end{pmatrix} = 0, \tag{2.53}$$

$$I\ddot{\theta} - f_1 Y_1 + f_2 Y_2 + l_1\lambda_1 - l_2\lambda_2 = 0. \tag{2.54}$$

Since it follows from the meaning of the constraints expressed by Equations (2.47) and (2.50) that

$$\begin{cases} \frac{\mathrm{d}Q_i}{\mathrm{d}t} = \dot{q}_i \frac{\partial Q_i}{\partial q_i} + \dot{x}\frac{\partial Q_i}{\partial x} + \dot{y}\frac{\partial Q_i}{\partial y} + \dot{\theta}\frac{\partial Q_i}{\partial \theta} \\ \frac{\mathrm{d}R_i}{\mathrm{d}t} = \dot{q}_i \frac{\partial R_i}{\partial q_i} + \dot{x}\frac{\partial R_i}{\partial x} + \dot{y}\frac{\partial R_i}{\partial y} + \dot{\theta}\frac{\partial R_i}{\partial \theta} \end{cases} \quad i = 1, 2 \tag{2.55}$$

the sum of multiplications of Equation (2.52) by $\dot{q}_i$ $(i = 1, 2)$ and Equation (2.54) by $\dot{\theta}$ and inner product between Equation (2.53) and $(\dot{x}, \dot{y})^{\mathrm{T}}$ yields

$$\frac{\mathrm{d}}{\mathrm{d}t}K = \sum_{i=1,2} \dot{q}_i u_i. \tag{2.56}$$

Now, we are ready to discuss how to design a controller that can stabilise both translational and rotational motions of the object by balancing forces and torques affecting the object. Obviously from Equations (2.53) and (2.54), the wrench vectors exerted on the object are described as follows:

$$\boldsymbol{w}_{f1} = f_1 \begin{pmatrix} -\cos\theta \\ \sin\theta \\ -Y_i \end{pmatrix}, \quad \boldsymbol{w}_{f2} = f_2 \begin{pmatrix} \cos\theta \\ -\sin\theta \\ Y_2 \end{pmatrix},$$
$$\boldsymbol{w}_{\lambda 1} = \lambda_1 \begin{pmatrix} \sin\theta \\ \cos\theta \\ l_1 \end{pmatrix}, \quad \boldsymbol{w}_{\lambda 2} = \lambda_2 \begin{pmatrix} \sin\theta \\ \cos\theta \\ -l_2 \end{pmatrix}. \tag{2.57}$$

In order that the sum of these four wrenches become zero, it is necessary and sufficient that

$$f_1 = f_2 = f_d, \quad \lambda_1 + \lambda_2 = 0, \quad -f_d(Y_1 - Y_2) + \lambda_1(l_1 + l_2) = 0, \tag{2.58}$$

where f_d must be some positive constant. One possible solution to the simultaneous conditions of Equation (2.58) is to control the overall system motion so as to let $f_i \to f_d$, $Y_1 - Y_2 \to 0$ and $\lambda_i \to 0$ $(i = 1, 2)$. Following this observation, we are able to devise the following control signal:

$$u_i = -c_i\dot{q}_i - (-1)^i f_d \left\{ J_{0i}^{\mathrm{T}}(q_i) \begin{pmatrix} \cos\theta \\ -\sin\theta \end{pmatrix} - \frac{r_i}{r_1 + r_2}(Y_1 - Y_2) \right\} \tag{2.59}$$

where f_0 can be chosen as being equal to f_d or any other positive constant. Substituting Equation (2.59) into Equation (2.52) yields

$$I_i\ddot{q}_i + c_i\dot{q}_i - (-1)^i \left\{ \Delta f_i J_{0i}^{\mathrm{T}}(q_i) \begin{pmatrix} \cos\theta \\ -\sin\theta \end{pmatrix} + \frac{r_i f_d}{r_1 + r_2}(Y_1 - Y_2) \right\}$$
$$-\lambda_i \left\{ J_{0i}^{\mathrm{T}}(q_i) \begin{pmatrix} \sin\theta \\ \cos\theta \end{pmatrix} - r_i \right\} = 0, \tag{2.60}$$

where $\Delta f_i = f_i - f_0$ $(i = 1, 2)$. We conveniently rewrite Equations (2.53) and (2.54) in the following equivalent formulae:

$$M\begin{pmatrix}\ddot{x}\\ \ddot{y}\end{pmatrix} - (\Delta f_1 - \Delta f_2)\begin{pmatrix}\cos\theta\\ -\sin\theta\end{pmatrix} + (\lambda_1 + \lambda_2)\begin{pmatrix}\sin\theta\\ \cos\theta\end{pmatrix} = 0, \quad (2.61)$$

$$I\ddot{\theta} - \Delta f_1 Y_1 + \Delta f_2 Y_2 - f_d(Y_1 - Y_2) + l_1\lambda_1 - l_2\lambda_2 = 0. \quad (2.62)$$

Similarly to the derivation of Equation (2.56) by referring to Equation (2.55), the sum of the multiplications of Equation (2.60) by $\dot{q}_i$ for $i = 1, 2$, Equation (2.62) by $\dot{\theta}$, and the inner product between Equation (2.61) and $(\dot{x}, \dot{y})^{\mathrm{T}}$ takes the form

$$\frac{\mathrm{d}}{\mathrm{d}t}K + \sum_{i=1,2}\left\{c_i\dot{q}_i^2 - (-1)^i\frac{r_i f_d}{r_1 + r_2}\dot{q}_i(Y_1 - Y_2)\right\} - f_d\dot{\theta}(Y_1 - Y_2) = 0. \quad (2.63)$$

Since from Equations (2.48) and (2.49) it follows that

$$\begin{aligned}\dot{Y}_1 - \dot{Y}_2 &= -r_1\dot{\phi}_1 + r_2\dot{\phi}_2 = -r_1(\dot{\theta} - \dot{q}_1) + r_2(-\dot{\theta} - \dot{q}_2)\\ &= -(r_1 + r_2)\dot{\theta} + (r_1\dot{q}_1 - r_2\dot{q}_2)\end{aligned} \quad (2.64)$$

Equation (2.63) can be reduced to

$$\frac{\mathrm{d}}{\mathrm{d}t}E(\boldsymbol{X}, \dot{\boldsymbol{X}}) = -\sum_{i=1,2} c_i\dot{q}_i^2, \quad (2.65)$$

where $\boldsymbol{X} = (q_1, q_2, x, y, \theta)^{\mathrm{T}}$,

$$P = \frac{f_d}{2(r_1 + r_2)}(Y_1 - Y_2)^2, \quad (2.66)$$

$$\begin{aligned}E(\boldsymbol{X}, \dot{\boldsymbol{X}}) &= K + P\\ &= \sum_{i=1,2}\frac{I_i}{2}\dot{q}_i^2 + \frac{M}{2}(\dot{x}^2 + \dot{y}^2) + \frac{I}{2}\dot{\theta}^2 + \frac{f_d}{2(r_1 + r_2)}(Y_1 - Y_2)^2, \end{aligned} \quad (2.67)$$

and K is the total kinetic energy already given by Equation (2.43). Equation (2.65) can be interpreted as stating that the time rate of the total energy $E(\boldsymbol{X})$ is equal to the instantaneous energy dissipation rate. Hence, we call the scalar function P the artificial potential. It is important to note that the closed-loop dynamics of Equations (2.60–2.62) is equivalent to Lagrange's equation of motion for the Lagrangian

$$L = K - P + \sum_{i=1,2}\{\Delta f_i Q_i + \lambda_i R_i\}. \quad (2.68)$$

Note that the overall fingers–object system of Figure 2.9 has a single DOF because the system has five independent position variables $\boldsymbol{X} = (q_1, q_2, x, y, \theta)^{\mathrm{T}}$ but they are subject to four independent holonomic constraints. Therefore P is positive definite with respect to $\boldsymbol{X}$ under the four constraints and therefore the total energy $E(\boldsymbol{X}, \dot{\boldsymbol{X}})$ is positive definite for the state variables $(\boldsymbol{X}, \dot{\boldsymbol{X}})$ under the following eight constraints

$$\begin{cases} Q_i = 0, \quad R_i = 0 \\ \dot{Q}_i = \dot{\boldsymbol{X}}^{\mathrm{T}} \dfrac{\partial Q_i}{\partial \boldsymbol{X}} = 0, \quad \dot{R}_i = \dot{\boldsymbol{X}}^{\mathrm{T}} \dfrac{\partial R_i}{\partial \boldsymbol{X}} = 0 \end{cases} \quad i = 1, 2. \tag{2.69}$$

Hence, due to Dirichlet's theorem of stability, the equilibrium state $(\boldsymbol{X}_\infty, \dot{\boldsymbol{X}}_\infty = 0)$ that satisfies $Y_1 - Y_2 = 0$ at $\boldsymbol{X} = \boldsymbol{X}_\infty$ is stable.

Now, we show that the equilibrium state $(\boldsymbol{X}_\infty, 0)$ is asymptotically stable for the system of Equations (2.60–2.62), *i.e.*, there exists a positive number $\delta > 0$ such that any solution $(\boldsymbol{X}(t), \dot{\boldsymbol{X}}(t))$ of Equations (2.60–2.62) subject to constraints (2.69) starting from an arbitrary initial state $(\boldsymbol{X}(0), \dot{\boldsymbol{X}}(0))$ satisfying $E(\boldsymbol{X}(0), \dot{\boldsymbol{X}}(0)) \leq \delta$ converges asymptotically to the equilibrium state $(\boldsymbol{X}_\infty, 0)$ as $t \to \infty$. For the sake of convenience for proving this, we rewrite the closed-loop dynamics of Equations (2.60–2.62) into the single matrix–vector form:

$$H\ddot{\boldsymbol{X}} + C\dot{\boldsymbol{X}} - A\Delta\boldsymbol{\lambda} - \frac{f_d}{r_1 + r_2}(Y_1 - Y_2)\boldsymbol{e} = 0 \tag{2.70}$$

where

$$H = \begin{pmatrix} I_1 & 0 & 0 & 0 & 0 \\ 0 & I_2 & 0 & 0 & 0 \\ 0 & 0 & M & 0 & 0 \\ 0 & 0 & 0 & M & 0 \\ 0 & 0 & 0 & 0 & I \end{pmatrix}, \quad \Delta\boldsymbol{\lambda} = \begin{pmatrix} \Delta f_1 \\ \Delta f_2 \\ \lambda_1 \\ \lambda_2 \end{pmatrix}, \quad \boldsymbol{e} = \begin{pmatrix} -r_1 \\ r_2 \\ 0 \\ 0 \\ r_1 + r_2 \end{pmatrix}, \tag{2.71}$$

$$A = \begin{pmatrix} -J_{01}^{\mathrm{T}}\boldsymbol{r}_X & 0 & J_{01}^{\mathrm{T}}\boldsymbol{r}_Y - r_1 & 0 \\ 0 & J_{02}^{\mathrm{T}}\boldsymbol{r}_X & 0 & J_{02}^{\mathrm{T}}\boldsymbol{r}_Y - r_2 \\ \boldsymbol{r}_X & -\boldsymbol{r}_X & -\boldsymbol{r}_Y & -\boldsymbol{r}_Y \\ Y_1 & -Y_2 & -l_1 & l_2 \end{pmatrix}, \tag{2.72}$$

$$C = \begin{pmatrix} \begin{matrix} c_1 & 0 \\ 0 & c_2 \end{matrix} & 0_{2\times 3} \\ 0_{3\times 2} & 0_{3\times 3} \end{pmatrix}, \quad \boldsymbol{r}_X = \begin{pmatrix} \cos\theta \\ -\sin\theta \end{pmatrix}, \quad \boldsymbol{r}_Y = \begin{pmatrix} \sin\theta \\ \cos\theta \end{pmatrix}. \tag{2.73}$$

Obviously, the 5×4 matrix A is of full rank [*i.e.*, rank $(A) = 4$] when the pair of fingers in contact with the object takes an ordinary position like that shown in Figure 2.9. The proof should go in the follows steps.

1) According to the energy relation of Equation (2.68), $E(\boldsymbol{X}(t), \dot{\boldsymbol{X}}(t)) \leq E(\boldsymbol{X}(0), \dot{\boldsymbol{X}}(0)) \leq \delta$ for any $t > 0$. Hence, $\dot{\boldsymbol{X}}(t)$ is uniformly bounded and $Y_1 - Y_2$ is also bounded, in particular

$$|Y_1(t) - Y_2(t)| \leq \sqrt{\frac{2(r_1 + r_2)\delta}{f_d}}. \tag{2.74}$$

2) Next, note that

$$\begin{aligned} 0 = \dot{R}_1 - \dot{R}_2 &= (\dot{x}_{01} - \dot{x}_{02}) \sin\theta + (\dot{y}_{01} - \dot{y}_{02}) \cos\theta \\ &\quad -\dot{\theta}(l_1 + l_2 + r_1 + r_2) + (r_1 + r_2)\dot{\theta} - r_1\dot{q}_1 - r_2\dot{q}_2 \end{aligned} \tag{2.75}$$

from which it follows that

$$\dot{\theta} = \frac{1}{l_1 + l_2} \left\{-r_1\dot{q}_1 + r_2\dot{q}_2 + (\dot{x}_{01} - \dot{x}_{02}) \sin\theta + (\dot{y}_{01} - \dot{y}_{02}) \cos\theta\right\}. \tag{2.76}$$

Hence, $|\dot{\theta}|$ is bounded. Similarly, it follows from differentiations of Q_1 and R_1 with respect to t that

$$R_\theta^{\mathrm{T}} \begin{pmatrix} \dot{x} \\ \dot{y} \end{pmatrix} - R_\theta^{\mathrm{T}} \begin{pmatrix} \dot{x}_{01} \\ \dot{y}_{01} \end{pmatrix} + \begin{pmatrix} Y_1\dot{\theta} \\ r_1\dot{q}_1 - l_1\dot{\theta} \end{pmatrix} = 0, \tag{2.77}$$

where

$$R_\theta = \begin{pmatrix} \boldsymbol{r}_X & \boldsymbol{r}_Y \end{pmatrix} = \begin{pmatrix} \cos\theta & \sin\theta \\ -\sin\theta & \cos\theta \end{pmatrix}. \tag{2.78}$$

Since R_θ is an orthogonal matrix, $R_\theta^{-1} = R_\theta^{\mathrm{T}}$ and therefore Equation (2.77) is reduced to

$$\begin{pmatrix} \dot{x} \\ \dot{y} \end{pmatrix} = \begin{pmatrix} \dot{x}_{01} \\ \dot{y}_{01} \end{pmatrix} - R_\theta \begin{pmatrix} Y_1\dot{\theta} \\ r_1\dot{q}_1 - l_1\dot{\theta} \end{pmatrix} \tag{2.79}$$

from which $(\dot{x}, \dot{y})^{\mathrm{T}}$ is also bounded.

3) Note that multiplication of Equation (2.70) from the left by $A^{\mathrm{T}}H^{-1}$ yields

$$\begin{aligned} &-\dot{A}^{\mathrm{T}}\dot{\boldsymbol{X}} + A^{\mathrm{T}}H^{-1}C\dot{\boldsymbol{X}} - A^{\mathrm{T}}H^{-1}A\Delta\boldsymbol{\lambda} \\ &\qquad -\frac{f_d}{r_1 + r_2}(Y_1 - Y_2)A^{\mathrm{T}}H^{-1}\boldsymbol{e} = 0, \end{aligned} \tag{2.80}$$

where we used the relation

$$0 = \frac{\mathrm{d}}{\mathrm{d}t}(A^{\mathrm{T}}\dot{\boldsymbol{X}}) = A^{\mathrm{T}}\ddot{\boldsymbol{X}} + \dot{A}^{\mathrm{T}}\dot{\boldsymbol{X}}. \tag{2.81}$$

Multiplying Equation (2.80) by $(A^{\mathrm{T}}H^{-1}A)^{-1}$ from the left yields

$$\Delta\boldsymbol{\lambda} = (A^{\mathrm{T}}H^{-1}A)^{-1}\Big\{ -\dot{A}^{\mathrm{T}}\dot{\boldsymbol{X}} + A^{\mathrm{T}}H^{-1}C\dot{\boldsymbol{X}} - \frac{f_d}{r_1+r_2}(Y_1 - Y_2)A^{\mathrm{T}}H^{-1}\boldsymbol{e}\Big\}. \tag{2.82}$$

Since A is of full rank at the equilibrium state $\boldsymbol{X} = \boldsymbol{X}_\infty$, it is possible to choose $\delta > 0$ small enough that A is nondegenerate for all $\boldsymbol{X}$ satisfying $E(\boldsymbol{X}, \dot{\boldsymbol{X}}) \leq \delta$ together with constraints of Equation (2.69). Hence, $\Delta\boldsymbol{\lambda}$ is also bounded.

4) Since $Y_1 - Y_2$, $\dot{\boldsymbol{X}}$, and $\Delta\boldsymbol{\lambda}$ are all uniformly bounded, $\ddot{\boldsymbol{X}}$ must be uniformly bounded according to Equation (2.70). This implies that $\dot{\boldsymbol{X}}$ is uniformly continuous. In particular, $\dot{q}_1$ and $\dot{q}_2$ are uniformly continuous and also belong to $L^2(0,\infty)$ from the energy relation of Equation (2.65). Thus, on account of Lemma 2 of Appendix A, $\dot{q}_1(t)$ and $\dot{q}_2(t)$ converge to zero as $t \to \infty$. Then, according to Equations (2.76) and (2.79), $\dot{\theta}(t) \to 0$ as $t \to \infty$ and subsequently $\dot{x}(t)$ and $\dot{y}(t)$ must converge to zero as $t \to \infty$.

5) Since $\dot{\boldsymbol{X}}$ and $Y_1 - Y_2$ are uniformly continuous in t, $\ddot{\boldsymbol{X}}(t)$ is also uniformly continuous. Since $\ddot{\boldsymbol{X}}$ is uniformly continuous and $\dot{\boldsymbol{X}}(t) \to 0$ as $t \to \infty$, Lemma A of Appendix A implies that $\ddot{\boldsymbol{X}}(t) \to 0$ as $t \to \infty$.

6) Thus, from Equation (2.70) it follows that

$$-[A, \boldsymbol{e}]\begin{bmatrix} \Delta\boldsymbol{\lambda} \\ -f_d(Y_1 - Y_2)/(r_1 + r_2) \end{bmatrix} \to 0 \quad \text{as} \quad t \to \infty. \tag{2.83}$$

Since the 5×5-matrix $[A, \boldsymbol{e}]$ is nonsingular obviously, Equation (2.83) implies

$$Y_1(t) - Y_2(t) \to 0, \quad f_i(t) \to f_d, \quad \lambda_i(t) \to 0, \quad i = 1, 2 \tag{2.84}$$

as $t \to \infty$.

Thus, the proof of the asymptotic stability of the equilibrium state $(\boldsymbol{X}_\infty, 0)$ that satisfies $Y_1 - Y_2 = 0$ together with $f_i = f_d$ and $\lambda_i = 0$ has been completed.

However, there still remain uncertainties in the mathematical rigour of the proof and the practical effectiveness of the control scheme given by Equation (2.59). In fact, we could not argue how rapidly the solution trajectory $(\boldsymbol{X}(t), \dot{\boldsymbol{X}}(t))$ of Equation (2.70) converges to the equilibrium state $(\boldsymbol{X}_\infty, 0)$. We could not find out how large a neighbourhood of the equilibrium point $(\boldsymbol{X}_\infty, 0)$ characterised by $E(\boldsymbol{X}(0), \dot{\boldsymbol{X}}(0)) \leq \delta$ with $\delta > 0$ can be selected. If any solution trajectories starting from any initial state inside such a neighbourhood of $(\boldsymbol{X}_\infty, 0)$ converge asymptotically to $(\boldsymbol{X}_\infty, 0)$, the neighbourhood is called an attractor of the equilibrium point $(\boldsymbol{X}_\infty, 0)$. In the process of proving the asymptotic convergence, we needed to show that boundedness of the finger angular velocities $\dot{q}_1$ and $\dot{q}_2$ implies that of the velocities of the object variables (x, y, θ) based on Equations (2.76) and (2.79) owing to the four contact constraints. However, Equation (2.76) shows that the attractor of $(\boldsymbol{X}_\infty, 0)$ may not be selected large enough if the object width $l_1 + l_2$ is very small. Another problem is whether both contacts between the finger-ends and

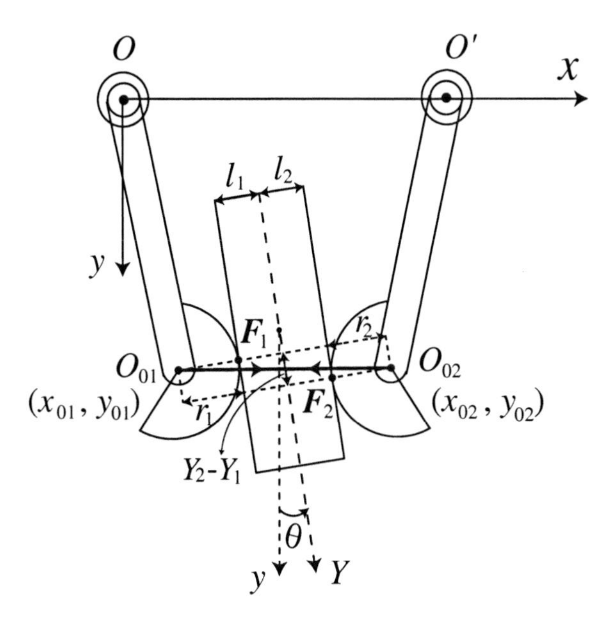

Fig. 2.10. A pair of single-DOF robot fingers grasping a 2-D object with parallel flat sides

object surfaces are maintained during the motion of the overall system. In the next section, a more mathematically rigorous treatment of the problem will be presented by introducing a far more important class of coordinated motor control signals for stable grasping.

2.6 Blind Grasping and Robustness Problems

Consider again a pair of dual single-DOF robot fingers contacting a 2-D rectangular object as shown in Figure 2.10. Now, let us consider the following class of control signals:

$$u_i = -c_i \dot{q}_i + (-1)^i \frac{f_d}{r_1 + r_2} J_{0i}^{\mathrm{T}}(q_i) \begin{pmatrix} x_{01} - x_{02} \\ y_{01} - y_{02} \end{pmatrix}, \quad i = 1, 2. \tag{2.85}$$

The first term on the right-hand side indicates damping injection for finger joint motion and the second term is introduced to exert an approximated opposition force, denoted by $\boldsymbol{F}_i$ in Figure 2.10, that presses the object coordinatedly from the left by the left finger ($i = 1$) and from the right by the right one ($i = 2$). If in the construction of the control signal of Equation (2.85) any information about the object is not available, *i.e.*, the positions of the contact points O_1 and O_2 are uncertain, we must use the information only about finger kinematics and measured data on finger joint angles. Hence we assume that the Jacobian matrices $J_{0i}(q_i)$ for $i = 1, 2$ are known and the orientation vector $(x_{01} - x_{02}, y_{01} - y_{02})^{\mathrm{T}}$ that has the same direction of $\overline{O_{01}O_{02}}$ is also known and available for the construction of the control signal. In other words, it is expected that the direction of the opposition force, coincident with the line $\overline{O_1O_2}$ but unknown, can be well approximated by the known axis of $\overline{O_{01}O_{02}}$.

Next, note that multiplication of Equation (2.85) by $\dot{q}_i$ and summing the resutant equations for $i = 1, 2$ yields

$$\sum_{i=1,2} \dot{q}_i u_i$$

$$= - \sum_{i=1,2} c_i \dot{q}_i^2 - \frac{\mathrm{d}}{\mathrm{d}t} \left[\frac{f_d}{2(r_1 + r_2)} \left\{ (x_{01} - x_{02})^2 + (y_{01} - y_{02})^2 \right\} \right]. \quad (2.86)$$

Then, it is easy to show (see Figure 2.10) that

$$(x_{01} - x_{02})^2 + (y_{01} - y_{02})^2 = (Y_1 - Y_2)^2 + l_w^2, \quad (2.87)$$

$$l_w = r_1 + r_2 + l_1 + l_2. \quad (2.88)$$

On the other hand, it follows from Equations (2.47) and (2.48) that

$$R_\theta^{\mathrm{T}} \begin{pmatrix} x_{01} - x_{02} \\ y_{01} - y_{02} \end{pmatrix} = \begin{pmatrix} -l_w \\ Y_1 - Y_2 \end{pmatrix} \quad (2.89)$$

from which it follows that

$$\begin{pmatrix} x_{01} - x_{02} \\ y_{01} - y_{02} \end{pmatrix} = R_\theta \begin{pmatrix} -l_w \\ Y_1 - Y_2 \end{pmatrix} = -l_w \boldsymbol{r}_X + (Y_1 - Y_2) \boldsymbol{r}_Y. \quad (2.90)$$

Thus, the control signals u_i $(i = 1, 2)$ can be recast into the form

$$u_i = -c_i \dot{q}_i - (-1)^i f_0 J_{0i}^{\mathrm{T}}(q_i) \boldsymbol{r}_X + \frac{f_d}{r_1 + r_2} (Y_1 - Y_2)(-1)^i \left\{ J_{0i}^{\mathrm{T}}(q_i) \boldsymbol{r}_Y - r_i \right\}$$
$$+ (-1)^i \frac{r_i f_d}{r_1 + r_2} (Y_1 - Y_2), \quad (2.91)$$

where

$$f_0 = \left(1 + \frac{l_1 + l_2}{r_1 + r_2} \right) f_d. \quad (2.92)$$

Substituting Equation (2.91) into Equation (2.52) yields

$$I_i \ddot{q}_i - (-1)^i \Delta f_i J_{0i}^{\mathrm{T}}(q_i) \boldsymbol{r}_X - \Delta \lambda_i \left\{ J_{0i}^{\mathrm{T}}(q_i) \boldsymbol{r}_Y - r_i \right\}$$
$$- (-1)^i \frac{r_i f_d}{r_1 + r_2} (Y_1 - Y_2) = 0, \qquad i = 1, 2, \quad (2.93)$$

where

$$\Delta f_i = f_i - f_0, \quad \Delta\lambda_i = \lambda_i + (-1)^i \frac{f_d}{r_1 + r_2}(Y_1 - Y_2), \quad i = 1, 2. \tag{2.94}$$

Note that Equation (2.93) becomes the same as Equation (2.60) if $\Delta\lambda_i$ is replaced with λ_i. Equations (2.53) and (2.54) can be rewritten as

$$M \begin{pmatrix} \ddot{x} \\ \ddot{y} \end{pmatrix} - (\Delta f_1 - \Delta f_2)\, \boldsymbol{r}_X + (\Delta\lambda_1 + \Delta\lambda_2)\, \boldsymbol{r}_Y = 0, \tag{2.95}$$

$$I\ddot{\theta} - \Delta f_1 Y_1 + \Delta f_2 Y_2 + \Delta\lambda_1 l_1 - \Delta\lambda_2 l_2 - f_d(Y_1 - Y_2) = 0. \tag{2.96}$$

These are also the same as Equation (2.61) and Equation (2.62), respectively, if $\Delta\lambda_i$ is replaced by λ_i for $i = 1, 2$. Thus, similarly to derivation of Equation (2.70), Equations (2.93), (2.95) and (2.96) can be recast in the vector-matrix equation:

$$H\ddot{\boldsymbol{X}} + C\dot{\boldsymbol{X}} - A\Delta\boldsymbol{\lambda} - \frac{f_d}{r_1 + r_2}(Y_1 - Y_2)\boldsymbol{e} = 0, \tag{2.97}$$

where

$$\Delta\boldsymbol{\lambda} = (\Delta f_1, \Delta f_2, \Delta\lambda_1, \Delta\lambda_2)^{\mathrm{T}}. \tag{2.98}$$

Note that this $\Delta\boldsymbol{\lambda}$ differs slightly from that in Equation (2.71). Similarly, by taking inner product between Equation (2.97) and $\dot{\boldsymbol{X}}$ or substituting Equation (2.86) into Equation (2.56), we obtain

$$\frac{\mathrm{d}}{\mathrm{d}t} E(\boldsymbol{X}, \dot{\boldsymbol{X}}) = - \sum_{i=1,2} c_i \dot{q}_i^2, \tag{2.99}$$

where E is given by Equation (2.67). This relation is the same as Equation (2.65).

It can easily be reconfirmed that the closed-loop dynamics of Equations (2.93–2.96) is derived as Lagrange's equation of motion for the Lagrangian

$$L = K - P + \sum_{i=1,2} \{\Delta f_i Q_i + \Delta\lambda_i R_i\}, \tag{2.100}$$

where Q_i and R_i ($i = 1, 2$) are defined by (2.47) and (2.50), K is given by (2.43) and P is the same artificial potential as given in (2.66). This means that the same argument developed in the previous section for proving the convergence of solution trajectories of Equation (2.70) can apply to Equations (2.93), (2.95) and (2.96) or Equation (2.97). Then, it is concluded that

$$Y_1(t) - Y_2(t) \to 0, \quad \Delta f_i(t) \to 0, \quad \Delta\lambda_i(t) \to 0, \quad i = 1, 2 \tag{2.101}$$

as $t \to \infty$. This means that

Table 2.2. Physical parameters

$m_{11} = m_{21}$	link mass	$0.025[\mathrm{kg}]$
$I_{11} = I_{21}$	inertia moment	$3.333 \times 10^{-6}[\mathrm{kg} \cdot \mathrm{m}^2]$
$l_{11} = l_{21}$	link length	$0.040[\mathrm{m}]$
$r_1 = r_2$	radius	$0.01[\mathrm{m}]$
M	object mass	$0.027[\mathrm{kg}]$
h	object height	$0.025[\mathrm{m}]$
$w = (l_1 + l_2)$	object width	$0.03[\mathrm{m}]$
I	object inertia moment	$3.431 \times 10^{-6}[\mathrm{kg} \cdot \mathrm{m}^2]$
$l_1 = l_2$	object length	$0.015[\mathrm{m}]$

Table 2.3. Parameters of the control signals

f_d	internal force	1.0[N]
$c_1 = c_2$	damping coefficient	0.002[msN]
γ_f	CSM gain	1500.0
γ_λ	CSM gain	3000.0

$$f_i(t) \to f_0, \quad \lambda_i(t) \to 0, \quad i = 1, 2 \tag{2.102}$$

as $t \to \infty$. This concludes that solution trajectories of Equation (2.97) converge asymptotically to the equilibrium state.

In order to see how rapidly such solution trajectories converge to the equilibrium state, we will show results of numerical simulation conducted on the basis of physical models of robot fingers and a 2-D rectangular object with regular size and weight as shown in Table 2.2. In this simulation, Baumgarte's constraint stabilisation method (CSM) is employed by introducing a class of over-damped second-order differential equation with gains γ_f and γ_λ to approximate the holonomic constraints well. The details of the CSM method will be given in Section 4.5. By using the control gains given in Table 2.3 and starting from the initial state with $\dot{\boldsymbol{X}}(0) = 0$ given in Table 2.3, we can obtain a numerical solution to Equation (2.97). In Figure 2.11 we show the transient behaviours of the key physical variables involved in Equation (2.97). As can be seen from Figure 2.11, $Y_1 - Y_2$, f_i and λ_i $(i = 1, 2)$ converge asymptotically to each expected constant value. As a matter of course, the rotational angle θ of the object also converges asymptotically to a certain constant. Furthermore, all the transient behaviours of the physical variables suggest that the speed of all these convergences must be exponential in t. In the next section, we shall confirm this exponential convergence of the solution trajectories in a rigorous mathematical way.

Before closing the section we will show another coordinated control signal for establishing force/torque balance for the physical setup shown in Figure 2.10. This is of the form

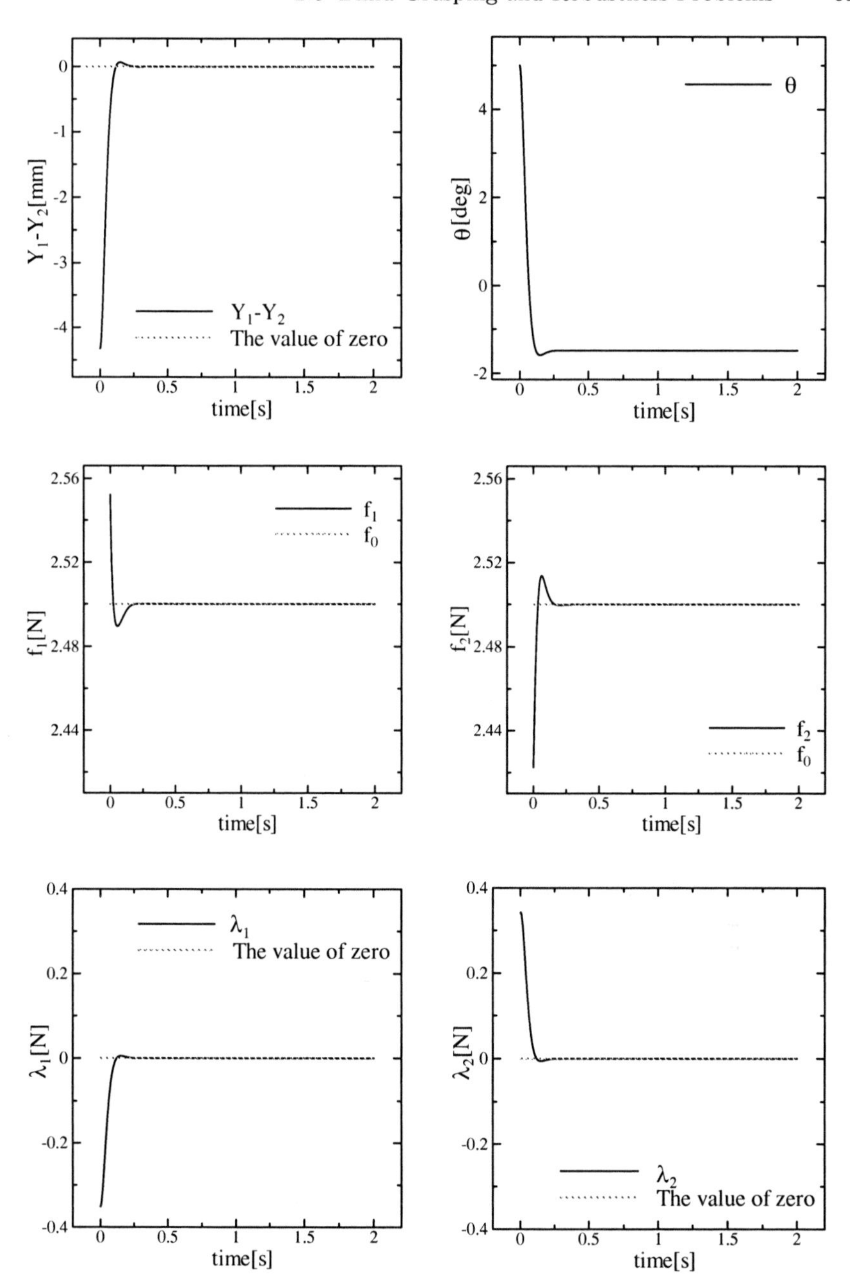

Fig. 2.11. The transient responses of physical variables along a solution to Equation (2.97) when the control signals of Equation (2.85) are exerted on finger joints

$$u_i = -c_i\dot{q}_i + (-1)^i \frac{f_d}{r_1 + r_2} J_{0i}^{\mathrm{T}}(q_i) \begin{pmatrix} x_{01} - x_{02} \\ y_{01} - y_{02} \end{pmatrix} - r_i \hat{N}_i, \quad i = 1, 2 \quad (2.103)$$

where

$$\begin{aligned} \hat{N}_i(t) &= \hat{N}_i(0) + \gamma_i^{-1} \int_0^t r_i \dot{q}_i(\tau)\,\mathrm{d}\tau \\ &= \hat{N}_i(0) + (r_i/\gamma_i)\{q_i(t) - q_i(0)\}, \quad i = 1, 2. \end{aligned} \quad (2.104)$$

The closed-loop dynamics obtained by substituting Equation (2.103) into Equation (2.52) can be written as the vector–matrix form

$$H\ddot{\boldsymbol{X}} + C\dot{\boldsymbol{X}} - A\Delta\boldsymbol{\lambda} - \frac{f_d}{r_1 + r_2}(Y_1 - Y_2)\boldsymbol{e} + \sum_{i=1,2} r_i \hat{N}_i \boldsymbol{e}_i = 0, \quad (2.105)$$

where

$$\boldsymbol{e}_1 = (1, 0, 0, 0, 0)^{\mathrm{T}}, \quad \boldsymbol{e}_2 = (0, 1, 0, 0, 0)^{\mathrm{T}}. \quad (2.106)$$

Then, it is easy to see that taking inner product between Equation (2.105) and $\dot{\boldsymbol{X}}$ yields

$$\frac{\mathrm{d}}{\mathrm{d}t} E_N(\boldsymbol{X}, \dot{\boldsymbol{X}}) = -\sum_{i=1,2} c_i \dot{q}_i^2, \quad (2.107)$$

where

$$E_N(\boldsymbol{X}, \dot{\boldsymbol{X}}) = K + P_N, \quad (2.108)$$

$$P_N = \frac{f_d}{2(r_1 + r_2)}(Y_1 - Y_2)^2 + \sum_{i=1,2} \frac{\gamma_i}{2} \hat{N}_i^2. \quad (2.109)$$

Applying a similar argument to that given in verifying the convergence of solution trajectories to Equation (2.97), we can conclude that $\dot{\boldsymbol{X}}$ and $\ddot{\boldsymbol{X}}$ tend to vanish as $t \to \infty$ and thereby as $t \to \infty$

$$A\Delta\boldsymbol{\lambda} + \frac{f_d}{r_1 + r_2}(Y_1 - Y_2)\boldsymbol{e} - \sum_{i=1,2} r_i \hat{N}_i \boldsymbol{e}_i \to 0. \quad (2.110)$$

Since matrix $[A, \boldsymbol{e}]$ is of 5×5 and non-singular at a regular position of the fingers–object setup shown in Figure 2.10, it is expected from Equation (2.110) that as $t \to \infty$ physical variables Δf_i, $\Delta\lambda_i$ $(i = 1, 2)$, $Y_1 - Y_2$ and $\hat{N}_i$ $(i = 1, 2)$ may converge to some constant respectively. More explicitly, let us consider the minimisation problem:

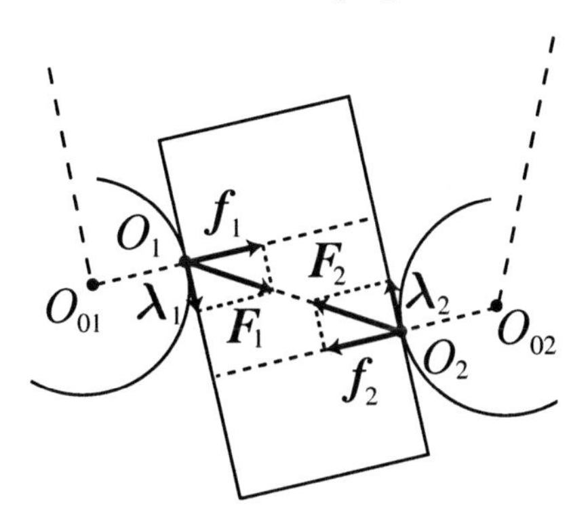

Fig. 2.12. Pressing forces $\boldsymbol{F}_1$ and $\boldsymbol{F}_2$ on the object must be collinear and oppositely directed

Table 2.4. Parameters of the control signals

f_d	internal force	1.0[N]
$c_1 = c_2 = c_m$	damping coefficient	0.002[msN]
$\gamma_i (i = 1, 2)$	regressor gain	0.001
$\hat{N}_i(0)(i = 1, 2)$	initial estimate value	0.0
γ_f	CSM gain	1500.0
γ_λ	CSM gain	3000.0

$$\begin{cases} \text{Minimise } P_N = \dfrac{f_d}{2(r_1 + r_2)}(Y_1 - Y_2)^2 + \dfrac{\gamma_1}{2}\hat{N}_1^2 + \dfrac{\gamma_2}{2}\hat{N}_2^2 \\ \text{under the constraints} \\ \qquad Q_1 = 0, \quad Q_2 = 0, \quad R_1 = 0, \quad R_2 = 0 \end{cases}$$

Then, the solution $\boldsymbol{X} = \boldsymbol{X}^*$ that minimises P_N under the above constraints must satisfy the equations

$$A\Delta\boldsymbol{\lambda} + \frac{f_d}{r_1 + r_2}(Y_1 - Y_2)\boldsymbol{e} = r_1\hat{N}_1\boldsymbol{e}_1 + r_2\hat{N}_2\boldsymbol{e}_2. \tag{2.111}$$

The minimising position state $\boldsymbol{X} = \boldsymbol{X}^*$ actually happens in a physical state as shown in Figure 2.12 where the pressing force $\boldsymbol{F}_1$ to the object from the left finger and $\boldsymbol{F}_2$ from the right finger must be collinear and oppositely directed. Before asertaining this observation theoretically, we show how fast a solution to Equation (2.105) tends to the position $\boldsymbol{X} = \boldsymbol{X}^*$ that attains force/torque balance with the aid of computer simulation. Again, numerical simulation for the closed-loop dynamics of Equation (2.105) has been carried out by using the same physical parameters of the fingers–object system given in Table 2.2 and the control gains given in Table 2.4. We show the transient responses of the key physical variables in Figure 2.13. Apparently from the last two graphs

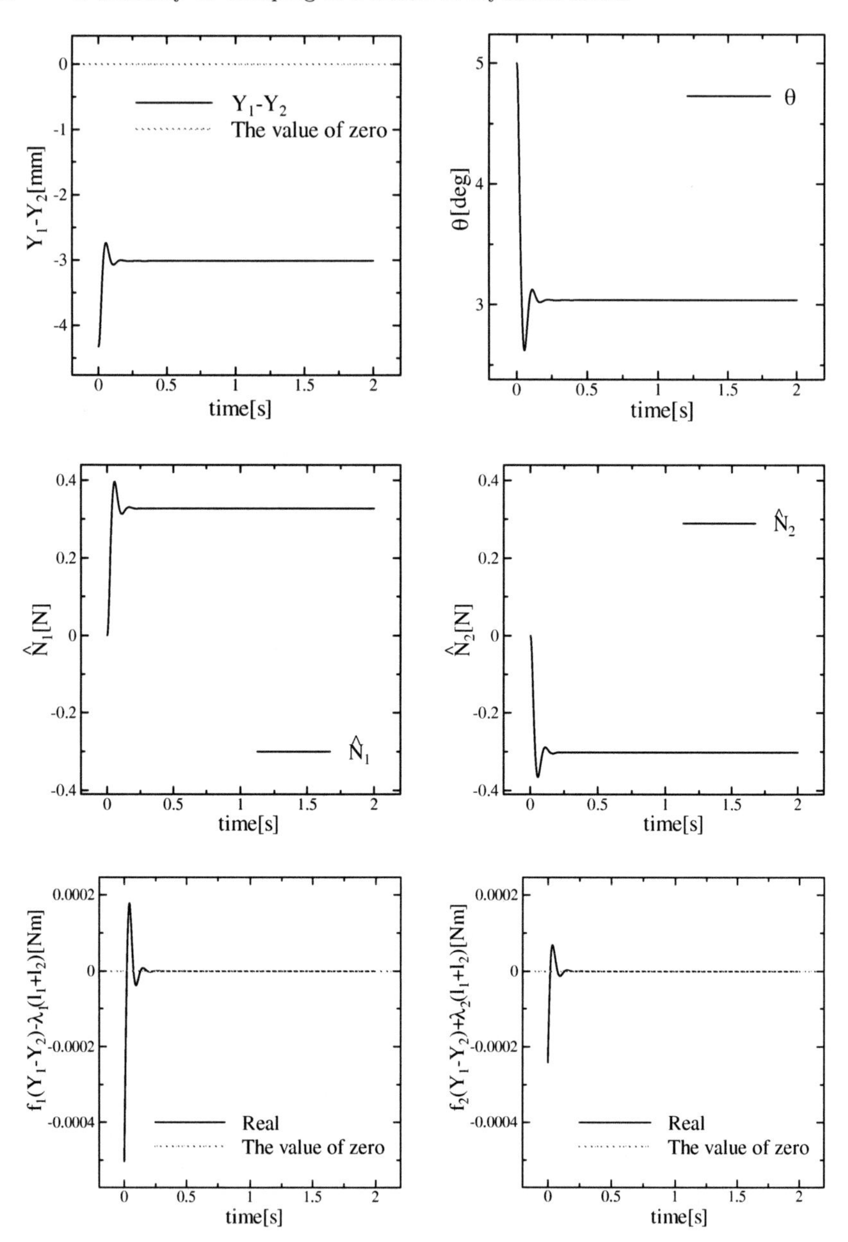

Fig. 2.13. The transient responses of physical variables along a solution to the closed-loop equation when the control signals of Equation (2.103) are used

of the figure, the solution converges to satisfy

$$f_i(Y_1 - Y_2) + (-1)^i \lambda_i (l_1 + l_2) = 0, \quad i = 1, 2, \tag{2.112}$$

which shows that as $t \to \infty$ force/torque balance is established.

Let us now find a solution satisfying Equation (2.111). Since x and y components of Equation (2.105) are the same as Equation (2.95), it should follow that

$$\Delta f_1 = \Delta f_2, \quad \Delta\lambda_1 = -\Delta\lambda_2. \tag{2.113}$$

Substituting this into Equation (2.96), we obtain

$$-\Delta f_1(Y_1 - Y_2) + \Delta\lambda_1(l_1 + l_2) - f_d(Y_1 - Y_2) = 0. \tag{2.114}$$

Subsequently, substituting Equation (2.94) into this equation yields

$$f_1(Y_1 - Y_2) + \lambda_1(l_1 + l_2) = 0, \tag{2.115}$$

which together with Equation (2.113) implies Equation (2.112). Thus, once f_1 is determined, the other magnitudes of the constraint forces f_2, λ_1, and λ_2 can be determined through Equations (2.113) and (2.115). The remaining six optimal values for the five components of $\boldsymbol{X}$ and f_1 can be determined by six equations, which are 1) the first two components of Equation (2.105), where f_2 and λ_2 are substituted by f_1 and $-\lambda_1$, respectively, and again λ_1 is substituted by $-f_1(Y_1 - Y_2)/(l_1 + l_2)$ owing to Equation (2.115) and 2) the four constraint equations $Q_i = 0$ and $R_i = 0$ for $i = 1, 2$. Finally, we remark that the last term $-r_i\hat{N}_i$ of the control signal of Equation (2.103) plays a role of saving abundant joint movements from initial angles.

In view of these theoretical arguments and simulation results shown in Figure 2.13, $Y_1 - Y_2$ may not tend to zero as $t \to \infty$ and therefore the λ_i $(i = 1, 2)$ also do not vanish with increasing t. This means that, around the state of force/torque balance of the fingers–object system (see Figure 2.9 or 2.10), the control torque inputs of Equation (2.103) generated at the finger joints are transmitted to the fingertips so as to withstand the reaction forces $-\boldsymbol{F}_i$ that are exerted on the contact points O_i for $i = 1, 2$. In other words, in this case non-zero tangential forces λ_i at the contact points to the object should be sustained by finger joint actuators even when the fingers–object state converges approximately to a still state attaining force/torque balance.

One of the advantages of using the coordinated control signal of Equation (2.103) is that it may be robust against the geometrical shapes of objects. In fact, the signals of Equation (2.103) can be constructed without knowing the geometrical shape of the object surface. In the next chapter, we shall discuss stability problems of the closed-loop dynamics when the same control signals as Equation (2.103) are used for 2-D objects with non-parallel but flat surfaces. Robustness problems of the control signals of Equation (2.103) for a general class of 2-D objects with smooth convex sides remain unsolved.

2.7 Exponential Convergence to Force/Torque Balance

Some of the arguments of the last two sections can be brought out more clearly and rigorously by discussing the speed of convergences of solution trajectories of the closed-loop dynamics toward the equilibrium point satisfying the balance of forces and torques acting on the object.

We consider the closed-loop Equation (2.97) of motion of the fingers–object system depicted in Figure 2.9 when the co-ordinated control signals of Equation (2.85) are used. First, we introduce a scale factor $r > 0$ and transform x and y to $\bar{x}$ and $\bar{y}$ in such a way that

$$\bar{x} = r^{-1}x, \quad \bar{y} = r^{-1}y. \tag{2.116}$$

Then, by defining $\bar{M} = r^2 M$, we see that

$$\frac{1}{2}M\left(\dot{x}^2 + \dot{y}^2\right) = \frac{1}{2}\bar{M}\left(\dot{\bar{x}}^2 + \dot{\bar{y}}^2\right). \tag{2.117}$$

The reason why such a scale transformation is required is that Equation (2.61) expresses translational motion of the obejct on the basis of physical units [m] but Equations (2.60) and (2.62) express the rotational motion based on physical units [radian]. This was caused by the adoption of the generalized position coordinates $\boldsymbol{X} = (q_1, q_2, x, y, \theta)^{\mathrm{T}}$ mixed with physical units [m] and [radian]. This also causes imbalance among the eigenvalues of the inertia matrix H defined in Equation (2.71). Therefore, once Equation (2.61) for the translational motion of the object is rewritten by this scale transformation as

$$\bar{M}\begin{pmatrix}\ddot{\bar{x}} \\ \ddot{\bar{y}}\end{pmatrix} - r\left(\Delta f_1 - \Delta f_2\right)\boldsymbol{r}_X + r\left(\lambda_1 + \lambda_2\right)\boldsymbol{r}_Y = 0 \tag{2.118}$$

Equation (2.70) can be written in the form

$$\bar{H}\ddot{\bar{\boldsymbol{X}}} + C\dot{\bar{\boldsymbol{X}}} - \bar{A}\Delta\boldsymbol{\lambda} - \frac{f_d}{r_1 + r_2}(Y_1 - Y_2)\boldsymbol{e} = 0, \tag{2.119}$$

where

$$\left\{\begin{aligned}
&\bar{\boldsymbol{X}} = \left(q_1, q_2, r^{-1}x, r^{-1}y, \theta\right), \quad \bar{\boldsymbol{r}}_X = r\begin{pmatrix}\cos\theta \\ -\sin\theta\end{pmatrix}, \\
&\bar{H} = \begin{pmatrix} I_1 & 0 & 0 & 0 & 0 \\ 0 & I_2 & 0 & 0 & 0 \\ 0 & 0 & r^2M & 0 & 0 \\ 0 & 0 & 0 & r^2M & 0 \\ 0 & 0 & 0 & 0 & I \end{pmatrix}, \quad \bar{\boldsymbol{r}}_Y = r\begin{pmatrix}\sin\theta \\ \cos\theta\end{pmatrix}, \\
&\bar{A} = \begin{pmatrix} -J_{01}^{\mathrm{T}}\boldsymbol{r}_X & 0 & J_{01}^{\mathrm{T}}\boldsymbol{r}_Y - r_1 & 0 \\ 0 & J_{02}^{\mathrm{T}}\boldsymbol{r}_X & 0 & J_{02}^{\mathrm{T}}\boldsymbol{r}_Y - r_2 \\ \bar{\boldsymbol{r}}_X & -\bar{\boldsymbol{r}}_X & -\bar{\boldsymbol{r}}_Y & -\bar{\boldsymbol{r}}_Y \\ Y_1 & -Y_2 & -l_1 & l_2 \end{pmatrix}.
\end{aligned}\right. \tag{2.120}$$

It is also necessary to introduce the pseudo-inverse of 4×5 matrix $\bar{A}^{\mathrm{T}}$, which is defined as

$$\left(\bar{A}^{\mathrm{T}}\right)^{+} = \bar{A}\left(\bar{A}^{\mathrm{T}}\bar{A}\right)^{-1}. \tag{2.121}$$

Evidently it follows that

$$\dot{\boldsymbol{X}}^{\mathrm{T}}\left(\bar{A}^{\mathrm{T}}\right)^{+} = \dot{\boldsymbol{X}}^{\mathrm{T}}\bar{A}\left(\bar{A}^{\mathrm{T}}\bar{A}\right)^{-1} = 0. \tag{2.122}$$

Let us define another important 5×5 matrix

$$P = I_5 - \left(\bar{A}^{\mathrm{T}}\right)^{+}\bar{A}^{\mathrm{T}} = I_5 - \bar{A}\left(\bar{A}^{\mathrm{T}}\bar{A}\right)^{-1}\bar{A}^{\mathrm{T}}. \tag{2.123}$$

Then, it is easy to see that

$$P\bar{A} = 0_{5\times 4} \tag{2.124}$$

and it follows that

$$P^{\mathrm{T}} = P, \quad PP = P, \quad \boldsymbol{e}^{\mathrm{T}}P\boldsymbol{e} \leq \boldsymbol{e}^{\mathrm{T}}\boldsymbol{e} \tag{2.125}$$

for any five-dimensional vector $\boldsymbol{e}$.

Now we are in a position to prove the exponential convergence of a solution to Equation (2.119). First, note that taking inner product between Equation (2.119) and $P\boldsymbol{e}(Y_1 - Y_2)$ yields

$$\boldsymbol{e}^{\mathrm{T}}P\bar{H}\ddot{\boldsymbol{X}}(Y_1 - Y_2) + \boldsymbol{e}^{\mathrm{T}}PC\dot{\boldsymbol{X}}(Y_1 - Y_2) - \frac{f_d(Y_1 - Y_2)^2}{r_1 + r_2}\boldsymbol{e}^{\mathrm{T}}P\boldsymbol{e} = 0 \tag{2.126}$$

from which it follows that

$$\begin{aligned}&\frac{\mathrm{d}}{\mathrm{d}t}\left\{-\boldsymbol{e}^{\mathrm{T}}P\bar{H}\dot{\boldsymbol{X}}(Y_1 - Y_2)\right\}\\&\quad = -\frac{f_d(Y_1 - Y_2)^2}{r_1 + r_2}\boldsymbol{e}^{\mathrm{T}}P\boldsymbol{e} + \boldsymbol{e}^{\mathrm{T}}PC\dot{\boldsymbol{X}}(Y_1 - Y_2) + h(\dot{\boldsymbol{X}}),\end{aligned} \tag{2.127}$$

where

$$h(\dot{\boldsymbol{X}}) = -\boldsymbol{e}^{\mathrm{T}}\left\{\dot{P}\bar{H}\dot{\boldsymbol{X}}(Y_1 - Y_2) + P\bar{H}\dot{\boldsymbol{X}}(\dot{Y}_1 - \dot{Y}_2)\right\}. \tag{2.128}$$

Here, $\boldsymbol{e}$ signifies the five-dimensional vector defined in Equation (2.71). Since in general it follows that for any $\gamma > 0$

$$|ab| \leq \frac{1}{2}\left(\gamma a^2 + (1/\gamma)b^2\right) \tag{2.129}$$

we see that

$$\boldsymbol{e}^{\mathrm{T}}PC\dot{\boldsymbol{X}}(Y_1 - Y_2) \leq \frac{f_d(Y_1 - Y_2)^2}{2(r_1 + r_2)}\boldsymbol{e}^{\mathrm{T}}P\boldsymbol{e} + \frac{r_1 + r_2}{2f_d}\left(c_1^2\dot{q}_1^2 + c_2^2\dot{q}_2^2\right). \tag{2.130}$$

Now we assume that at the equilibrium state $(\bar{\boldsymbol{X}}_\infty, 0)$ satisfying $Y_1 - Y_2 = 0$, $f_i = f_d$, $\lambda_i = 0$ $(i = 1, 2)$ as discussed in the paragraph containing Equation (2.84) and its subsequent paragraphs, matrix $\bar{A}$ is of full rank (non-degenerate) and the matrix $[\bar{A}, \boldsymbol{e}]$ is non-singular. Then, obviously at $\bar{\boldsymbol{X}} = \bar{\boldsymbol{X}}_\infty$, $\boldsymbol{e}^{\mathrm{T}} P \boldsymbol{e}$ does not vanish. Let us define

$$0 < \gamma_e = \frac{\boldsymbol{e}^{\mathrm{T}} P \boldsymbol{e}}{\boldsymbol{e}^{\mathrm{T}} \boldsymbol{e}} = \left\{ 1 - \frac{\boldsymbol{e}^{\mathrm{T}} \left(\bar{A}^{\mathrm{T}}\right)^{+} \bar{A}^{\mathrm{T}} \boldsymbol{e}}{\boldsymbol{e}^{\mathrm{T}} \boldsymbol{e}} \right\}, \tag{2.131}$$

which is evaluated at $\bar{\boldsymbol{X}} = \bar{\boldsymbol{X}}_\infty$. At this stage, we note that the Lyapunov function $E(\boldsymbol{X}, \dot{\boldsymbol{X}})$ [or equivalently, $E(\bar{\boldsymbol{X}}, \dot{\bar{\boldsymbol{X}}})$] is positive definite with respect to $(\bar{\boldsymbol{X}}, \dot{\bar{\boldsymbol{X}}})$ under the constraints $Q_i = 0$, $R_i = 0$, $\dot{Q}_i = 0$ and $\dot{R}_i = 0$ $(i = 1, 2)$ in a neighbourhood of $\bar{\boldsymbol{X}} = \bar{\boldsymbol{X}}_\infty$. Hence, it is possible to choose $\delta > 0$ such that, at any $(\bar{\boldsymbol{X}}, \dot{\bar{\boldsymbol{X}}})$ satisfying

$$E\left(\bar{\boldsymbol{X}}, \dot{\bar{\boldsymbol{X}}}\right) = E\left(\boldsymbol{X}, \dot{\boldsymbol{X}}\right) < \delta \tag{2.132}$$

and constraints $Q_i = 0$ and $R_i = 0$ $(i = 1, 2)$, $\boldsymbol{e}^{\mathrm{T}} P \boldsymbol{e}$ also satisfies

$$\boldsymbol{e}^{\mathrm{T}} P \boldsymbol{e} / \boldsymbol{e}^{\mathrm{T}} \boldsymbol{e} \geq \frac{1}{2} \gamma_e \tag{2.133}$$

and in addition that the matrix $\bar{A}$ is non-degenerate. Further, we choose f_d and c_i $(i = 1, 2)$ so that they satisfy

$$\frac{(r_1 + r_2) c_i}{f_d \gamma_e \boldsymbol{e}^{\mathrm{T}} \boldsymbol{e}} \leq \frac{1}{2} \qquad i = 1, 2. \tag{2.134}$$

Then, by substituting inequality (2.130) into (2.127) and referring to Equations (2.133) and (2.134), we obtain

$$\begin{aligned} &\frac{\mathrm{d}}{\mathrm{d}t} \left\{ -\frac{2}{\gamma_e \boldsymbol{e}^{\mathrm{T}} \boldsymbol{e}} \boldsymbol{e}^{\mathrm{T}} P \bar{H} \dot{\bar{\boldsymbol{X}}} (Y_1 - Y_2) \right\} \\ &\qquad \leq -\frac{f_d (Y_1 - Y_2)^2}{2(r_1 + r_2)} + \frac{1}{2} \left(c_1 \dot{q}_1^2 + c_2 \dot{q}_2^2 \right) + \frac{2}{\gamma_e \boldsymbol{e}^{\mathrm{T}} \boldsymbol{e}} h(\dot{\bar{\boldsymbol{X}}}). \end{aligned} \tag{2.135}$$

Now, we define

$$V_\alpha = E\left(\bar{\boldsymbol{X}}, \dot{\bar{\boldsymbol{X}}}\right) - \frac{2\alpha}{\gamma_e \boldsymbol{e}^{\mathrm{T}} \boldsymbol{e}} \boldsymbol{e}^{\mathrm{T}} P \bar{H} \dot{\bar{\boldsymbol{X}}} (Y_1 - Y_2), \tag{2.136}$$

where α is a positive parameter such that $0 < \alpha \leq 1$. Obviously, V_α is a quadratic function of $\bar{\boldsymbol{X}}$, $\dot{\bar{\boldsymbol{X}}}$ and $Y_1 - Y_2$ and, according to Equations (2.99) and (2.135), the time derivative $\dot{V}_\alpha$ becomes

$$\begin{aligned} \frac{\mathrm{d}}{\mathrm{d}t} V_\alpha \leq &-\left(1 - \frac{\alpha}{2}\right) \left(c_1 \dot{q}_1^2 + c_2 \dot{q}_2^2 \right) \\ &\qquad - \frac{\alpha f_d}{2(r_1 + r_2)} (Y_1 - Y_2)^2 + \alpha \bar{h}(\dot{\bar{\boldsymbol{X}}}). \end{aligned} \tag{2.137}$$

where

$$\bar{h}(\dot{\boldsymbol{X}}) = \frac{2}{\gamma_e \boldsymbol{e}^{\mathrm{T}} \boldsymbol{e}} h(\dot{\boldsymbol{X}})$$
$$= -\frac{2}{\gamma_e \boldsymbol{e}^{\mathrm{T}} \boldsymbol{e}} \boldsymbol{e}^{\mathrm{T}} \left\{ \dot{P} \bar{H} \dot{\boldsymbol{X}} (Y_1 - Y_2) + P \bar{H} \dot{\boldsymbol{X}} (\dot{Y}_1 - \dot{Y}_2) \right\}. \quad (2.138)$$

Note that this can be regarded as a quadratic function of $\dot{\boldsymbol{X}}$ and further $|Y_1 - Y_2|$ is at least of $O(r_1 + r_2)$. Hence, there is a constant β of $O(1)$ such that

$$\left| \bar{h}(\dot{\boldsymbol{X}}) \right| \leq \frac{\beta}{\gamma_e} K(\dot{\boldsymbol{X}}) = \frac{\beta}{\gamma_e} K(\dot{\boldsymbol{X}}), \quad (2.139)$$

where K denotes the kinetic energy defined by Equation (2.43). To simplify the mathematical argument, we assume at this stage that $c_1 = c_2 = c_{\max}$ and $r_1 = r_2 = r_m$. Furthermore, we assume that the object width $l_1 + l_2$ is not so small relative to r_i $(= r_m)$. Then, as discussed around the derivation of Equations (2.76) and (2.79), it is possible to confirm that

$$\begin{cases} \dot{\theta}^2 \leq \beta_\theta \left(\dot{q}_1^2 + \dot{q}_2^2 \right) \\ r^{-2} \left(\dot{x}^2 + \dot{y}^2 \right) \leq \beta_0 \left(\dot{q}_1^2 + \dot{q}_2^2 \right), \end{cases} \quad (2.140)$$

where β_θ and β_0 are positive constants of numerical order $O(1)$, and the scale factor r can be selected around $r = 0.01$–0.02 in relation to the physical parameters given in Tables 2.2 and 2.3. Then, the total kinetic energy must be of order

$$K(\dot{\boldsymbol{X}}) = \gamma \left(\dot{q}_1^2 + \dot{q}_2^2 \right) \quad (2.141)$$

with a positive constant γ that is of $O(10^{-5})$. Thus, referring to Equations (2.139) and (2.141), we can conclude that

$$\frac{\mathrm{d}}{\mathrm{d}t} V_\alpha \leq -\left(1 - \frac{\alpha}{2} - \frac{\alpha \beta \gamma}{\gamma_e c_m} \right) \left(c_m \dot{q}_1^2 + c_m \dot{q}_2^2 \right)$$
$$- \frac{\alpha f_d}{2(r_1 + r_2)} (Y_1 - Y_2)^2. \quad (2.142)$$

We now assume that γ_e defined by Equation (2.131) is large enough to satisfy $\gamma_e \geq 0.2$ and the finger joint damping coefficient c_m $(= c_1 = c_2)$ is chosen around $c_m = 0.001$–0.003 [Nms]. Then, $\beta\gamma / \gamma_e c_m$ must be smaller than $1/4$ and therefore inequality (2.142) can be reduced to, in reference to Equation (2.141), the following fundamental inequality:

$$\frac{\mathrm{d}}{\mathrm{d}t} V_\alpha \leq -\alpha E \left(\bar{\boldsymbol{X}}, \dot{\boldsymbol{X}} \right). \quad (2.143)$$

Next, we evaluate the upper bound of scalar function V_α in the following way:

$$V_\alpha \leq E + \frac{\alpha}{\gamma_e \boldsymbol{e}^{\mathrm{T}} \boldsymbol{e}} \left\{ \eta \boldsymbol{e}^{\mathrm{T}} P \boldsymbol{e} (Y_1 - Y_2)^2 + \eta^{-1} \dot{\boldsymbol{X}}^{\mathrm{T}} \bar{H} \bar{H} \dot{\boldsymbol{X}} \right\}. \tag{2.144}$$

Since $\boldsymbol{e}^{\mathrm{T}}\boldsymbol{e} = (r_1 + r_2)^2 + r_1^2 + r_2^2 = 6r_m^2$ and $\boldsymbol{e}^{\mathrm{T}} P \boldsymbol{e} / \boldsymbol{e}^{\mathrm{T}} \boldsymbol{e} \leq 1$, choosing

$$\eta = \gamma_e f_d / 8(r_1 + r_2) = \gamma_e f_d / 16 r_m \tag{2.145}$$

yields

$$V_\alpha \leq E + \frac{\alpha f_d}{8(r_1 + r_2)} (Y_1 - Y_2)^2 + \frac{8(r_1 + r_2)\alpha}{\gamma_e^2 f_d \boldsymbol{e}^{\mathrm{T}} \boldsymbol{e}} \lambda_M(\bar{H}) K(\dot{\boldsymbol{X}}), \tag{2.146}$$

where $\lambda_M(\bar{H})$ denotes the maximum eigenvalue of $\bar{H}$. Since $\lambda_M(\bar{H})$ is of numerical order $O(10^{-6})$ and

$$\frac{8(r_1 + r_2)}{\gamma_e^2 f_d \boldsymbol{e}^{\mathrm{T}} \boldsymbol{e}} = \frac{2 \times 10^2}{3 r_m f_d} = O(10^4) \tag{2.147}$$

as far as f_d is of $O(1)$ in [N], Equation (2.146) can be reduced to

$$V_\alpha \leq E + \frac{\alpha}{4} E = \left(1 + \frac{\alpha}{4}\right) E \left(\boldsymbol{X}, \dot{\boldsymbol{X}}\right). \tag{2.148}$$

From the same argument, it also follows that

$$V_\alpha \geq \left(1 - \frac{\alpha}{4}\right) E \left(\boldsymbol{X}, \dot{\boldsymbol{X}}\right), \tag{2.149}$$

that is,

$$\left(1 - \frac{\alpha}{4}\right) E \leq V_\alpha \leq \left(1 + \frac{\alpha}{4}\right) E \tag{2.150}$$

as far as $0 < \alpha \leq 1$. Thus, it follows from Equations (2.143) and (2.150) that

$$\frac{\mathrm{d}}{\mathrm{d}t} V_\alpha(t) \leq -\frac{4\alpha}{4 + \alpha} V_\alpha(t). \tag{2.151}$$

In particular, if we choose $\alpha = 1.0$, then it follows from Equations (2.150) and (2.151) that

$$E\left(\boldsymbol{X}(t), \dot{\boldsymbol{X}}(t)\right) \leq \frac{5}{3} E\left(\boldsymbol{X}(0), \dot{\boldsymbol{X}}(0)\right) \mathrm{e}^{-0.8t}. \tag{2.152}$$

In conclusion, it has been proved that any solution $(\boldsymbol{X}(t), \dot{\boldsymbol{X}}(t))$ starting from an arbitrary initial state satisfying $E(\boldsymbol{X}(0), \dot{\boldsymbol{X}}(0)) < (3/5)\delta$ and constraints $Q_i = 0$, $R_i = 0$, $\dot{Q}_i = 0$ and $\dot{R}_i = 0$ for $i = 1, 2$ remains in the neighbourhood $M_1(\boldsymbol{X}_\infty) = \{(\boldsymbol{X}, \dot{\boldsymbol{X}}) : E(\boldsymbol{X}, \dot{\boldsymbol{X}}) < \delta \text{ and } Q_i = 0, R_i = 0, \dot{Q}_i = 0, \dot{R}_i = 0 \text{ for } i = 1, 2\}$ of $(\boldsymbol{X}_\infty, 0)$ and converges exponentially to the equilibrium point $(\boldsymbol{X}_\infty, 0)$.

The proof presented is based upon the numerical orders of the physical parameters of the fingers and object given in Table 2.2 and the control gains given in Table 2.3. That is, the proof is not generic but context dependent as is usually the case in human dexterity seen in our everyday life, as discussed in Section 1.4. Notwithstanding such context dependency, a similar mathematical argument can be developed for stability proof of grasping by similar fingers–object mechanisms with different numerical orders. However, it should be remarked that a synergistic choice for c_i (the damping factors for finger joints) and f_d satisfying Equation (2.134) is vital to regulate the exponential speed of convergence of solutions to the closed-loop dynamics. Furthermore, it should be noted that, if the object width $l_1 + l_2$ becomes small relative to the radius of the finger-end spheres, then the attracter region of convergence must be shrunk and the parameter α in the definition of the function V_α should be chosen considerably less than 1, that is, $0 < \alpha << 1.0$. Fortunately, however, human finger-ends are not rigid but rather soft and deformable. In Chapter 6 we will show that the visco-elastic properties of the finger-end material widen such attrator regions even if the object is very thin and light like a credit card or a paper name card.

Finally, the exponential convergence of solutions to the closed-loop dynamics in the case that the control signals of Equation (2.103) are used will be treated in Chapter 4 as a case of more general problems of grasping under the effect of gravity and the condition that the object has non-parallel flat surfaces.

3

Testbed Problems to Control a 2-D Object Through Rolling Contact

This chapter poses a class of testbed problems of control for dynamic grasping or immobilisation of a 2-D rigid object, in order to gain physical insight into stable grasping in a dynamic sense. Such problems may play a principal role in understanding the important concept of sensory feedback stabilisation similarly to the inverted pendulum on a cart problem that played an essential role in the history of stabilisation and control of mechanical systems. The simplest problem is to stabilise or immobilise the rotational motion of a 2-D object with a flat side surface by using a multi-joint robot finger where the object can only rotate around a single fixed axis. It is assumed that the rotational motion of the object pivoted around the fixed axis is frictionless, the finger-end is hemispherical, and therefore that rolling between the finger-end and object surface is induced without incurring any slip. Lagrange's equation of motion for such a testbed finger–object system is derived together with two constraints: the point contact constraint and the rolling contact constraint. It is shown that there arises a rolling constraint force tangential to both the finger-end sphere and the object surface and originating at the contact point. By taking advantage of induced tangential constraint force, rotational motion of the object can be immobilised or torque-balanced by a single finger with at least two joints in a dynamic manner, although the object can be controlled indirectly through only constraint forces. This can be regarded as like an extended version of the feedback control of inverted pendulums. Through studying these fundamental testbed problems, the basic concepts of the key mathematics including constraint manifold, tangent space, Riemannian distance and Morse function are introduced, which will play crucial roles in subsequent chapters where the dynamics of grasping and its stabilisation by using a pair of multi-joint fingers are analyzed.

3.1 Stabilisation of Motion of a 2-D Object Through Rolling Contact

Consider a simple control problem of how to stop the rotational motion of a 2-D object pivoted at a fixed point $O_m(x_m, y_m)$ by using a single-finger robot with three joints as shown in Figure 3.1. The robot finger can be regarded as an idealized physical model of a human index finger shown in Figure 3.2. Differently from the human finger, we assume that the finger-end is hemispherical and rigid. It is assumed that all the rotational axes of the three finger joints and the object rotational axis have a direction in z perpendicular to the xy-plane and therefore the overall motion of the finger–object system is confined to the horizontal plane equivalent to the xy-plane. Further, we assume that rotational motion of the object around $O_m(x_m, y_m)$ is free and frictionless and that the translational motion of the object is pinned at the point $O_m(x_m, y_m)$. In the following, the effect of gravity is reasonably ignored, because only the motion of the system in the horizontal plane is considered.

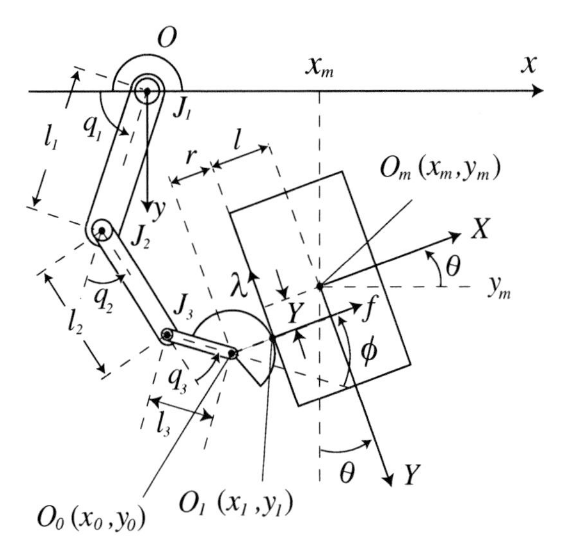

Fig. 3.1. Manipulation of an object by a three-DOF robot finger with a hemispherical end. The object is pinned to the horizontal xy-plane but its rotational motion around O_m is free and frictionless

Now, let us derive the equation of motion of this finger–object system as a Lagrange equation. To do this, denote the kinetic energy of the system by

$$K = \frac{1}{2}\dot{q}^{\mathrm{T}} H(q)\dot{q} + \frac{1}{2} I\dot{\theta}^2, \tag{3.1}$$

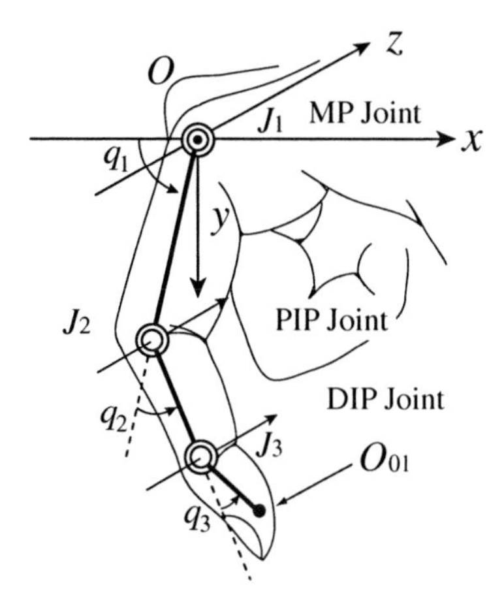

Fig. 3.2. Skeletal mechanism of a human-like index finger

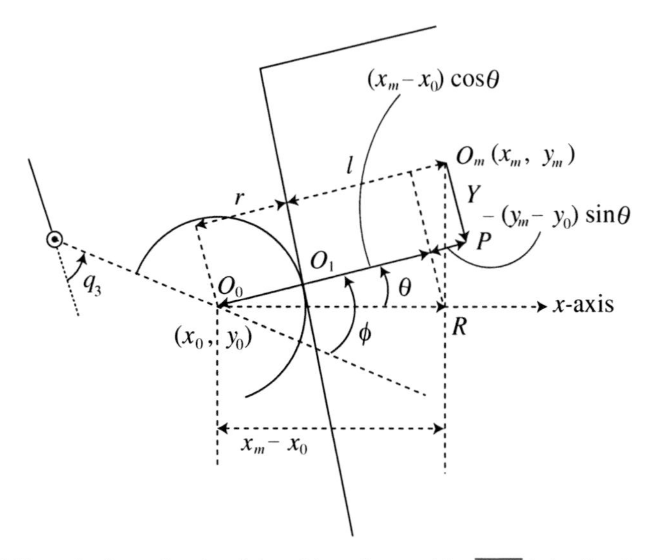

Fig. 3.3. When the lateral axis of the object denoted by $\overline{O_0P}$ is inclined to the x-axis by θ [radian], the length of $\overline{O_0P}$ ($|\overline{O_0P}| = r + l$) can be expressed as $(x_m - x_0)\cos\theta$ $-(y_m - y_0)\sin\theta$. Note that $|\overline{O_mP}| = Y$ and $|\overline{O_0R}| = x_m - x_0$

where $q = (q_1, q_2, q_3)^{\mathrm{T}}$, $H(q)$ stands for the inertia matrix of the robot finger and I the inertia moment of the object around the z-axis at the fixed point O_m. Obviously there arises a contact constraint reflecting the physical situation that the finger-end is in contact with the object surface. This contact constraint can be expressed as the following equation:

$$Q = -(r + l) + (x_m - x_0)\cos\theta - (y_m - y_0)\sin\theta = 0 \tag{3.2}$$

as seen from Figure 3.3, where $(x_0, y_0)^{\mathrm{T}}$ stands for the position of the centre of the finger-end sphere, r is its radius, and l is the one-side width of the object as shown in Figure 3.1. On the other hand, rolling contact is defined as a physical condition that the velocity of the contact point on the finger-end sphere is equivalent to that on the object surface, that is, the rolling contact constraint is expressed as

$$\frac{\mathrm{d}}{\mathrm{d}t}(r\phi) = -\frac{\mathrm{d}}{\mathrm{d}t}Y, \tag{3.3}$$

where ϕ denotes the angle as specified in Figure 3.1 and Y the Y-component of the contact point described by the coordinates (X, Y) attached and fixed to the object as shown in Figure 3.1. As observed from Figures 3.1 and 3.3, it follows that

$$Y = (x_0 - x_m)\sin\theta + (y_0 - y_m)\cos\theta \tag{3.4}$$

$$\phi = \pi + \theta - q_1 - q_2 - q_3 = \pi + \theta - q^{\mathrm{T}}\boldsymbol{e}, \tag{3.5}$$

where $\boldsymbol{e} = (1, 1, 1)^{\mathrm{T}}$ and θ denotes the angle of inclination of the object to the x-axis as defined in Figure 3.1. The zero-relative-velocity constraint of rolling contact expressed as Equation (3.3) can apparently be integrated in t, which reduces it to

$$R = c_0 + Y + r\phi = 0, \tag{3.6}$$

where c_0 is a constant of integration. We assume that no contact slip arises between the finger-end and object surfaces. Hence the constant c_0 in Equation (3.6) should be regarded as being fixed as long as the contact is maintained and does not slip. Then, it is possible to introduce Lagrange's multipliers f and λ for the constraint Equations (3.2) and (3.6), respectively, and define the Lagrangian

$$L = K + fQ + \lambda R. \tag{3.7}$$

Then, applying the variational principle for the Lagrangian in the form

$$\int_{t_1}^{t_2} \left\{\delta(K + fQ + \lambda R) + u^{\mathrm{T}}\delta q\right\} \mathrm{d}t = 0 \tag{3.8}$$

we obtain Lagrange's equation of motion [see Equations (2.39) and (2.40)]:

$$\left\{H(q)\frac{\mathrm{d}}{\mathrm{d}t} + \frac{1}{2}\dot{H}(q) + S(q, \dot{q})\right\}\dot{q} - f\frac{\partial Q}{\partial q} - \lambda\frac{\partial R}{\partial q} = u \tag{3.9}$$

$$I\ddot{\theta} - f\frac{\partial Q}{\partial \theta} - \lambda\frac{\partial R}{\partial \theta} = 0 \tag{3.10}$$

As discussed in Section 1.9, $S(q, \dot{q})$ is homogeneous in $\dot{q}$ and skew-symmetric. Equations (3.9) and (3.10) can be described in detail by using the Jacobian matrix

$$J(q) = \frac{\partial(x_0, y_0)^{\mathrm{T}}}{\partial q^{\mathrm{T}}} = \begin{pmatrix} \frac{\partial x_0}{\partial q_1} & \frac{\partial x_0}{\partial q_2} & \frac{\partial x_0}{\partial q_3} \\ \frac{\partial y_0}{\partial q_1} & \frac{\partial y_0}{\partial q_2} & \frac{\partial y_0}{\partial q_3} \end{pmatrix} \tag{3.11}$$

and calculating the gradients of Q and R in q or θ, respectively, as follows:

$$\begin{cases} \frac{\partial Q}{\partial q} = -J^{\mathrm{T}}(q)\boldsymbol{r}_X, & \boldsymbol{r}_X = \begin{pmatrix} \cos\theta \\ -\sin\theta \end{pmatrix}, \\ \frac{\partial R}{\partial q} = J^{\mathrm{T}}(q)\boldsymbol{r}_Y - r\boldsymbol{e}, & \boldsymbol{r}_Y = \begin{pmatrix} \sin\theta \\ \cos\theta \end{pmatrix}, \\ \frac{\partial Q}{\partial \theta} = Y, & \frac{\partial R}{\partial \theta} = -l. \end{cases} \tag{3.12}$$

Note that $\boldsymbol{r}_X$ stands for the unit vector of the X-axis and $\boldsymbol{r}_Y$ for that of the Y-axis attached to the object with the fixed origin $O_m(x_m, y_m)$. The equation of motion of the object described by (3.10) can be written in detail as

$$I\ddot{\theta} - fY + \lambda l = 0. \tag{3.13}$$

This equation evidently shows that the Lagrange multiplier f is regarded as the contact force that is pressing the object in the direction normal to the object surface and another multiplier λ as the rolling constraint force arising at the contact point in the common direction tangential to both the finger-end sphere and the object surface (see Figure 3.1). In order to stop rotational motion of the object, two rotational moments fY and λl around the origin O_m must become equal. Before discussing control problems for stopping motion of the object and/or controlling its rotational angle, we show an important physical law called "passivity" concerning the input u and the output $\dot{q}$ of Lagrange's equation described by Equations (3.9) and (3.10). This is shown by taking inner products between $\dot{q}$ and Equation (3.9) and between $\dot{\theta}$ and Equation (3.10) and summing these results, from which it follows that

$$\frac{\mathrm{d}}{\mathrm{d}t}\{K\} = \dot{q}^{\mathrm{T}} u. \tag{3.14}$$

This implies

$$\int_0^t \dot{q}^{\mathrm{T}}(\tau)\, u(\tau)\, \mathrm{d}\tau = K(t) - K(0) \geq -K(0), \tag{3.15}$$

which is called the passivity, where $K(t)$ denotes the value of the kinetic energy K at time t. In the derivation of Equation (3.14), the velocity constraints

$\dot{Q} = \dot{q}^{\mathrm{T}}(\partial Q/\partial q) + \dot{\theta}(\partial Q/\partial \theta) = 0$ and $\dot{R} = \dot{q}^{\mathrm{T}}(\partial R/\partial q) + \dot{\theta}(\partial R/\partial \theta) = 0$ were employed.

Next consider the problem of how to stop the motion of an object by designing a control input u in the motion equation of the finger described by Equation (3.9). It should be remarked that the motion of the object can be controlled indirectly through the contact constraint forces f and λ as seen in Equation (3.13). First, we intend not only to stabilise the motion of the object but also to control the object rotational angle θ toward $\theta = \theta_d$. Assume that not only the finger joint angle $q_i(t)$ $(i = 1, 2, 3)$ but also $\theta(t)$ and $Y(t)$ can be measured. Then, it is possible to consider the following control signal:

$$u = -C\dot{q} - f_d\left(\frac{\partial Q}{\partial q}\right) - f_d Y \boldsymbol{e} - \beta\Delta\theta\left(\frac{\partial R}{\partial q}\right), \tag{3.16}$$

where $\Delta\theta = \theta - \theta_d$ and $C = \mathrm{diag}(c_1, c_2, c_3)$ with $c_i > 0$. It should be noted that the inner product between $\dot{q}$ and u of Equation (3.16) yields

$$\begin{aligned}\dot{q}^{\mathrm{T}}u &= -\dot{q}^{\mathrm{T}}C\dot{q} + f_d\dot{\theta}(\partial Q/\partial\theta) - f_d Y\dot{q}^{\mathrm{T}}\boldsymbol{e} + \beta\Delta\theta\cdot\dot{\theta}(\partial R/\partial\theta)\\ &= -\dot{q}^{\mathrm{T}}C\dot{q} + f_d Y(\dot{\theta} - \dot{q}\boldsymbol{e}) - \frac{\mathrm{d}}{\mathrm{d}t}\frac{\beta l}{2}\Delta\theta^2\\ &= -\dot{q}^{\mathrm{T}}C\dot{q} - \frac{\mathrm{d}}{\mathrm{d}t}\left\{\frac{1}{2}\left(\frac{f_d}{r}Y^2 + \beta l\Delta\theta^2\right)\right\}.\end{aligned} \tag{3.17}$$

Substituting this equality into Equation (3.14) yields

$$\frac{\mathrm{d}}{\mathrm{d}t}E = -\dot{q}^{\mathrm{T}}C\dot{q}, \tag{3.18}$$

where

$$E = \frac{1}{2}\left\{\dot{q}^{\mathrm{T}}H(q)\dot{q} + I\dot{\theta}^2 + \frac{f_d}{r}Y^2 + \beta l\Delta\theta^2\right\}. \tag{3.19}$$

On the other hand, substituting u from Equation (3.16) into Equation (3.9) leads to

$$\left\{H(q)\frac{\mathrm{d}}{\mathrm{d}t} + \frac{1}{2}\dot{H}(q) + S(q,\dot{q}) + C\right\}\dot{q} - \Delta f\frac{\partial Q}{\partial q} - \Delta\lambda\frac{\partial R}{\partial q} + f_d Y\boldsymbol{e} = 0, \tag{3.20}$$

where $\Delta f = f - f_d$ and $\Delta\lambda = \lambda - \beta\Delta\theta$. Equation (3.10) or (3.13) can be rewritten in the form

$$I\ddot{\theta} - \Delta f Y + \Delta\lambda l - f_d Y + \beta l\Delta\theta = 0. \tag{3.21}$$

In fact, the sum of the inner products between $\dot{q}$ and Equation (3.20) and between $\dot{\theta}$ and Equation (3.21) is reduced exactly to Equation (3.18).

We are now in a position to discuss the stability of an equilibrium state of the system composed of Equations (3.20) and (3.21), which should satisfy

$$\begin{cases} \Delta\theta = 0, \quad Y = 0, \quad \Delta f = 0, \quad \Delta\lambda = 0 \\ \dot{q} = 0, \quad \dot{\theta} = 0. \end{cases} \tag{3.22}$$

The state of the system denoted by

$$\boldsymbol{X} = (q^{\mathrm{T}}, \theta)^{\mathrm{T}}, \qquad \dot{\boldsymbol{X}} = (\dot{q}^{\mathrm{T}}, \dot{\theta})^{\mathrm{T}} \tag{3.23}$$

can be regarded as belonging to $(\boldsymbol{X}, \dot{\boldsymbol{X}}) \in R^8$, but it is subject to two holonomic constraints described by Equations (3.2) and (3.6). Hence, the system has two DOFs and its state must lie on the following four-dimensional constraint manifold:

$$TB_4 = \left\{ (\boldsymbol{X}, \dot{\boldsymbol{X}}) : Q = 0, R = 0, \dot{Q} = 0, \dot{R} = 0 \right\}. \tag{3.24}$$

At the same time, the position $\boldsymbol{X}$ to be considered constitutes the two-dimensional manifold

$$CM_2 = \{ \boldsymbol{X} : Q = 0, R = 0 \}, \tag{3.25}$$

which is considered to be embedded in the four-dimensional configuration space $CS^4 = \left\{ (q^{\mathrm{T}}, \theta)^{\mathrm{T}} \right\} = \{ \boldsymbol{X} \}$. To develop the mathematical argument in a more rigorous way, we rewrite Equations (3.20) and (3.21) in the following single vector–matrix form:

$$\tilde{H}\ddot{\boldsymbol{X}} + \left(\frac{1}{2}\dot{\tilde{H}} + \tilde{S} + \tilde{C} \right) \dot{\boldsymbol{X}} - A\boldsymbol{\lambda} + B\boldsymbol{\eta} = 0, \tag{3.26}$$

where

$$\begin{cases} \boldsymbol{\lambda} = \begin{pmatrix} \Delta f \\ \Delta\lambda \end{pmatrix}, \qquad \boldsymbol{\eta} = \begin{pmatrix} Y \\ \Delta\theta \end{pmatrix}, \\ A = \begin{pmatrix} \dfrac{\partial Q}{\partial q} & \dfrac{\partial R}{\partial q} \\ Y & -l \end{pmatrix}, \quad B = \begin{pmatrix} f_d e & 0_{3\times 1} \\ -f_d & \beta l \end{pmatrix}, \\ \tilde{H} = \begin{pmatrix} H(q) & 0_{3\times 1} \\ 0_{1\times 3} & I \end{pmatrix}, \quad \tilde{S} = \begin{pmatrix} S(q, \dot{q}) & 0_{3\times 1} \\ 0_{1\times 3} & 0 \end{pmatrix}, \quad \tilde{C} = \begin{pmatrix} C & 0_{3\times 1} \\ 0_{1\times 3} & 0 \end{pmatrix}. \end{cases} \tag{3.27}$$

Since the holonomic constraints of Equations (3.2) and (3.6) imply that

$$\dot{\boldsymbol{X}}^{\mathrm{T}} A = \left(\dot{q}^{\mathrm{T}} \frac{\partial Q}{\partial q} + \dot{\theta} \frac{\partial R}{\partial \theta}, \quad \dot{q}^{\mathrm{T}} \frac{\partial R}{\partial q} + \dot{\theta} \frac{\partial R}{\partial \theta} \right) = 0, \tag{3.28}$$

multiplication of $A^{\mathrm{T}}\tilde{H}^{-1}$ by Equation (3.26) from the left yields

$$A^{\mathrm{T}}\ddot{\boldsymbol{X}} + A^{\mathrm{T}}\tilde{H}^{-1} \left\{ \left(\frac{1}{2}\dot{\tilde{H}} + \tilde{S} + \tilde{C} \right) \dot{\boldsymbol{X}} + B\boldsymbol{\eta} \right\} - A^{\mathrm{T}}\tilde{H}^{-1}A\boldsymbol{\lambda} = 0. \tag{3.29}$$

Since $A^{\mathrm{T}}\ddot{\boldsymbol{X}} = -\dot{A}^{\mathrm{T}}\dot{\boldsymbol{X}}$ according to Equation (3.28), it follows from Equation (3.29) that

$$\boldsymbol{\lambda} = \left(A^{\mathrm{T}}\tilde{H}^{-1}A\right)^{-1}\left[-\dot{A}^{\mathrm{T}}\dot{\boldsymbol{X}} + A^{\mathrm{T}}\tilde{H}^{-1}\left\{\left(\frac{1}{2}\dot{\tilde{H}} + \tilde{S} + \tilde{C}\right)\dot{\boldsymbol{X}} + B\boldsymbol{\eta}\right\}\right]. \quad (3.30)$$

It is obvious from this that, if $\dot{\boldsymbol{X}} \to 0$ as $t \to \infty$ and the 4×2 matrix A is non-degenerate during motion of the system, then as $t \to \infty$

$$\boldsymbol{\lambda} \to \left(A^{\mathrm{T}}\tilde{H}^{-1}A\right)^{-1} A^{\mathrm{T}}\tilde{H}^{-1}(B\boldsymbol{\eta}). \quad (3.31)$$

At this stage it is quite important to note that the closed-loop equation of motion of the system described by Equation (3.29) can be regarded as Lagrange's equation of motion with the external damping force $C\dot{q}$ and the Lagrangian

$$L = K(\boldsymbol{X}, \dot{\boldsymbol{X}}) - P(\boldsymbol{X}) + \Delta f Q + \Delta\lambda R. \quad (3.32)$$

where $K(\boldsymbol{X}, \dot{\boldsymbol{X}})$ is described in detail by Equation (3.1) and $P(\boldsymbol{X})$ is the artificial potential function given by

$$P(\boldsymbol{X}) = \frac{1}{2}\left\{\frac{f_d}{r}Y^2 + \beta l \Delta\theta^2\right\} \quad (3.33)$$

[see Equation (3.19)] and the total energy can be expressed as

$$E(\boldsymbol{X}, \dot{\boldsymbol{X}}) = K(\boldsymbol{X}, \dot{\boldsymbol{X}}) + P(\boldsymbol{X}) \quad (3.34)$$

Actually, it is possible to see from Equations (3.20) and (3.21) that Equation (3.26) is equivalent to

$$\frac{\mathrm{d}}{\mathrm{d}t}\left(\frac{\partial L}{\partial \dot{\boldsymbol{X}}}\right) - \frac{\partial L}{\partial \boldsymbol{X}} = -\tilde{C}\dot{\boldsymbol{X}}. \quad (3.35)$$

Since the position state $\boldsymbol{X}$ should lie on the two-dimensional constraint manifold CM_2, the scalar function $P(\boldsymbol{X})$ must be positive definite on CM_2 in the vicinity of $Y = 0$ and $\theta = \theta_d$ that corresponds to the condition of Equation (3.22). Therefore, it is reasonable to suppose that there exists a position state $\boldsymbol{X} = \boldsymbol{X}^*$ that attains the minimum of $P(\boldsymbol{X})$ and assume that at that point $\boldsymbol{X} = \boldsymbol{X}^*$ the 4×2 matrix A is non-degenerate. Such a physical situation actually can happen when in Figure 3.1 the rolling contact position $O_1(x_1, y_1)$ moves upward on the object surface with decreasing Y and increasing θ toward θ_d, where θ_d is given around $\pi/4$. It is easy to check that the 4×2 matrix A is degenerate if and only if the Jacobian matrix $J(q)$ is degenerte, *i.e.*, $\mathrm{rank}\{J(q)\} = 1$. This singularity of $J(q)$ arises if and only if $q_2 = q_3 = 0$ in a region of q such that $0 < q_1 \leq \pi/2$, $0 \leq q_2 \leq \pi/2$ and $0 \leq q_3 \leq \pi/2$. Then, it is possible to see that such an optimal state $\boldsymbol{X} = \boldsymbol{X}^*$ that minimises

$P(\boldsymbol{X})$ is unique in a neighbourhood of $\boldsymbol{X}^*$ on CM_2. Thus, the problem is to prove whether any solution of the closed-loop Equation (3.29) starting from $(\boldsymbol{X}(0), \dot{\boldsymbol{X}}(0))$ in a neighbourhood of $(\boldsymbol{X}^*, 0)$ on the four-dimensional manifold TB_4 defined by Equation (3.24) converges asymptotically to the equilibrium point $(\boldsymbol{X}^*, 0)$ that minimises the artificial potential $P(\boldsymbol{X})$ on the constraint manifold CM_2. In order to discuss this problem in a rigorous mathematical way, it is crucial to define the concept of neighbourhoods around the equilibrium point not only in TB_4 but also on CM_2.

3.2 Stability Problems under Redundancy of DOFs

If we do not care about controlling the pose of the object by specifying the desired orientation angle but concentrate only on stopping its rotational motion and immobilise it securely, then it is reasonable to consider a control signal of the form

$$u = -C\dot{q} - f_d\left(\frac{\partial Q}{\partial q}\right) - f_d Y \boldsymbol{e}. \tag{3.36}$$

Exertion of this control signal on the finger joints through joint actuators corresponds to the introduction of the artificial potential function

$$\tilde{P} = \frac{1}{2}(f_d/r)Y^2. \tag{3.37}$$

Note that minimisation of $\tilde{P}$ in $\boldsymbol{X}$ under the holonomic constraints of Equations (3.2) and (3.6) is equivalent to minimisation of the square of length $|\overline{O_0 O_m}|$ as shown in Figure 3.3, because

$$Y^2 + (r+l)^2 = (x_m - x_0)^2 + (y_m - y_0)^2. \tag{3.38}$$

Therefore, instead of the control signal of Equation (3.36), let us consider the signal

$$u = -C\dot{q} - (f_d/r)J^{\mathrm{T}}(q)\begin{pmatrix} x_0 - x_m \\ y_0 - y_m \end{pmatrix}. \tag{3.39}$$

It is interesting to note that this control signal can be constructed by easy calculation based on the knowledge of finger kinematics, the fixed point (x_m, y_m) and measurement data of joint angles together with appropriate choices for control gains $f_d > 0$ and $c_i > 0$ $(i = 1, 2, 3)$. Note again that in the construction of this control signal there is no need to know the object width l or measure Y or θ. At this stage it is convenient to introdue the orthogonal matrix parametrised by θ in such a form that

$$R_\theta = \begin{pmatrix} \cos\theta & -\sin\theta \\ \sin\theta & \cos\theta \end{pmatrix}. \tag{3.40}$$

Then, it is important to note that from Equations (3.2) and (3.4) the following relation follows:

$$\begin{pmatrix} r+l \\ -Y \end{pmatrix} = -R_\theta \begin{pmatrix} x_0 - x_m \\ y_0 - y_m \end{pmatrix}. \tag{3.41}$$

Since $R_\theta^{-1} = R_\theta^{\mathrm{T}}$, multiplication of $-R_\theta^{\mathrm{T}}$ to Equation (3.41) from the left yields

$$\begin{pmatrix} x_0 - x_m \\ y_0 - y_m \end{pmatrix} = -R_\theta^{\mathrm{T}} \begin{pmatrix} r+l \\ -Y \end{pmatrix} = -(r+l)\begin{pmatrix} \cos\theta \\ -\sin\theta \end{pmatrix} + Y \begin{pmatrix} \sin\theta \\ \cos\theta \end{pmatrix}. \tag{3.42}$$

Thus, by substituting this into Equation (3.39) and further substituting u from Equation (3.39) into Equation (3.9), we obtain the closed-loop dynamics of finger motion

$$\left(H\frac{\mathrm{d}}{\mathrm{d}t} + \frac{1}{2}\dot{H} + S + C \right)\dot{q} - \Delta f \left(\frac{\partial Q}{\partial q} \right) - \Delta\lambda \left(\frac{\partial R}{\partial q} \right) + f_d Y \boldsymbol{e} = 0, \tag{3.43}$$

where

$$\Delta f = f - \left(1 + \frac{l}{r} \right) f_d, \quad \Delta\lambda = \lambda - \frac{f_d}{r} Y. \tag{3.44}$$

It should be noted that Equation (3.10) can be rewritten in the form

$$I\ddot{\theta} - \Delta f \frac{\partial Q}{\partial \theta} - \Delta\lambda \frac{\partial R}{\partial \theta} - f_d Y = 0. \tag{3.45}$$

Similarly to the derivation of Equation (3.26), we rewrite Equations (3.43) and (3.45) in the following vector–matrix form:

$$\tilde{H}\ddot{\boldsymbol{X}} + \left(\frac{1}{2}\dot{\tilde{H}} + \tilde{S} + \tilde{C} \right) \dot{\boldsymbol{X}} - A\boldsymbol{\lambda} + (f_d Y / r)\boldsymbol{b} = 0, \tag{3.46}$$

where

$$\boldsymbol{\lambda} = (\Delta f, \Delta\lambda)^{\mathrm{T}}, \quad \boldsymbol{b} = (r\boldsymbol{e}^{\mathrm{T}}, -r)^{\mathrm{T}}, \tag{3.47}$$

and Δf and $\Delta\lambda$ are defined as in Equation (3.44). Note again that taking the inner product between $\dot{\boldsymbol{X}}(t)$ and Equation (3.46) leads to

$$\frac{\mathrm{d}}{\mathrm{d}t} E_0 = -\dot{\boldsymbol{X}}^{\mathrm{T}} \tilde{C} \dot{\boldsymbol{X}} = -\dot{q}^{\mathrm{T}} C \dot{q}, \tag{3.48}$$

where

$$E_0 = K + \tilde{P} = K + \frac{1}{2}(f_d / r) Y^2 \tag{3.49}$$

and K is the kinetic energy as defined in Equation (3.1).

Now, differently from the previous case where the Lyapunov relation expressed by Equation (3.18) holds with a positive definite function E defined by Equation (3.34), Equation (3.48) does not have any meaning as Lyapunov's relation because E_0 is no longer positive definite in TB_4. Nevertheless, it is fortunate to see through the relation of Equation (3.48) that the scalar function $E_0(\boldsymbol{X}, \dot{\boldsymbol{X}})$ is non-increasing with increasing t and hence $\dot{\boldsymbol{X}}$ and Y must be bounded uniformly in t. Furthermore, the constraint force $\boldsymbol{\lambda}$ also becomes bounded uniformly in t according to Equation (3.30) [where $B\boldsymbol{\eta}$ must be replaced with $(f_d/r)\boldsymbol{b}$], provided that A is non-degenerate during the motion of the overall system (this condition will be analysed later in a rigorous way). Thus, it follows from Equation (3.46) that $\ddot{\boldsymbol{X}}$ becomes uniformly bounded. This means that $\dot{\boldsymbol{X}}(t)$ becomes uniformly continuous in t. On the other hand, Equation (3.48) implies that

$$\int_0^\infty \dot{q}^{\mathrm{T}}(t) C \dot{q}(t)\, \mathrm{d}t \le E_0(\boldsymbol{X}(0), \dot{\boldsymbol{X}}(0)) = E_0(0). \tag{3.50}$$

That is, $\dot{q}(t) \in L^2(0, \infty)$. Hence, owing to Lemma 2 (see Appendix A), $\dot{q}(t)$ must converge to zero as $t \to \infty$. In order to verify the convergence of $\dot{\theta}(t)$ to zero as $t \to \infty$, it is necessary to use the constraint Equations (3.2) and (3.6). Indeed, differentiation of R defined by Equation (3.6) with respect to t leads to

$$\begin{aligned} 0 = \dot{Y} + r\dot{\phi} &= \dot{x}_0 \sin\theta + \dot{y}_0 \cos\theta \\ &\quad - \dot{\theta}\left\{(x_0 - x_m)\cos\theta - (y_0 - y_m)\sin\theta\right\} + r\left(\dot{\theta} - \dot{q}^{\mathrm{T}}\boldsymbol{e}\right) \\ &= \dot{q}^{\mathrm{T}}\left\{J^{\mathrm{T}}(q)\boldsymbol{r}_Y - r\boldsymbol{e}\right\} - \dot{\theta}(r + l) + r\dot{\theta} \\ &= \dot{q}^{\mathrm{T}}\left\{J^{\mathrm{T}}(q)\boldsymbol{r}_Y - r\boldsymbol{e}\right\} - l\dot{\theta} \end{aligned} \tag{3.51}$$

from which it follows that

$$\dot{\theta} = l^{-1}\dot{q}^{\mathrm{T}}\left\{J^{\mathrm{T}}(q)\boldsymbol{r}_Y - r\boldsymbol{e}\right\}. \tag{3.52}$$

This shows that $\dot{\theta}(t) \to 0$ as $t \to \infty$ provided that l is not too small to be compared with r (the radius of the finger-end sphere). Thus, it is verified that $\dot{\boldsymbol{X}}(t) \to 0$ as $t \to \infty$. Since $\dot{\boldsymbol{X}}(t)$ is also continuous uniformly in t, Lemma 2 implies that $\ddot{\boldsymbol{X}}(t) \to 0$ as $t \to \infty$. Thus, it can be concluded from Equation (3.43) that as $t \to \infty$

$$\Delta f\left(\frac{\partial Q}{\partial q}\right) + \Delta\lambda\left(\frac{\partial R}{\partial q}\right) - f_d Y\boldsymbol{e} \to 0. \tag{3.53}$$

This expression can be written in the following vector–matrix form:

$$D(q, \theta)\bar{\boldsymbol{\lambda}} \to 0, \tag{3.54}$$

where

$$\begin{cases} D(q,\theta) = (-J^{\mathrm{T}}(q)\boldsymbol{r}_X,\ J^{\mathrm{T}}(q)\boldsymbol{r}_Y - r\boldsymbol{e},\ -r\boldsymbol{e}) \\ \qquad \bar{\boldsymbol{\lambda}} = (\Delta f,\ \Delta\lambda,\ (f_d/r)Y)^{\mathrm{T}} \end{cases} \tag{3.55}$$

Since the 3×3 matrix $D(q,\theta)$ can be resolved into the multiplication of two matrices such that

$$D(q,\theta) = \left(J^{\mathrm{T}}(q), r\boldsymbol{e}\right) \begin{pmatrix} -\boldsymbol{r}_X & \boldsymbol{r}_Y & 0_{2\times 1} \\ 0 & -1 & -1 \end{pmatrix} \tag{3.56}$$

$D(q,\theta)$ is non-singular if and only if the matrix $(J^{\mathrm{T}}(q), r\boldsymbol{e})$ is non-singular. The Jacobian matrix $J(q)$ defined by Equation (3.11) can be calculated as follows [for brevity, $s_1 = \sin q_1$, $s_{12} = \sin(q_1 + q_2)$, $c_{12} = \cos(q_1 + q_2)$, *etc.*]:

$$\begin{cases} x_0 = -(l_1 c_1 + l_2 c_{12} + l_3 c_{123}) \\ y_0 = l_1 s_1 + l_2 s_{12} + l_3 s_{123} \end{cases} \tag{3.57}$$

$$J(q) = \begin{pmatrix} l_1 s_1 + l_2 s_{12} + l_3 s_{123} & l_2 s_{12} + l_3 s_{123} & l_3 s_{123} \\ l_1 c_1 + l_2 c_{12} + l_3 c_{123} & l_2 c_{12} + l_3 c_{123} & l_3 c_{123} \end{pmatrix}. \tag{3.58}$$

Therefore, $J(q)$ is degenerate if and only if $q_2 = q_3 = 0$. That is, singularity of the pose of the finger arises when it stretches straight so that $q_2 = 0$ and $q_3 = 0$ (see Figure 3.1). Referring to this analysis of the structure of $D(q,\theta)$ in Equation (3.55), we can see that $D(q,\theta)$ is non-singular and therefore the convergence of $D(q,\theta)\bar{\boldsymbol{\lambda}}$ to 0 as $t \to \infty$ shown in Equation (3.54) implies

$$\bar{\boldsymbol{\lambda}} = (\Delta f, \Delta\lambda, (f_d/r)Y)^{\mathrm{T}} \to 0 \quad \text{as} \quad t \to \infty \tag{3.59}$$

provided that during motion of the finger–object system the finger does not take the singular pose. Equation (3.59) implies that

$$f(t) \to \left(1 + \frac{l}{r}\right) f_d, \quad Y(t) \to 0, \quad \lambda(t) \to 0 \quad \text{as} \quad t \to \infty. \tag{3.60}$$

The argument of convergence of the physical variables $f(t)$, $\lambda(t)$ and $Y(t)$ along the solution trajectory $(\boldsymbol{X}(t), \dot{\boldsymbol{X}}(t))$ to the closed-loop Equation (3.46) is validated mathematically under the condition that the finger pose must not be singular during its motion and the finger-end maintains contact with the object, *i.e.*, $f(t) > 0$ for all $t > 0$. Indeed, we cannot yet discuss anything about convergences of position variables $\theta(t)$ and $q_i(t)$ $(i = 1,2,3)$ as t tends to infinity. We could show that $\dot{q}(t) \in L^2(0,\infty)$ and, as $t \to \infty$, $\dot{q}(t) \to 0$ and $\dot{\theta}(t) \to 0$ but not verify whether $\dot{q}(t) \in L^1(0,\infty)$ and $\dot{\theta}(t) \in L^1(0,\infty)$ or not. Rather, there may arise self-motion of the system owing to the redundancy of DOFs, that is, persistent rotational movements of the finger joints q_1, q_2, q_3, and object orientation angle θ even if $\theta - \sum_i q_i$ is kept constant [see Equations (3.4–3.6)]. Furthermore, there arises the possibility that the finger may approach its sigular pose during its motion. All these problems can be treated in a rigorous way by using the Riemannian metrics and distance introduced on the constraint manifold CM_2.

3.3 Riemannian Distance and Stability on a Manifold

In order to treat the stability of motion with physical interaction between the robotic finger and rigid object under DOF redundancy as shown in Figure 3.1 in a rigorous mathematical way, it is necessary and appropriate to introduce the concept of Riemannian distance on the two-dimensional manifold CM_2 defined by Equation (3.25) and embedded in the configuration space $R^4 = \{(q^{\mathrm{T}}, \theta)^{\mathrm{T}} = \boldsymbol{X}\}$. For two given points $\boldsymbol{X}_1$, $\boldsymbol{X}_2 \in CM_2$, the Riemannian distance between $\boldsymbol{X}_1$ and $\boldsymbol{X}_2$ is defined as

$$R(\boldsymbol{X}_1, \boldsymbol{X}_2) = \min_{\boldsymbol{X}(t)} \int_0^1 \sqrt{\sum_{i,j} \frac{1}{2} h_{ij}(\boldsymbol{X}(t)) \dot{X}_i(t) \dot{X}_j(t)}\, \mathrm{d}t, \tag{3.61}$$

where the minimisation is taken over all smooth curves $\boldsymbol{X}(t)$ parameterised by $t \in [0, 1]$ in such a way that $\boldsymbol{X}(0) = \boldsymbol{X}_1$, $\boldsymbol{X}(1) = \boldsymbol{X}_2$, constrained on $Q = 0$ and $R = 0$, and also with $\dot{\boldsymbol{X}}(t)$ constrained on $\dot{Q} = 0$ and $\dot{R} = 0$. Here, $h_{ij}(\boldsymbol{X})$ denotes the (i, j)-entry of the inertia matrix $\tilde{H}(\boldsymbol{X})$ defined in Equation (3.27). It is well known in differential geometry that the optimal curve that minimizes the integral of Equation (3.61) satisfies the Euler-Lagrange equation:

$$\sum_{j=1}^{4} h_{ij}(\boldsymbol{X}) \ddot{X}_j + \sum_{j,k=1}^{4} \Gamma_{jik} \dot{X}_j \dot{X}_k - A_i \Delta \boldsymbol{\lambda} = 0, \quad i = 1, \cdots, 4, \tag{3.62}$$

where A_i denotes the ith row of the 4×2-matrix A defined in Equation (3.27) and Γ_{ijk} is the Christoffel symbol of the first kind. Note that Equation (3.62) is equivalent to the following expression:

$$\tilde{H}\ddot{\boldsymbol{X}} + \left(\frac{1}{2}\dot{\tilde{H}} + \tilde{S}\right) \dot{\boldsymbol{X}} - A \Delta \boldsymbol{\lambda} = 0. \tag{3.63}$$

It should be remarked at this stage that:

1) The Riemannian distance $R(\boldsymbol{z}_1, \boldsymbol{z}_2)$ is invariant under any transformation $t = g(s)$ satisfying $\mathrm{d}t/\mathrm{d}s > 0$ and $T = g^{-1}(1)$ and $0 = g^{-1}(0)$, because it follows that

$$\begin{aligned}
&\int_0^T \sqrt{\sum_{i,j} \frac{1}{2} h_{ij}\left(\tilde{\boldsymbol{X}}(s)\right) \frac{\mathrm{d}\tilde{X}_i}{\mathrm{d}s} \frac{\mathrm{d}\tilde{X}_j}{\mathrm{d}s}}\, \mathrm{d}s \\
&= \int_0^T \sqrt{\sum_{i,j} \frac{1}{2} h_{ij}\left(\boldsymbol{X}(t)\right) \cdot \left(\frac{\mathrm{d}X_i}{\mathrm{d}t} \frac{\mathrm{d}t}{\mathrm{d}s}\right) \cdot \left(\frac{\mathrm{d}X_j}{\mathrm{d}t} \frac{\mathrm{d}t}{\mathrm{d}s}\right)}\, \mathrm{d}s \\
&= \int_0^1 \sqrt{\sum_{i,j} \frac{1}{2} h_{ij}\left(\boldsymbol{X}(t)\right) \frac{\mathrm{d}X_i}{\mathrm{d}t} \frac{\mathrm{d}X_j}{\mathrm{d}t}}\, \mathrm{d}t,
\end{aligned} \tag{3.64}$$

where $\tilde{\boldsymbol{X}}(s) = \boldsymbol{X}(g(s))$.

2) If the curve $\boldsymbol{X}(t)$ for $t \in [0,1]$ minimises the integral of Equation (3.61), then the quantity

$$L(\boldsymbol{X}, \dot{\boldsymbol{X}}) = \sqrt{\frac{1}{2}\sum_{i,j} h_{ij}(\boldsymbol{X})\dot{X}_i\dot{X}_j} \tag{3.65}$$

is constant for all $t \in [0,1]$. This follows from taking the inner product between $\dot{\boldsymbol{X}}(t)$ and Equation (3.63), which yields

$$\begin{aligned} 0 &= \dot{\boldsymbol{X}}^{\mathrm{T}}\left\{\tilde{H}\ddot{\boldsymbol{X}} + \left(\frac{1}{2}\dot{\tilde{\boldsymbol{H}}} + S\right) - A\Delta\boldsymbol{\lambda}\right\} \\ &= \frac{\mathrm{d}}{\mathrm{d}t}\left\{\frac{1}{2}\dot{\boldsymbol{X}}^{\mathrm{T}}\tilde{H}(\boldsymbol{X})\dot{\boldsymbol{X}}\right\}, \end{aligned} \tag{3.66}$$

that is, $(1/2)\dot{\boldsymbol{X}}^{\mathrm{T}}\tilde{H}(\boldsymbol{X})\dot{\boldsymbol{X}} = \text{const.}$

We shall now discuss the stability of an equilibrium state $(\boldsymbol{X}^*, 0)$, $\boldsymbol{X}^* = \left((q^*)^{\mathrm{T}}, \theta^*\right)^{\mathrm{T}}$, such that

$$0 = Y = -c_0 - r\phi = -c_0 - r\left\{\pi + \theta^* - \boldsymbol{e}^{\mathrm{T}}q^*\right\} \tag{3.67}$$

and the finger pose at $q = q^*$ is sufficiently distant from the singular pose in the sense of the Riemannian distance. To do this, it is necessary to define neighbourhoods of the equilibrium pose $\boldsymbol{X} = \boldsymbol{X}^*$ on the manifold CM_2 and at the same time those of the equilibrium state $(\boldsymbol{X}^*, 0)$ on TB_4 defined by Equation (3.24), which can be also represented as a point in the Enclidean state space $R^8 = \{(\boldsymbol{X}, \dot{\boldsymbol{X}})\}$. The former can be defined naturally by using the Riemannian distance in such a way that

$$N^2(\boldsymbol{X}^*; r_0) = \{\boldsymbol{X} : R(\boldsymbol{X}, \boldsymbol{X}^*) < r_0, \boldsymbol{X} \in CM_2\}. \tag{3.68}$$

The latter can be defined by introducing two positive parameters ρ and r_0 in such a way that

$$\begin{aligned} N^4\left\{(\boldsymbol{X}^*, 0); \rho, r_0\right\} = \Big\{&(\boldsymbol{X}, \dot{\boldsymbol{X}}) : E_0(\boldsymbol{X}, \dot{\boldsymbol{X}}) < \rho^2 \\ &\text{and} \quad R(\boldsymbol{X}, \boldsymbol{X}^*) < r_0, \quad (\boldsymbol{X}, \boldsymbol{X}^*) \in TB_4\Big\}, \end{aligned} \tag{3.69}$$

where $E_0 = K + \tilde{P}$ is described in detail as

$$E_0(\boldsymbol{X}, \dot{\boldsymbol{X}}) = \frac{1}{2}\dot{\boldsymbol{X}}^{\mathrm{T}}\tilde{H}(\boldsymbol{X})\dot{\boldsymbol{X}} + \frac{1}{2}(f_d/r)Y^2. \tag{3.70}$$

The reason we use the inequality $E_0(\boldsymbol{X}, \dot{\boldsymbol{X}}) < \rho^2$ instead of local coordinates on the manifold TB_4 (which is equivalent to the tangent bundle, the set of all tangent spaces $T_2(\boldsymbol{X}) = \{\dot{\boldsymbol{X}} : \dot{\boldsymbol{X}}^{\mathrm{T}}(\partial Q/\partial \boldsymbol{X}) = 0 \text{ and } \dot{\boldsymbol{X}}^{\mathrm{T}}(\partial R/\partial \boldsymbol{X}) = 0\}$ over all $\boldsymbol{X} \in CM_2$) will be cleared later. The reason why we use the inequality

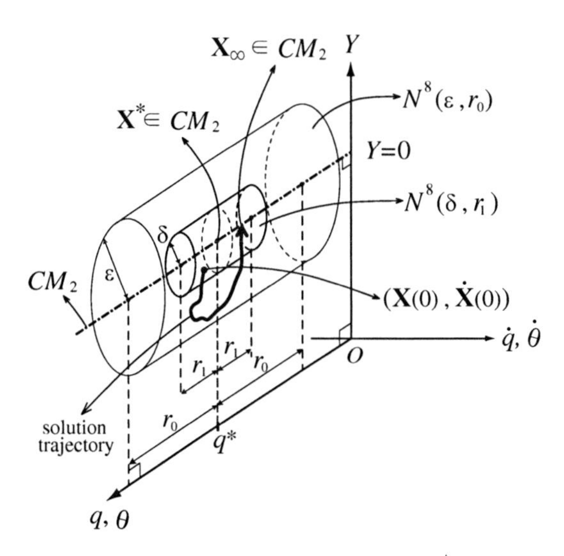

Fig. 3.4. Definitions of stability on a manifold and transferability to a subset of the equilibrium manifold

$R(\boldsymbol{X}, \boldsymbol{X}^*) < r_0$ in the definition of N^4 is only to avoid the occurence of the singular pose of the finger at any instant. Therefore, once r_0 is chosen appropriately for the given $\boldsymbol{X}^*$ so that there does not exist any singular pose inside $N^2(\boldsymbol{X}^*; r_0)$, the constant r_0 can be treated as fixed. Now we introduce the definition of stability of the given equilibrium state $(\boldsymbol{X}^*, 0)$ on the manifold TB_4 regarding the closed-loop dynamics of Equation (3.46).

Definition 3.1. If for an arbitrarily given $\varepsilon > 0$ there exist a constant $\delta > 0$ depending on ε and another constant $r_1 > 0$ independent of ε and less than r_0 such that a solution trajectory $(\boldsymbol{X}(t), \dot{\boldsymbol{X}}(t))$ of the closed-loop dynamics of Equation (3.46) starting from any initial state $(\boldsymbol{X}(0), \dot{\boldsymbol{X}}(0))$ inside $N^4\{(\boldsymbol{X}^*, 0); \delta(\varepsilon), r_1\}$ remains in $N^4\{(\boldsymbol{X}^*, 0); \varepsilon, r_0\}$, then the equilibrium state $(\boldsymbol{X}^*, 0)$ is said to be stable on a manifold (see Figure 3.4).

Definition 3.2. If for a reference equilibrium state $(\boldsymbol{X}^*, 0) \in M_4$ there exist constants $\varepsilon_1 > 0$ and $r_1 > 0$ $(r_1 < r_0)$ such that any solution of the closed-loop dynamics of Equation (3.46) starting from an arbitrary initial state in $N^4\{(\boldsymbol{X}^*, 0); \varepsilon_1, r_1\}$ remains in $N^4\{(\boldsymbol{X}^*, 0); \varepsilon_1, r_0\}$ and converges asymptotically as $t \to \infty$ to some point of the one-dimensional equilibrium manifold

$$EM_1 = \left\{(\boldsymbol{X}, 0) : E_0(\boldsymbol{X}, \dot{\boldsymbol{X}}) = 0\right\} \tag{3.71}$$

then the neighbourhood $N^4\{(\boldsymbol{X}^*, 0); \varepsilon_1, r_1\}$ is said to be treansferable to a subset of EM_1 containing the reference point $(\boldsymbol{X}^*, 0)$.

The intuitive idea behind introduction of the two neighbourhoods $N^2(\boldsymbol{X}^*; r_0)$ of Equation (3.68) and $N^4\{(\boldsymbol{X}^*, 0); \rho, r_0\}$ of Equation (3.69) is that the function $E_0(\boldsymbol{X}(t), \dot{\boldsymbol{X}}(t))$ of t evaluated along a solution $(\boldsymbol{X}(t), \dot{\boldsymbol{X}}(t))$ to the closed-loop dynamics of Equation (3.46) starting from an arbitrary initial state $(\boldsymbol{X}(0), \dot{\boldsymbol{X}}(0))$ in $N^4\{(\boldsymbol{X}^*, 0) : \varepsilon_1, r_1\}$ with some $\varepsilon_1 > 0$ and $r_1 > 0$ converges exponentially to zero, *i.e.*, there exist some positive constants α^* and β of $O(1)$ such that

$$E_0(\boldsymbol{X}(t), \dot{\boldsymbol{X}}(t)) \leq \beta E_0(\boldsymbol{X}(0), \dot{\boldsymbol{X}}(0)) \mathrm{e}^{-\alpha^* t} \tag{3.72}$$

though $E_0(\boldsymbol{X}, \dot{\boldsymbol{X}})$ is not a Lyapunov function for Equation (3.46). As to the derivation of the inequality above we will discuss this in detail based on the physical circumstances of the problem in the next section. Instead, we shall close this section by showing how to choose the constants ε_1 and r_1 in $N^4\{(\boldsymbol{X}^*, 0) : \varepsilon_1, r_1\}$ to prove the stability of the reference equilibrium state $(\boldsymbol{X}^*, 0)$ and the transferability of $N^4\{(\boldsymbol{X}^*, 0); \varepsilon_1, r_1\}$ to a subset of EM_1. Indeed, it is possible to choose $\varepsilon_1 > 0$ and $r_1 > 0$ so that

$$r_1 \leq \frac{r_0}{2}, \quad \varepsilon_1 < \min\left\{\frac{\alpha^* r_0}{4\sqrt{\beta}}, \ \rho\right\}. \tag{3.73}$$

Then, it is possible to see that for any $T > 0$

$$\begin{aligned} R(\boldsymbol{X}(T), \boldsymbol{X}^*) &\leq R(\boldsymbol{X}(T), \boldsymbol{X}(0)) + R(\boldsymbol{X}(0), \boldsymbol{X}^*) \\ &\leq R(\boldsymbol{X}(T), \boldsymbol{X}(0)) + r_1 \leq R(\boldsymbol{X}(T), \boldsymbol{X}(0)) + \frac{r_0}{2} \end{aligned} \tag{3.74}$$

along the solution $(\boldsymbol{X}(t), \dot{\boldsymbol{X}}(t))$ to Equation (3.46) starting from $(\boldsymbol{X}(0), \dot{\boldsymbol{X}}(0))$ in $N^4\{(\boldsymbol{X}^*, 0); \varepsilon_1, r_1\}$. Since the position trajectory $\boldsymbol{X}(t)$ starting from $\boldsymbol{X}(0)$ at $t = 0$ and reaching $\boldsymbol{X}(T)$ at $t = T$ under the constraints of Equations (3.2) and (3.6) is a special curve, it follows from the meaning of Riemannian distance that

$$\begin{aligned} R(\boldsymbol{X}(T), \boldsymbol{X}(0)) &\leq \int_0^T \sqrt{\frac{1}{2}\dot{\boldsymbol{X}}^{\mathrm{T}}(t)\tilde{H}(\boldsymbol{X}(t))\dot{\boldsymbol{X}}(t)}\ \mathrm{d}t \\ &\leq \int_0^T \sqrt{E_0(\boldsymbol{X}(t), \dot{\boldsymbol{X}}(t))}\ \mathrm{d}t \leq \int_0^T \sqrt{\beta E_0(\boldsymbol{X}(0), \dot{\boldsymbol{X}}(0))}\ \mathrm{e}^{-(\alpha^*/2)t}\mathrm{d}t \\ &\leq \frac{2\sqrt{\beta}}{\alpha^*}\sqrt{E_0(0)} < \frac{r_0}{2}, \end{aligned} \tag{3.75}$$

where $E_0(0)$ denotes $E_0(\boldsymbol{X}(0), \dot{\boldsymbol{X}}(0))$ for abbreviation, and the last inequality follows from Equation (3.73). Then, substituting Equation (3.75) into Equation (3.74) yields

$$R(\boldsymbol{X}(T), \boldsymbol{X}^*) < \frac{r_0}{2} + \frac{r_0}{2} = r_0. \tag{3.76}$$

At the same time, Equation (3.73) together with the Lyapunov-like relation of Equation (3.48) implies that $E_0(T) = E_0(\boldsymbol{X}(T), \boldsymbol{X}^*(T)) < E_0(0) \leq \varepsilon_1^2 \leq \rho^2$. This concludes that the trajectory $(\boldsymbol{X}(t), \dot{\boldsymbol{X}}(t))$ remains in $N^4\{(\boldsymbol{X}^*, 0); \rho, r_0\}$

Table 3.1. Scales of a finger and object

	Symbol	Numerical order	
link length	l_i $(i = 1\text{–}3)$	$0.02 \sim 0.05$	[m]
radius of finger-end sphere	r	$0.01 \sim 0.03$	[m]
link inertia momnet	I_i $(i = 1, 2, 3)$	$0.2 \times 10^{-7} \sim 1.0 \times 10^{-6}$	[Nm]
object width	l	$0.005 \sim 0.05$	[m]
object inertia moment	I	$0.2 \times 10^{-7} \sim 5.0 \times 10^{-6}$	[Nm]

and converge exponentially to some point on EM_1. As for the proof of stability on a manifold, if we choose $\delta(\varepsilon)$ for an arbitrarily given $\varepsilon > 0$ in such a way that

$$\delta(\varepsilon) < \min\left\{\frac{\alpha^* r_0}{4\sqrt{\beta}},\ \rho,\ \varepsilon\right\} \tag{3.77}$$

then the trajectory $(\boldsymbol{X}(t), \dot{\boldsymbol{X}}(t))$ starting from an arbitrarily given initial state $(\boldsymbol{X}(0), \dot{\boldsymbol{X}}(0))$ in $N^4\{(\boldsymbol{X}^*, 0); \delta(\varepsilon), r_1\}$ remains in $N^4\{(\boldsymbol{X}^*, 0); \varepsilon, r_0\}$. This shows the stability on a manifold of the reference equilibrium point $(\boldsymbol{X}^*, 0) \in EM_1$.

3.4 Exponential Convergence for Stabilisation of Rotational Moments

We shall discuss the speed of convergence of the Lyapunov function $E_0(\boldsymbol{X}(t), \dot{\boldsymbol{X}}(t))$ evaluated along the trajectory of a solution to the closed-loop dynamics of Equation (3.46). The convergence speed is, as a matter of course, dependent on the scale of the concerned finger and object and the choice of control gains c_i $(i = 1, 2, 3)$ and f_d, where we set the damping matrix C in Equation (3.36) as $C = \mathrm{diag}(c_1, c_2, c_3)$. As discussed in Section 1.5, we consider a somewhat broader class of robot fingers and rigid objects with ordinary scales shown in Table 3.1. The control gains c_i $(i = 1, 2, 3)$ and f_d are given as in Table 3.2. In this section, we further assume that

A$_1$) $r/l = 0.3 \sim 3.0$

A$_2$) $\dfrac{\max\{c_i\}}{2rf_d} = 0.05 \sim 0.2$ [s]

The case when the object width l is very small relative to the radius of the finger-end sphere will be treated in Sections 6.3 and 6.7 that are concerned with the case of soft and deformable finger-ends.

Suppose now that a concerned equilibrium point $(\boldsymbol{X}^*, 0)$ lies on the set of equilibrium manifold EM_1 and there exists a neighbourhood $N^2(\boldsymbol{X}^*; r_0)$

Table 3.2. Control gains

	Symbol	Numerical Order	
damping gain	c_i $(i = 1, 2, 3)$	0.001–0.005	[Nms]
pressing force	f_d	0.05–2.0	[N]

such that, at any $\boldsymbol{X} \in N^2(\boldsymbol{X}^*; r_0)$, the Jacobian matrix $J(\boldsymbol{X})$ $(= J(q), \boldsymbol{X} = (q^{\mathrm{T}}, \theta)^{\mathrm{T}})$ is non-degenerate. As discussed in Section 3.2, this means that the two column vectors $\boldsymbol{a}_1 = (\partial Q/\partial q^{\mathrm{T}}, Y)^{\mathrm{T}}$ and $\boldsymbol{a}_2 = (\partial R/\partial q^{\mathrm{T}}, -l)$ of matrix A and the vector $\boldsymbol{b} = r(\boldsymbol{e}^{\mathrm{T}}, -1)^{\mathrm{T}}$ are independent of each other. To gain mathematical insight into this mutual independence between the column vectors of A and the vector $\boldsymbol{b}$, we introduce the following two-dimensional subspaces at the point $\boldsymbol{X}$ in the configuration space R^4:

$$\mathrm{I_m}(A) = \{A\boldsymbol{r} : \boldsymbol{r} \in R^2\},$$
$$\mathrm{Ker}(A^{\mathrm{T}}) = \{\boldsymbol{s}^{\mathrm{T}} : A^{\mathrm{T}}\boldsymbol{s} = 0, \boldsymbol{s} \in R^4\}.$$

The former $\mathrm{I_m}(A)$ is called the image space of $A(\boldsymbol{X})$ at the point $\boldsymbol{X}$ and the latter $\mathrm{Ker}(A^{\mathrm{T}})$ the kernel space of $A^{\mathrm{T}}(\boldsymbol{X})$. The configuration space R^4 can be expressed as the direct sum of the two subspaces in the following way:

$$R^4 = \mathrm{I_m}(A)(+)\mathrm{Ker}(A^{\mathrm{T}}). \tag{3.78}$$

In other words, the constant vector $\boldsymbol{b} = r(\boldsymbol{e}^{\mathrm{T}}, -1)^{\mathrm{T}}$ can be expressed as

$$\boldsymbol{b} = \boldsymbol{b}_I(\boldsymbol{X}) + \boldsymbol{b}_K(\boldsymbol{X}), \tag{3.79}$$

where $\boldsymbol{b}_I(\boldsymbol{X}) \in \mathrm{I_m}(A(\boldsymbol{X}))$ and $\boldsymbol{b}_K(\boldsymbol{X}) \in \mathrm{Ker}(A^{\mathrm{T}}(\boldsymbol{X}))$ and $\boldsymbol{b}_I$ and $\boldsymbol{b}_K$ are mutually orthogonal, that is, $\boldsymbol{b}_I^{\mathrm{T}}\boldsymbol{b}_K = 0$. Independence of $\boldsymbol{b}$ from the column vectors of $A(\boldsymbol{X})$ implies that $\boldsymbol{b}_K(\boldsymbol{X}) \neq 0$, that is, the constant vector $\boldsymbol{b}$ does not lie on the image space of A. Further, the image component of $\boldsymbol{b}$ concerning matrix $A(\boldsymbol{X})$ can be calculated by the pseudo-inverse of A^{T} as follows:

$$\boldsymbol{b}_I(\boldsymbol{X}) = A(\boldsymbol{X})\left(A^{\mathrm{T}}(\boldsymbol{X})A(\boldsymbol{X})\right)^{-1}A^{\mathrm{T}}(\boldsymbol{X})\boldsymbol{b} = A(\boldsymbol{X})(A^{\mathrm{T}})^{+}\boldsymbol{b}, \tag{3.80}$$

where

$$(A^{\mathrm{T}})^{+} = (A^{\mathrm{T}}A)^{-1}A^{\mathrm{T}}. \tag{3.81}$$

Then, we can calculate the kernel component $\boldsymbol{b}_K(\boldsymbol{X})$ by the formula:

$$\boldsymbol{b}_K(\boldsymbol{X}) = \left\{I_4 - A(\boldsymbol{X})(A(\boldsymbol{X})^{\mathrm{T}})^{+}\right\}\boldsymbol{b}, = P_A(\boldsymbol{X})\boldsymbol{b}, \tag{3.82}$$

where $P_A(\boldsymbol{X})$ is called the projection matrix of A onto the kernel space. We further remark that the tangent space $TM_2(\boldsymbol{X}) = \{\dot{\boldsymbol{X}} = (\dot{q}^{\mathrm{T}}, \dot{\theta}) : \dot{Q} =$

0 and $\dot{R} = 0\}$ is equivalent to the kernel space $\mathrm{Ker}(A^{\mathrm{T}})$ of $A(\boldsymbol{X})$ in this problem. Since $A(\boldsymbol{X})$ is non-degenerate at any $\boldsymbol{X}$ in the neighbourhood $N^2(\boldsymbol{X}^*; r_0)$ and therefore $\boldsymbol{b}_K(\boldsymbol{X}) \neq 0$, it can be reasonably expected that

$$\frac{\boldsymbol{b}_K^{\mathrm{T}}(\boldsymbol{X})\boldsymbol{b}_K(\boldsymbol{X})}{\boldsymbol{b}^{\mathrm{T}}\boldsymbol{b}} \geq \gamma > 0, \quad \forall \boldsymbol{X} \in N^2(\boldsymbol{X}^*; r_0) \tag{3.83}$$

for some $\gamma > 0$. In particular at $\boldsymbol{X} = \boldsymbol{X}^*$, we denote

$$\gamma^* = \boldsymbol{b}_K^{\mathrm{T}}(\boldsymbol{X}^*)\boldsymbol{b}_K(\boldsymbol{X}^*)/\|\boldsymbol{b}\|^2 \tag{3.84}$$

and assume that $\gamma^* > 0.2$. Then, it is possible to find some number $\bar{r}_0$ ($\leq r_0$) such that

$$\mathrm{A}_3) \quad \frac{\boldsymbol{b}_K^{\mathrm{T}}(\boldsymbol{X})\boldsymbol{b}_K(\boldsymbol{X})}{\|\boldsymbol{b}\|^2} \geq \frac{2}{3}\gamma^* \qquad \forall \boldsymbol{X} \in N^2(\boldsymbol{X}^*; \bar{r}_0)$$

Choose such a number $\bar{r}_0$ as large as possible.

We are now in a position to prove the exponential convergence of trajectories $(\boldsymbol{X}(t), \dot{\boldsymbol{X}}(t))$ of a solution to the closed-loop dynamics of Equation (3.46) starting from an arbitrary initial state $(\boldsymbol{X}(0), \dot{\boldsymbol{X}}(0)) \in N^4\{(\boldsymbol{X}^*, 0); \delta, r_1\}$ with $\delta > 0$ and r_1. The two parameters δ and r_1 that specified the scale of $N^4\{(\boldsymbol{X}^*, 0); \delta, r_1\}$ will be determined necessarily in the following argument. First, consider a scalar quantity

$$V = \left(\tilde{H}\dot{\boldsymbol{X}}\right)^{\mathrm{T}} P_A \boldsymbol{b} \left(Y/\gamma^*\|\boldsymbol{b}\|^2\right) \tag{3.85}$$

and define

$$W_\alpha = E_0(\boldsymbol{X}, \dot{\boldsymbol{X}}) + \alpha V \tag{3.86}$$

with some positive parameter $\alpha > 0$ less than or equal to 1.0. Referring to the inequality

$$\beta x^2 + (1/\beta)y^2 \geq 2xy \geq -\beta x^2 - (1/\beta)y^2 \tag{3.87}$$

for any $\beta > 0$, we see that

$$\begin{aligned} V &\geq -\frac{r}{f_d(\gamma^*)^2\|\boldsymbol{b}\|^2}\dot{\boldsymbol{X}}^{\mathrm{T}}\tilde{H}\tilde{H}\dot{\boldsymbol{X}} - \frac{f_d\boldsymbol{b}^{\mathrm{T}}P_A\boldsymbol{b}}{4r\|\boldsymbol{b}\|^2}Y^2 \\ &\geq -\frac{1}{4f_d(\gamma^*)^2 r}\dot{\boldsymbol{X}}^{\mathrm{T}}\tilde{H}\tilde{H}\dot{\boldsymbol{X}} - \frac{f_d}{4r}Y^2 \\ &\geq -\frac{1}{2}E_0(\boldsymbol{X}, \dot{\boldsymbol{X}}) \end{aligned} \tag{3.88}$$

since $\|\boldsymbol{b}\|^2 = 4r^2$, $\boldsymbol{b}^{\mathrm{T}}P_A\boldsymbol{b} \leq \|\boldsymbol{b}\|^2$, $\gamma^* \geq 0.2$ and the maximum eigenvalue of matrix $\tilde{H}(\boldsymbol{X})$ is at most of numerical order $O(10^{-6})$ (also see Table 3.1). Next, let us differentiate V in time t. This results in

$$\dot{V} = \left(\tilde{H}\ddot{\boldsymbol{X}}\right)^{\mathrm{T}} P_A \boldsymbol{b} \left(Y/\gamma^* \|\boldsymbol{b}\|^2\right) + \left(\tilde{H}\dot{\boldsymbol{X}}\right)^{\mathrm{T}} P_A \boldsymbol{b} \left(\dot{Y}/\gamma^* \|\boldsymbol{b}\|^2\right)$$
$$+ \left\{\left(\dot{\tilde{H}}\dot{\boldsymbol{X}}\right)^{\mathrm{T}} P_A + \left(\tilde{H}\dot{\boldsymbol{X}}\right)^{\mathrm{T}} \dot{P}_A\right\} \left(Y/\gamma^* \|\boldsymbol{b}\|^2\right). \tag{3.89}$$

Then, substituting Equation (3.46) into the first term of the right-hand side yields

$$\dot{V} = -\frac{\boldsymbol{b}^{\mathrm{T}} P_A \boldsymbol{b}}{\gamma^* \|\boldsymbol{b}\|^2} \cdot \frac{f_d Y^2}{r} - \left(\tilde{C}\dot{\boldsymbol{X}}\right)^{\mathrm{T}} P_A \boldsymbol{b} \cdot \frac{Y}{\gamma^* \|\boldsymbol{b}\|^2} + h\left(\boldsymbol{X}, \dot{\boldsymbol{X}}\right), \tag{3.90}$$

where

$$h(\boldsymbol{X}, \dot{\boldsymbol{X}}) = \left[\left\{\left(\frac{1}{2}\dot{\tilde{H}} - S\right)\dot{\boldsymbol{X}}\right\}^{\mathrm{T}} P_A + \left(\tilde{H}\dot{\boldsymbol{X}}\right)^{\mathrm{T}} \dot{P}_A\right\} \frac{Y}{\gamma^* \|\boldsymbol{b}\|^2} \boldsymbol{b}$$
$$+ \left(\tilde{H}\dot{\boldsymbol{X}}\right)^{\mathrm{T}} P_A \frac{\dot{Y}}{\gamma^* \|\boldsymbol{b}\|^2} \boldsymbol{b} \tag{3.91}$$

and we refer to $P_A A = 0$. Note that $h(\boldsymbol{X}, \dot{\boldsymbol{X}})$ is quadratic in $\dot{\boldsymbol{X}}$. By comparing the numerical order of $\tilde{H}$ with that of C, it is possible to verify that

$$\|h(\boldsymbol{X}, \dot{\boldsymbol{X}})\| \leq \frac{1}{4} \dot{\boldsymbol{X}} \tilde{C} \dot{\boldsymbol{X}} \tag{3.92}$$

under the contact constraints of Equations (3.2) and (3.6), and in particular, the relationship expressed by Equation (3.52). Next note that

$$-\left(\tilde{C}\dot{\boldsymbol{X}}\right)^{\mathrm{T}} P_A \boldsymbol{b} \frac{Y}{\gamma^* \|\boldsymbol{b}\|^2} \leq \frac{r}{f_d \gamma^* \|\boldsymbol{b}\|^2} \dot{\boldsymbol{X}}^{\mathrm{T}} \tilde{C} \tilde{C} \dot{\boldsymbol{X}} + \frac{f_d Y^2}{4r} \cdot \frac{\boldsymbol{b}^{\mathrm{T}} P_A \boldsymbol{b}}{\gamma^* \|\boldsymbol{b}\|^2}$$
$$\leq \frac{\max\{c_i\}}{2r f_d} \cdot \frac{1}{2\gamma^*} \dot{\boldsymbol{X}}^{\mathrm{T}} \tilde{C} \dot{\boldsymbol{X}} + \frac{f_d Y^2}{4r} \cdot \frac{\boldsymbol{b}^{\mathrm{T}} P_A \boldsymbol{b}}{\gamma^* \|\boldsymbol{b}\|^2}. \tag{3.93}$$

According to the assumption A_2, we obtain

$$-\left(\tilde{C}\dot{\boldsymbol{X}}\right)^{\mathrm{T}} P_A \boldsymbol{b} \frac{Y}{\gamma^* \|\boldsymbol{b}\|^2} \leq \frac{1}{2} \dot{\boldsymbol{X}}^{\mathrm{T}} \tilde{C} \dot{\boldsymbol{X}} + \frac{f_d Y^2}{4r} \cdot \frac{\boldsymbol{b}^{\mathrm{T}} P_A \boldsymbol{b}}{\gamma^* \|\boldsymbol{b}\|^2}. \tag{3.94}$$

Substituting inequalities (3.92) and (3.94) into Equation (3.90) yields

$$\dot{V} \leq \frac{3}{4}\left\{\dot{\boldsymbol{X}}^{\mathrm{T}} \tilde{C} \dot{\boldsymbol{X}} - \frac{\boldsymbol{b}^{\mathrm{T}} P_A \boldsymbol{b}}{\gamma^* \|\boldsymbol{b}\|^2} \cdot \frac{f_d}{r} Y^2\right\}. \tag{3.95}$$

Bearing in mind that $P_A \boldsymbol{b} = \boldsymbol{b}_K(\boldsymbol{X})$ and referring to the assumption A_3, we conclude

$$\dot{V} \leq \frac{3}{4} \dot{\boldsymbol{X}}^{\mathrm{T}} \tilde{C} \dot{\boldsymbol{X}} - \frac{f_d}{2r} Y^2. \tag{3.96}$$

Table 3.3. Physical parameters of a three-DOF finger

m_1	link mass	0.045 [kg]
m_2	link mass	0.025 [kg]
m_3	link mass	0.015 [kg]
I_1	inertia moment	9.375×10^{-6}[kg · m^2]
I_2	inertia moment	3.333×10^{-6}[kg · m^2]
I_3	inertia moment	1.125×10^{-6}[kg · m^2]
l_1	link length	0.050 [m]
l_2	link length	0.040 [m]
l_3	link length	0.030 [m]
r	radius	0.010 [m]
M	object mass	0.009 [kg]
h	object length	0.050 [m]
w	object width	0.030 [m]
d	object depth	0.010 [m]
I	object inertia moment	3.000×10^{-6}[kg · m^2]
l	object width	0.020 [m]

Hence,

$$\begin{aligned}\dot{W}_\alpha &= \dot{E}_0 + \alpha\dot{V} \\ &\leq -\left(1-\frac{3\alpha}{4}\right)\dot{\boldsymbol{X}}^{\mathrm{T}}\tilde{C}\dot{\boldsymbol{X}} - \frac{\alpha f_d}{2r}Y^2. \end{aligned} \tag{3.97}$$

Since $0 < \alpha \leq 1$ and $(1/4)\dot{\boldsymbol{X}}^{\mathrm{T}}\tilde{C}\dot{\boldsymbol{X}} \geq K(\boldsymbol{X}, \dot{\boldsymbol{X}})$ due to the relation of Equation (3.52) derived from the contact constraints, we finally obtain

$$\dot{W}_\alpha \leq -\alpha E_0(\boldsymbol{X}, \dot{\boldsymbol{X}}). \tag{3.98}$$

On the other hand, we can obtain

$$V \leq \frac{1}{2}E_0(\boldsymbol{X}, \dot{\boldsymbol{X}})$$

similarly to derivation of Equation (3.88). Thus, it follows that

$$\left(1-\frac{\alpha}{2}\right)E_0(\boldsymbol{X}, \dot{\boldsymbol{X}}) \leq W_\alpha \leq \left(1+\frac{\alpha}{2}\right)E_0(\boldsymbol{X}, \dot{\boldsymbol{X}}) \tag{3.99}$$

and hence it follows from Equation (3.98) that

$$\dot{W}_\alpha \leq -\frac{\alpha}{1+(\alpha/2)}W_\alpha \tag{3.100}$$

and

Table 3.4. Parameters of the control signals

f_d	internal force	0.250 [N]
$c_1 = c_2 = c_3$	damping coefficient	0.001 [msN]
γ_f	CSM gain	1500.0
γ_λ	CSM gain	3000.0

$$E_0(\boldsymbol{X}(t), \dot{\boldsymbol{X}}(t)) \leq \frac{1+(\alpha/2)}{1-(\alpha/2)} E_0(\boldsymbol{X}(0), \dot{\boldsymbol{X}}(0)) \mathrm{e}^{-\alpha^* t}, \tag{3.101}$$

where $\alpha^* = \alpha/(1+\alpha/2)$. Note that α can be chosen maximally as $\alpha = 1.0$ in the concerned class of finger–object setups as shown in Table 3.1 under the choice of control gains of Table 3.2 satisfying A_3.

In order to confirm this theoretical proof of exponential convergence of the closed-loop dynamics to an equilibrium state, we show some results of computer simulation based on a physical model of the setup shown in Figure 3.1 with the physical parameters given in Table 3.3. The control gain f_d and damping coefficients c_i $(i = 1, 2, 3)$ for the ith diagonal entry of diagonal damping matrix C as well as the CSM parameters (see Section 4.2) are given in Table 3.4. We show transient responses of principal physical variables in Figure 3.5. As seen from Figure 3.5, $Y(t)$ converges to zero quickly as $t \to \infty$ and $\lambda(t)$ converges to zero with the same speed as $t \to \infty$. In Figure 3.5, we see also that p $(= q_{11} + q_{12} + q_{13})$ and q_{1i} $(i = 1, 2, 3)$ converge to some constant values exponentially as $t \to \infty$. Finally, note that the pressing force $f(t)$ in the normal direction to the object converges exponentially to the specified value $f_0 = (1 + l/r)f_d$.

3.5 Dynamic Force/Torque Balance Based upon Morse Theory

In the previous two sections, we showed that, even if the finger–object system has redundant DOFs, the control signal constructed without knowing the object kinematics or sensing the object orientation can sustain the object dynamically by balancing rotational moments of the object around the pivotal axis, though the equilibrium state that attains the force/torque balance for the overall system cannot be uniquely specified. In this section, we shall show another type of control signal that cannot only accomplish the state of zero moment for the object but also determine the final pose of the finger and object uniquely.

Now, in addition to the right-hand side of the control signal described by Equation (3.39), we append a signal that can be constructed as follows:

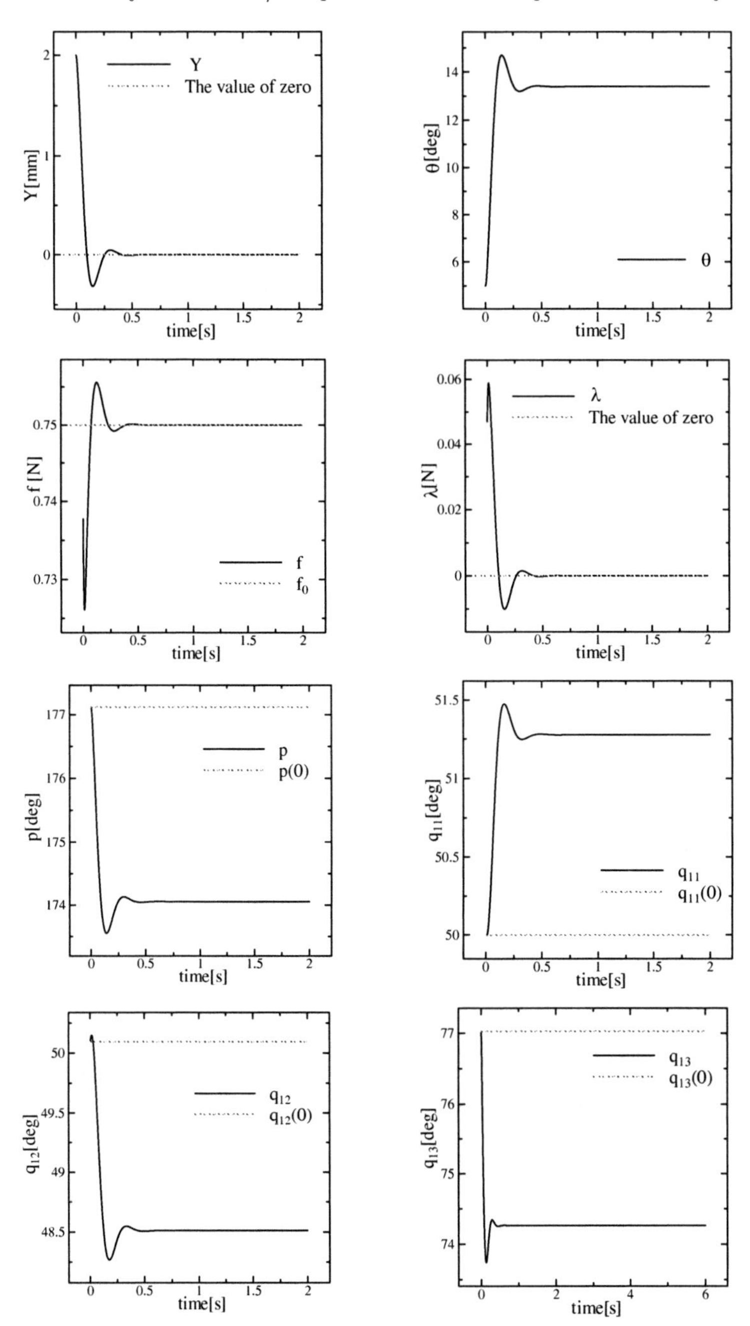

Fig. 3.5. Transient responses of physical variables in the closed-loop dynamics of Equation (3.46) when the control signal of Equation (3.39) is used

$$\hat{N}_0(t) = \hat{N}_0(0) + \gamma_0^{-1} \int_0^t r\dot{q}^{\mathrm{T}}(\tau) \boldsymbol{e} \, \mathrm{d}\tau$$
$$= \hat{N}_0(0) + (r/\gamma_0) \sum_{i=1}^{3} \{q_i(t) - q_i(0)\}, \tag{3.102}$$

where γ_0 is an appropriate positive control gain specified later. For convenience, we set $\hat{N}_0(0) = 0$. Note that the signal $\hat{N}_0(t)$ can also be constructed by measurement data of finger joint angles and the knowledge of the radius r of the finger-end sphere. As a total, the control signal is described as

$$u = -C\dot{q} - (f_d/r) J^{\mathrm{T}}(q) \begin{pmatrix} x_0 - x_m \\ y_0 - y_m \end{pmatrix} - r\hat{N}_0 \boldsymbol{e}. \tag{3.103}$$

Then, the closed-loop dynamics of motion of the finger/object system is given by

$$H\ddot{q} + \left(\frac{1}{2}\dot{H} + S\right)\dot{q} + C\dot{q} - \Delta f \left(\frac{\partial Q}{\partial q}\right) - \Delta\lambda \left(\frac{\partial R}{\partial q}\right) + r\Delta N_0 \boldsymbol{e} = 0, \tag{3.104}$$
$$I\ddot{\theta} - \Delta f Y + \Delta\lambda l + S_N = 0, \tag{3.105}$$

where

$$\begin{cases} \Delta N_0 = \hat{N}_0 - N_0, \quad N_0 = -\dfrac{f_d}{r} Y, \\ S_N = -f_d Y. \end{cases} \tag{3.106}$$

The sum of the inner products between $\dot{q}$ and Equation (3.104) and between $\dot{\theta}$ and Equation (3.105) is reduced to

$$\frac{\mathrm{d}}{\mathrm{d}t} E = -\dot{q}^{\mathrm{T}} C \dot{q}, \tag{3.107}$$

where

$$E = K + P, \qquad P = \frac{f_d}{2r} Y^2 + \frac{\gamma_0}{2} \hat{N}^2 \tag{3.108}$$

and K is the total kinetic energy expressed by Equation (3.1). For the sake of convenience, we call the scalar function P the artificial potential. It is in this case a quadratic function of Y and $\hat{N}$. At the same it is also a quadratic function of θ and p where

$$p = \sum_{i=1}^{3} q_i \tag{3.109}$$

since $Y = -c_0 - r\phi = -c_0 - r(\pi + \theta - p)$ according to Equations (3.5) and (3.6) and $\hat{N}(t) = (r/\gamma_0)\{p(t) - p(0)\}$ according to Equation (3.102).

Let us now confirm that the closed-loop dynamics of Equations (3.104) and (3.105) must be equivalent to Lagrange's equation of motion for the Lagrangian

$$L = K - P + \Delta f Q + \Delta\lambda R \tag{3.110}$$

with external dissipation $-C\dot{q}$. In fact,

$$\begin{aligned}\frac{\mathrm{d}}{\mathrm{d}t}\left(\frac{\partial L}{\partial \dot{q}}\right) - \frac{\partial L}{\partial q} &= H(q)\ddot{q} + \left\{\frac{1}{2}\dot{H} + S\right\}\dot{q} \\ &\quad -\Delta f \frac{\partial Q}{\partial q} - \Delta\lambda \frac{\partial R}{\partial q} + \frac{\partial P}{\partial q} = -C\dot{q}\end{aligned} \tag{3.111}$$

$$\frac{\mathrm{d}}{\mathrm{d}t}\left(\frac{\partial L}{\partial \dot{\theta}}\right) - \frac{\partial L}{\partial \theta} = I\ddot{\theta} - \Delta f \frac{\partial Q}{\partial \theta} - \Delta\lambda \frac{\partial R}{\partial \theta} + \frac{\partial P}{\partial \theta} = 0 \tag{3.112}$$

and

$$\begin{cases} \dfrac{\partial P}{\partial q} = \left(f_d Y + \gamma_0 \hat{N}_0 \cdot \dfrac{r}{\gamma_0}\right) \boldsymbol{e} = r\Delta N_0, \\ \dfrac{\partial P}{\partial \theta} = -f_d Y = S_N. \end{cases} \tag{3.113}$$

Therefore, it is expected that any solution $(\boldsymbol{X}, \dot{\boldsymbol{X}})$ $(= (q(t)^{\mathrm{T}}, \theta(t), \dot{q}^{\mathrm{T}}(t), \dot{\theta}(t))^{\mathrm{T}})$ to the closed-loop dynamics of Equations (3.111) and (3.112) converges asymptotically to the state $(\boldsymbol{X}^*, 0)$ that minimises the artificial potential $P(\boldsymbol{X})$. Nevertheless, it is important to remark that the dynamics of Equations (3.111) and (3.112) should be treated under the contact constraints of Equations (3.2) and (3.6). Minimisation of the artificial potential should also be executed under the same constraints. In other words, the scalar function $P(\boldsymbol{X})$ should be minimised on the two-dimensional constraint manifold $CM_2 = \{\boldsymbol{X} : Q = 0 \text{ and } R = 0\}$. Then, to find a point $\boldsymbol{X} = \boldsymbol{X}^*$ that minimises $P(\boldsymbol{X})$ so that $P(\boldsymbol{X}) \geq P(\boldsymbol{X}^*)$ in a neighbourhood of $\boldsymbol{X}^*$ in CM_2, it is reasonable to consider the scalar function

$$\tilde{P}(\boldsymbol{X}) = P(\boldsymbol{X}) + \Delta f Q(\boldsymbol{X}) + \Delta\lambda R(\boldsymbol{X}) \tag{3.114}$$

with Lagrange multipliers Δf and $\Delta\lambda$ corresponding to constraints $Q = 0$ and $R = 0$, respectively, and derive the gradient equation of $\tilde{P}$ in $\boldsymbol{X}$ $(= (q^{\mathrm{T}}, \theta)^{\mathrm{T}})$. This results in the following:

$$-A\boldsymbol{\lambda} + B\boldsymbol{\eta} = 0, \tag{3.115}$$

where $\boldsymbol{\lambda} = (\Delta f, \Delta\lambda)^{\mathrm{T}}$, $\boldsymbol{\eta} = (\Delta N_0, r_0^{-1} S_N)^{\mathrm{T}}$ and

$$A = \begin{pmatrix} \dfrac{\partial Q}{\partial q} & \dfrac{\partial R}{\partial q} \\ Y & -l \end{pmatrix}, \qquad B = \begin{pmatrix} r\boldsymbol{e} & 0_{3\times 1} \\ 0 & r_0 \end{pmatrix}, \tag{3.116}$$

where r_0 stands for a scale factor introduced for the purpose of balancing the numerical orders of the two column vectors of matrix B. Note that the coefficient matrix A in Equation (3.115) is equivalent to that defined by Equation (3.27). A point $\boldsymbol{X} = \boldsymbol{X}^*$ that satisfies Equation (3.115) with a certain vector $\boldsymbol{\lambda} = \boldsymbol{\lambda}^*$ is called the critical point. In this case, there exists a unique critical point $\boldsymbol{X}^*$ at which

$$\boldsymbol{\lambda} = 0_{2\times 1}, \qquad \boldsymbol{\eta} = 0_{2\times 1} \tag{3.117}$$

because the 4×4 matrix (A, B) is non-singular in a certain region on CM_2 as discussed in Section 3.3. Thus, at the critical point, $Y = 0$ according to $S_N = 0$ and $p = p(0)$, from which all values of q_i $(i = 1, 2, 3)$ and θ can be determined through constraint Equations (3.2) and (3.6).

Before concluding the argument, we must remark on some important properties of the Hessian matrix of the artificial potential $P(\boldsymbol{X})$ in $\boldsymbol{X}$, which is derived in the following way:

$$G = \left(\frac{\partial^2 P}{\partial X_i \partial X_j}\right) = r \begin{pmatrix} \left(f_d + \dfrac{r}{\gamma_0}\right) \boldsymbol{e}\boldsymbol{e}^{\mathrm{T}} & -f_d \boldsymbol{e} \\ -f_d \boldsymbol{e}^{\mathrm{T}} & f_d \end{pmatrix}. \tag{3.118}$$

This 4×4 matrix is non-negative definite but not positive definite. Apparently, G has the two positive eigenvalues and the other two are zero. To see this explicitly, consider the coordinate transformation

$$\begin{cases} \delta \boldsymbol{X} = \Gamma \delta \boldsymbol{Z}, \quad \Gamma = \begin{pmatrix} 1 & 0 & 0 & 0 \\ -1 & 1 & 0 & 0 \\ 0 & -1 & 1 & 0 \\ 0 & 0 & 0 & 1 \end{pmatrix}, \\ \delta \boldsymbol{X} = \boldsymbol{X} - \boldsymbol{X}^*, \quad \delta \boldsymbol{Z} = \boldsymbol{Z} - \boldsymbol{Z}^*, \end{cases} \tag{3.119}$$

where $\boldsymbol{X}^*$ denotes the critical point on CM_2, $\boldsymbol{Z}^* = \Gamma^{-1}\boldsymbol{X}^*$, and $\boldsymbol{Z} = \Gamma^{-1}\boldsymbol{X} = (q_1, q_1 + q_2, q_1 + q_2 + q_3, \theta)^{\mathrm{T}}$. Then, it is easy to see that

$$\begin{cases} \delta \boldsymbol{X}^{\mathrm{T}} G \delta \boldsymbol{X} = \delta \boldsymbol{Z}^{\mathrm{T}} \Gamma^{\mathrm{T}} G \Gamma \delta \boldsymbol{Z} = \delta \boldsymbol{Z}^{\mathrm{T}} \tilde{G} \delta \boldsymbol{Z} \\ \tilde{G} = \Gamma^{\mathrm{T}} G \Gamma = \begin{pmatrix} 0_{2\times 2} & 0_{2\times 2} \\ 0_{2\times 2} & r\tilde{G}_{22} \end{pmatrix}, \quad \tilde{G}_{22} = \begin{pmatrix} f_d + \dfrac{r}{\gamma_0} & -f_d \\ -f_d & f_d \end{pmatrix}. \end{cases} \tag{3.120}$$

This shows that the 2×2 sub-matrix $r\tilde{G}_{22}$ is positive definite with respect to the sub-vector $\delta \boldsymbol{Z}_2 = (\delta p, \delta\theta)^{\mathrm{T}}$ of $\delta \boldsymbol{Z} = (\delta \boldsymbol{Z}_1, \delta \boldsymbol{Z}_2)$. Further, this implies that $\delta \boldsymbol{Z}_2 = (\delta p, \delta\theta)^{\mathrm{T}}$ can play a role of local coordinates for the two-dimensional manifold CM_2 in the vicinity of the critical point $\boldsymbol{X} = \boldsymbol{X}^*$. More explicitly, let us consider a further coordinate transformation of the form:

$$\begin{cases} \delta \boldsymbol{Z}_2 = (\delta p, \delta\theta)^{\mathrm{T}} = \Gamma_{22}\delta\tilde{\boldsymbol{Z}}_2, \\ \Gamma_{22} = \begin{pmatrix} 1 & 0 \\ 1 & 1 \end{pmatrix}. \end{cases} \tag{3.121}$$

Then, $\delta\tilde{\boldsymbol{Z}}_2 = \Gamma_{22}^{-1}\delta\boldsymbol{Z}_2 = (\delta p, \delta\theta - \delta p)^{\mathrm{T}}$ and

$$\tilde{G}_2 = \Gamma_{22}^{\mathrm{T}} r\tilde{G}_{22}\Gamma_{22} = \begin{pmatrix} r^2/\gamma_0 & 0 \\ 0 & rf_d \end{pmatrix}. \tag{3.122}$$

Hence, it follows that

$$\begin{aligned} \frac{1}{2}\delta\tilde{\boldsymbol{Z}}_2^{\mathrm{T}}\tilde{G}_2\delta\tilde{\boldsymbol{Z}}_2 &= \frac{1}{2}\left\{\frac{r^2}{\delta_0}\delta p^2 + rf_d(\delta\theta - \delta p)^2\right\} \\ &= \frac{1}{2}\left\{\gamma_0(\delta\hat{N})^2 + (f_d/r)(\delta Y)^2\right\} \\ &= \delta\tilde{P}. \end{aligned} \tag{3.123}$$

According to the theory of calculus of variations in the large, that is called the Morse theory, the set of singular values (eigenvalues) of the Hessian with respect to certain local coordinates at the critical point on the manifold is called the Morse function, or equivalently the quadratic form of a diagonalised Hessian matrix is called the Morse function. In this case, the form

$$(r^2/\gamma_0)\delta p^2 + (rf_d)(\delta\theta - \delta p)^2 \tag{3.124}$$

is the Morse function at the critical point $\boldsymbol{X} = \boldsymbol{X}^*$ on the manifold CM_2. The Morse theory assures that if the Morse function is positive definite at the critical point then the concerned scalar function attains a locally unique minimum at the critical point. In this illustrative model for immobilisation of a two-dimensional pivoted object, most parts of the argument developed above may seem to be intelligible without taking advantage of the Morse theory, because the Hessian matrix of the potential becomes constant. However, such a rigorous argument will be crucial in later chapters where the stability analysis of pinching or precision prehension of 3-D objects by using a pair of multi-joint robot fingers with rigid or soft finger-ends with hemispherical shape will be presented.

3.6 Minimum DOF for Dynamic Immobilisation of a 2-D Pivoted Object

In the last section of this chapter, we shall present an answer to the question posed in Section 2.5: how many actuated joints (or DOFs) must a finger have to stop the rotational moment of the free pivotal motion of a 2-D rigid object hinged at a point on a horizontal plane.

Let us consider a robot finger with two joints as shown in Figure 2.8, whose finger-end is in contact with a 2-D rigid object. Contact constraints are given as in Equations (3.2) and (3.6). Then, the Lagrange equation of motion of the finger/object system is given by Equations (3.9) and (3.10) or Equations (2.39) and (2.40). As in the previous section, we consider the control input signal described by

$$u = -C\dot{q} - (f_d/r)J^{\mathrm{T}}(q)\begin{pmatrix} x_0 - x_m \\ y_0 - y_m \end{pmatrix} - r\hat{N}_0\boldsymbol{e}, \tag{3.125}$$

which is the same in form as Equation (3.103). However, in this case, $\boldsymbol{e} = (1,1)^{\mathrm{T}}$, $q = (q_1, q_2)^{\mathrm{T}}$, and therefore the Jacobian matrix $J(q)$ becomes a 2×2 matrix. Substituting Equation (3.125) into Equation (3.9), we obtain the same form of closed-loop dynamics in Equations (3.104) and (3.105). However, in this two-DOF finger case, the three two-dimensional vectors $\partial Q/\partial q$, $\partial R/\partial q$, and $\boldsymbol{e}$ are no more independent. Moreover, the constraint manifold $CM_1 = \{(q^{\mathrm{T}}, \theta) : Q = 0 \text{ and } R = 0\}$ is of one dimension and therefore either component of the coordinates $(p = q_1 + q_2, \theta)$ is dependent on the other one. Hence, minimisation of the scalar function $\tilde{P} = (1/2)\{(f_d/r)Y^2 + \gamma_0\hat{N}^2\}$ may certainly be attained at non-zero solutions $\Delta f = \Delta f_\infty$, $\Delta\lambda = \Delta\lambda_\infty$, $r^{-1}S_N = r^{-1}S_{N\infty}$, $\Delta N_0 = \Delta N_{0\infty}$ that satisfy

$$-A\boldsymbol{\lambda}_\infty + B\boldsymbol{\eta}_\infty = 0, \tag{3.126}$$

where $\boldsymbol{\lambda}_\infty = (\Delta f_\infty, \Delta\lambda_\infty)^{\mathrm{T}}$, $\boldsymbol{\eta}_\infty = (r^{-1}S_{N\infty}, \Delta N_{0\infty})$,

$$A = \begin{pmatrix} \dfrac{\partial Q}{\partial q} & \dfrac{\partial R}{\partial q} \\ Y & -l \end{pmatrix}, \quad B = \begin{pmatrix} r\boldsymbol{e} & 0_{2\times 1} \\ 0 & r \end{pmatrix}. \tag{3.127}$$

In particular, the final balanced rotational moment around the pivotal axis at O_m is expressed by Equation (3.112) as follows:

$$-\Delta f_\infty Y_\infty + \Delta\lambda_\infty l - f_d Y_\infty = 0. \tag{3.128}$$

Since f_∞ and λ_∞ are expressed from the definitions of Δf and $\Delta\lambda$ [see Equation (3.38)] as

$$f_\infty = \Delta f_\infty + \left(1 + \frac{l}{r}\right) f_d, \quad \lambda_\infty = \Delta\lambda_\infty + \frac{f_d}{r} Y_\infty. \tag{3.129}$$

Equation (3.128) must be equivalent to

$$-f_\infty Y_\infty + \lambda_\infty l = 0. \tag{3.130}$$

This shows that the rotational moments induced by the contact constraint force f_∞ with the direction indicated in Figure 2.8 and the rolling constraint

Table 3.5. DOF physical parameters

m_1	link mass	0.025 [kg]
m_2	link mass	0.015 [kg]
I_1	inertia moment	3.333×10^{-6}[kg · m^2]
I_2	inertia moment	1.125×10^{-6}[kg · m^2]
l_1	link length	0.040 [m]
l_2	link length	0.030 [m]
r	radius	0.010 [m]
M	object mass	0.009 [kg]
h	object length	0.050 [m]
w	object width	0.030 [m]
d	object depth	0.010 [m]
I	object inertia moment	3.000×10^{-6}[kg · m^2]
l	object width	0.020 [m]

Table 3.6. Parameters of the control signals

f_d	internal force	0.250 [N]
$c_1 = c_2$	damping coefficient	0.001 [msN]
γ_0	regressor gain	0.001 [m · rad/N]
$\hat{N}_0(0)$	initial estimate value	0.000 [N]
γ_f	CSM gain	1500.0
γ_λ	CSM gain	3000.0

force λ_∞ with the direction tangential to the object surface have been balanced.

All these theoretical predictions concerning the convergence of the closed-loop dynamics to the equilibrium point that satisfies force/torque balance can be confirmed through numerical simulation. In the simulation, a two-DOF robot finger and a 2-D rigid object with the physical parameters in Table 3.5 are used. The control gains f_d and γ_0 and damping coefficients c_i $(i = 1, 2)$ used in the simulation are given in Table 3.6. Figure 3.6 shows the transient responses of the physical variables when the control signal of Equation (3.125) is used. In this figure, α' is defined as

$$\alpha' = \tan^{-1}(\lambda/f). \tag{3.131}$$

As shown in Figure 3.7, at the convergent stage the net force vector $\boldsymbol{F}$ originating at the contact point P $(= (x, y))$ is directed toward the hinged point O_m $(= (x_m, y_m))$. As seen from Figure 3.7, α' must converge to the angle α specified in Figure 3.7 and defined by

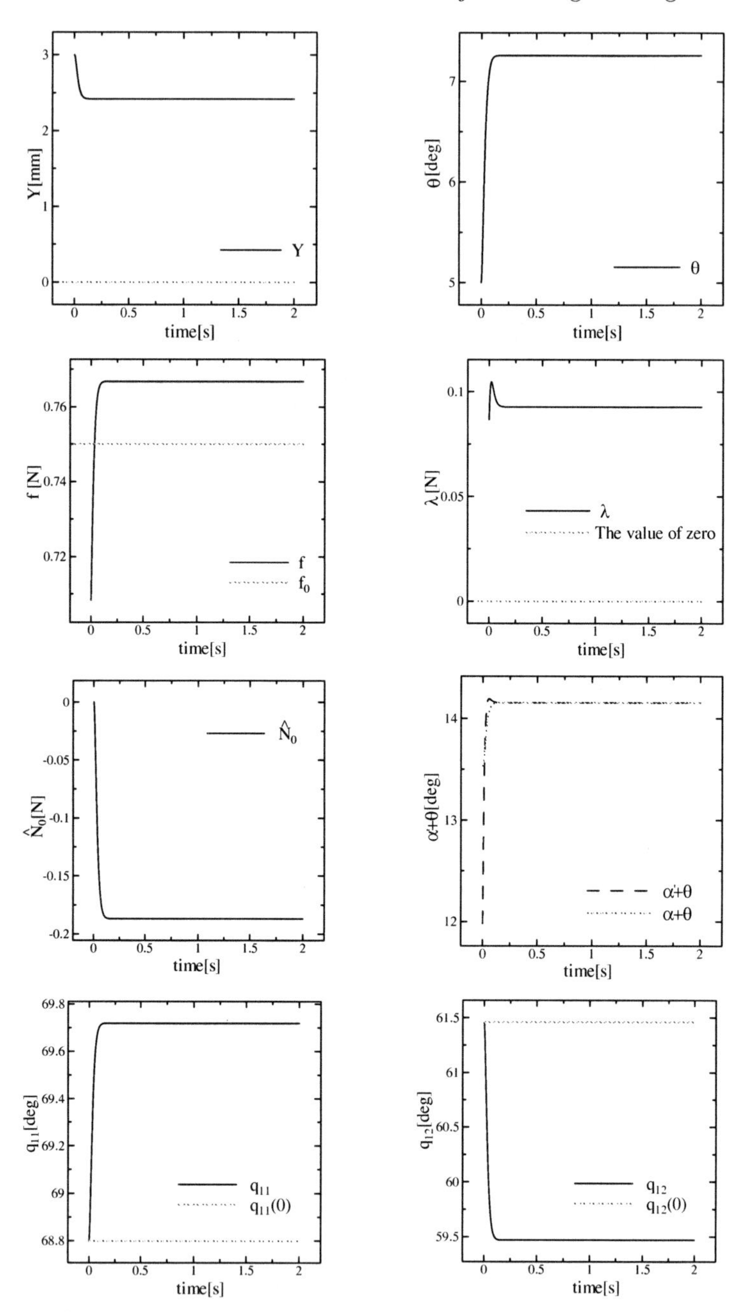

Fig. 3.6. The transient responses of the physical variables in the closed-loop dynamics of Equations (3.104) and (3.105) when the control signal of Equation (3.125) is used

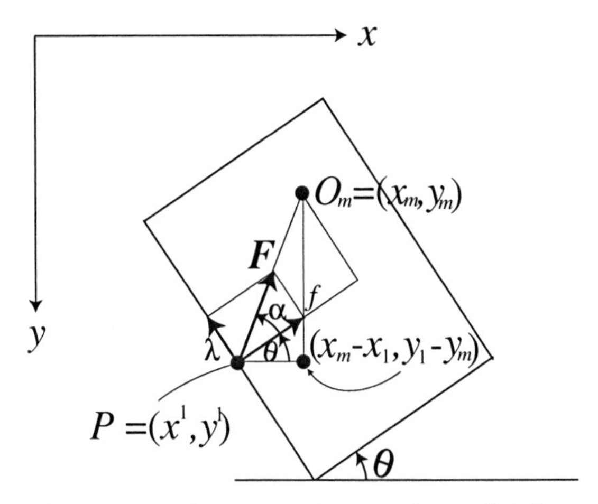

Fig. 3.7. Robot finger manipulating an object with parallel flat surfaces in a 2-D plane

$$\alpha = -\theta_\infty + \tan^{-1}\left(\frac{y_{1\infty} - y_m}{x_m - x_{1\infty}}\right), \tag{3.132}$$

where θ_∞ and $(x_{1\infty}, y_{1\infty})$ are constants to which $\theta(t)$ and $(x_1(t), y_1(t))$ converge, respectively, as $t \to \infty$. As discussed theoretically in the previous paragraph, $Y(t)$ converges to some constant value that is not necessarily zero. Therefore, as seen from Figure 3.6 $f(t)$ and $\lambda(t)$ converge to some constant values f_∞ and λ_∞ different from their corresponding values f_0 and $\lambda = 0$. It should be noted again that, from Equations (3.130), (3.131) and (3.132) and the convergence of α' to α as $t \to \infty$, α must be coincident with $\tan^{-1}(\lambda_\infty/f_\infty)$, which is also equal to $\tan^{-1}(Y_\infty/l)$.

For the sake of comparison of effects of the control signal of Equation (3.125) with those of Equation (3.39) in the case of two-DOF finger, we present Figure 3.8 that shows the transient responses of the principal physical variables of the closed-loop dynamics of Equation (3.46). In this simulation, the same values for f_d and c_i $(i = 1, 2)$ as given in Table 3.6 are employed. It is interesting to note that the speed of convergence to the equilibrium state $(Y = 0,\ \lambda = 0,\ f = (1 + l/r)f_d)$ becomes noticeably slower in comparison with the former case where the term $r\hat{N}_0\boldsymbol{e}$ is included in the control signal.

Even in the case of a robot finger with redundant joints like the three-DOF finger of Figure 3.1, is the performance of the closed-loop dynamics using the control signal of Equation (3.103) superior to that using the control of Equation (3.36) in which the term $r\hat{N}_0\boldsymbol{e}$ is missing? Unexpectedly and contrarily to the case of use of a non-redundant two-DOF finger, we observe from the computer simulation results that the addition of the term $r\hat{N}_0\boldsymbol{e}$ to the control signal rather degrades the control performance by slowing down the

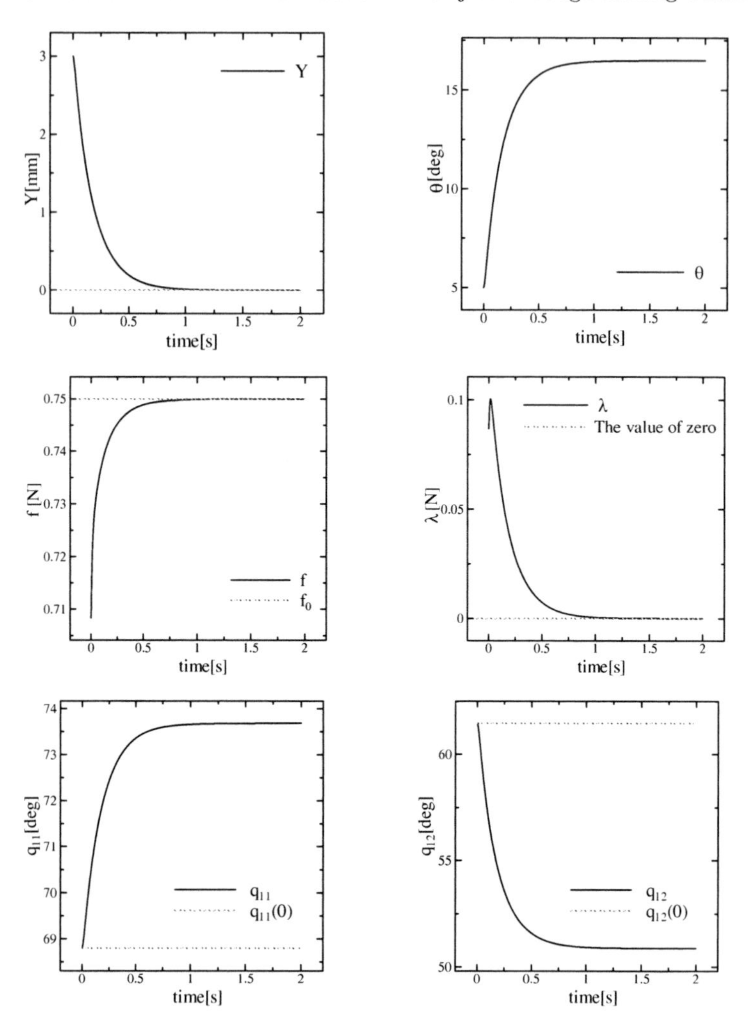

Fig. 3.8. The transient responses of the physical variables of the closed-loop dynamics when the term $r\hat{N}_0\boldsymbol{e}$ in Equation (3.125) is missing

speed of the convergence of the trajectory to the equilibrium point satisfying $Y_\infty = 0$ and $p_\infty = p(0)$. The reason is that in this redundant DOF case the physical variable p $(= q_1 + q_2 + q_3)$ should eventually converge and move back to the initial value $p_\infty = p(0)$ in order for Y and $\hat{N}_0$ to converge to zero as $t \to \infty$, as discussed in Section 3.5.

Finally we turn to the simplest case of immobilisation of rotational motion of the 2-D object by using a single-joint finger as shown in Figure 2.6. In this case, it is possible to stop the object motion only at its definite orientational angle specified by two constraint conditions expressed by Equations (3.2) and

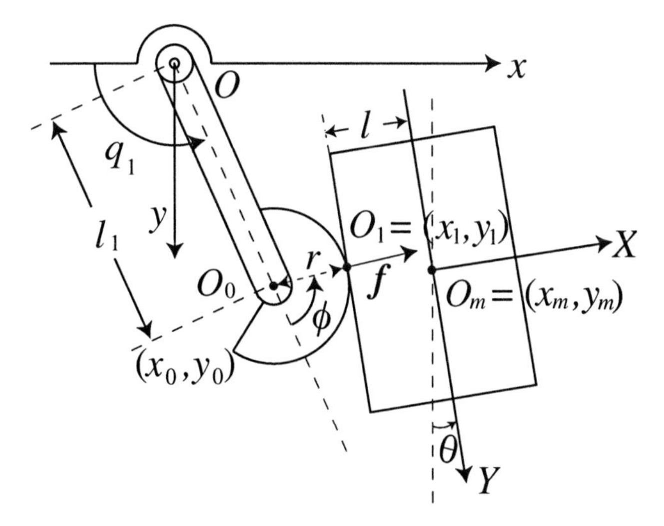

Fig. 3.9. When a finger-end sphere slowly contacts with the rigid object by increasing q_1, the finger motion is stacked at this contact position and immobilised in a static sense

(3.6). In fact, at the instant when the finger-end sphere contacts with the rigid object by slowly increasing q_1 (see Figure 3.9), the value of q_1 determines the value of Y according to Equation (3.38), from which the inclination angle θ of the object is determined through Equation (3.4). Then, the finger-end is stacked at this position and immobilised in a static sense. In other words, it is impossible to cause rolling between the finger-end sphere and the object surface and therefore it is impossible to restabilise the object motion once the contact is released by disturbances. Quite interestingly, however, a single-DOF finger whose finger-end is soft and visco-elastic works well in stabilisation of an equilibrium state of grasp in immobilising the object in a dynamic sense. This simplest testbed problem for stabilising rotational motion of a 2-D object by using a single-DOF robot finger with a soft fingertip will be treated in Section 6.2.

4

Two-dimensional Grasping by a Pair of Rigid Fingers

This chapter discusses a sensory-motor coordination control scheme that realises stable grasping (precision prehension) of rigid objects with parallel or non-parallel flat surfaces movable in a two-dimensional vertical plane by a pair of multi-joint robot fingers with hemispherical ends. The proposed control is composed of linear superposition of signals for gravity compensation for fingers, damping shaping, exertion of forces to the object from opposite directions, generation of moments for balancing of rotational moments, and regressors for estimating unknown steady-state terms, all of which neither need the knowledge of object parameters nor use any object sensing data. In other words, stable grasping can be realised in a blind manner by using measurement data of only finger joint angles without using force sensors or tactile sensing. Stability of pinching motion with convergence to the state of force/torque balance is defined on a constraint manifold based upon the Riemannian distance. Exponential convergence of the closed-loop dynamics to an equilibrium manifold satisfying the force/torque balance is verified by examining the Morse function derived from the Hessian matrix regarding the artificial potential. In the sequel, it is shown that the proposed coordinated control scheme does not need any alteration even if the object side surfaces are not parallel. It is also interesting to know eventually that the smallest number of total DOFs for pair of fingers to realize stable grasping in a blind manner in a veritical plane under the gravity effect is four, that is, each finger should have at least two joints.

4.1 Dynamics of the Physical Interaction Between Fingers and an Object

First, the dynamics of a pair of two-DOF finger robots and a 2-D object with flat surfaces are derived. The coordinates of the overall system is shown in Figure 4.1. The vector $\boldsymbol{q}_1 = (q_{11}, q_{12})^{\mathrm{T}}$ denotes the joint angles of the left-hand side finger and vector $\boldsymbol{q}_2 = (q_{21}, q_{22})^{\mathrm{T}}$ denotes those of the right finger.

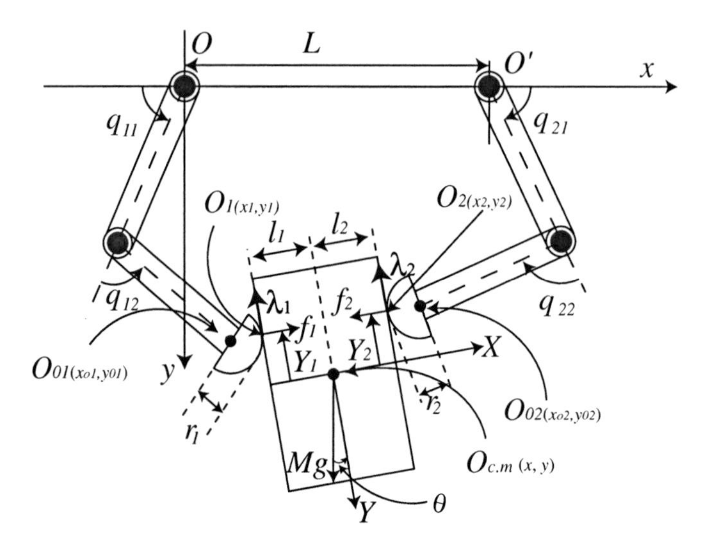

Fig. 4.1. A pair of robot fingers pinching a rigid object

The fingertips are rigid and hemispherically shaped with radius r_i. The centre of mass of the object is denoted by $O_{\mathrm{c.m.}}$. We define the vector $\boldsymbol{w} = (x, y, \theta)^{\mathrm{T}}$ in the coordinate system $\{O{-}xy\}$, where (x,y) denotes the position of the object centre of mass and θ is the rotational angle of the object. The symbols l_{ij}, m_{ij} and I_{ij} $(i, j = 1, 2)$ denote the length, mass and inertia moment of link j of finger i, respectively. The symbols M and I denote the mass and moment of inertia of the object, respectively, and L the distance between the origin$\{O\}$ of the left finger and the origin$\{O'\}$ of the right finger.

The distance Y_i between the contact point and the other surface point at which the straight line (the X-axis shown in Figure 4.1) from the centre of mass $O_{\mathrm{c.m.}}$ crosses the object surface perpendicularly is (see Figure 4.2):

$$Y_i = (x_{0i} - x)\sin\theta + (y_{0i} - y)\cos\theta, \quad i = 1, 2. \tag{4.1}$$

This quantity must be subject to the constraint that the fingertip does not slip on the object surface, or equivalently the velocity $-\mathrm{d}Y_i/\mathrm{d}t$ on the object surfaces equals that on its corresponding finger end $r_i\mathrm{d}\phi_i/\mathrm{d}t$, that is,

$$r_i \frac{\mathrm{d}}{\mathrm{d}t}\phi_i = -\frac{\mathrm{d}}{\mathrm{d}t}Y_i, \tag{4.2}$$

where ϕ_i is defined as

$$q_{i1} + q_{i2} + \phi_i = \pi - (-1)^i\theta, \quad i = 1, 2. \tag{4.3}$$

Throughout this book, we adopt the rule of defining the sign of an angle θ or q_{1j} $(j = 1, 2)$ to be positive when it directs counter-clockwise. However,

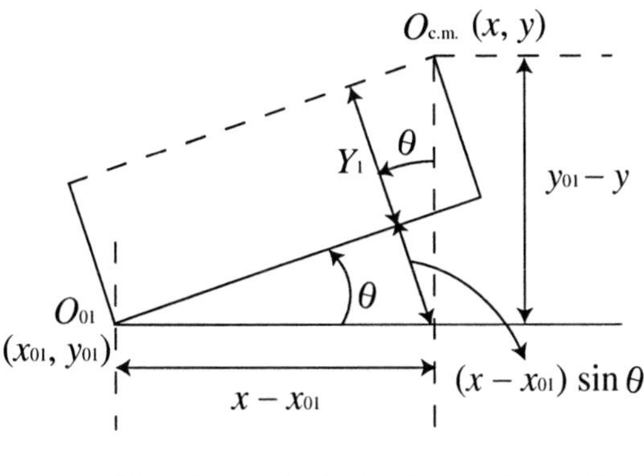

Fig. 4.2. Geometric relations among the centre O_{01} of fingertip sphere, object centre of mass $O_{\mathrm{c.m.}}$, θ and Y_1

exceptionally and for convenience, the joint angle q_{2j} ($j = 1, 2$) for the right-hand finger is defined to be positive if it directs clockwise. Then, integrating Equation (4.2) with respect to t leads to

$$Y_i = c_{0i} - r_i \left\{\pi - (-1)^i\theta - q_{11} - q_{12}\right\}, \quad i = 1, 2, \tag{4.4}$$

where c_{01} and c_{02} are constant. Then, substituting Equation (4.1) into (4.4) yields

$$\begin{aligned} R_i &= (x_{0i} - x)\sin\theta + (y_{0i} - y)\cos\theta - c_{01} + r_i\left\{\pi - (-1)^i\theta - q_{i1} - q_{i2}\right\} \\ &= 0, \qquad\qquad i = 1, 2. \end{aligned} \tag{4.5}$$

These constraints lead to the introduction of the scalar function R_0 in the following way by using the Lagrange multipliers λ_1 and λ_2:

$$R_0 = \lambda_1 R_1 + \lambda_2 R_2. \tag{4.6}$$

It is easy to calculate the partial differentials of $R_i (i = 1, 2)$ with respect to $\boldsymbol{q}_i$ as follows:

$$\frac{\partial R_i}{\partial \boldsymbol{q}_i} = J_{0i}^{\mathrm{T}} \begin{pmatrix} \sin\theta \\ \cos\theta \end{pmatrix} - r_i \boldsymbol{e}_i, \quad i = 1, 2 \tag{4.7}$$

where $\boldsymbol{e}_i = (1, 1)^{\mathrm{T}}$ and J_{0i} denotes the Jacobian matrix defined by

$$J_{0i}^{\mathrm{T}} = \begin{pmatrix} \dfrac{\partial x_{0i}}{\partial \boldsymbol{q}_i} & \dfrac{\partial y_{0i}}{\partial \boldsymbol{q}_i} \end{pmatrix}. \tag{4.8}$$

The constraint that the fingertip is stuck to the object is described by the following geometric relation:

$$Q_i = -(l_i + r_i) - (-1)^i\{(x - x_{0i})\cos\theta - (y - y_{0i})\sin\theta\} = 0, \quad i = 1, 2. \tag{4.9}$$

From a similar derivation to that of Equation (4.7), it follows that

$$\frac{\partial Q_i}{\partial \boldsymbol{q}_i} = (-1)^i J_{0i}^{\mathrm{T}} \begin{pmatrix} \cos\theta \\ -\sin\theta \end{pmatrix}, \quad i = 1, 2. \tag{4.10}$$

These constraints represent the scalar Q in the following way by using the Lagrange multipliers f_1 and f_2:

$$Q = f_1 Q_1 + f_2 Q_2. \tag{4.11}$$

Thus, the Lagrangian of the overall fingers–object system is composed in the form

$$L = K - (P_1 + P_2 - Mgy) + R_0 + Q, \tag{4.12}$$

where

$$K = \frac{1}{2}\sum_{i=1,2} \dot{\boldsymbol{q}}_i^{\mathrm{T}} H_i(\boldsymbol{q}_i)\dot{\boldsymbol{q}}_i + \frac{1}{2}(M\dot{x}^2 + M\dot{y}^2 + I\dot{\theta}^2) \tag{4.13}$$

and $H_i(\boldsymbol{q}_i)$ denotes the finger inertia matrix, P_i stands for the finger potential energy and K is the total kinetic energy. Then, by applying the variational principle to the form

$$\int_{t_0}^{t_1} \{\delta L + \boldsymbol{v}_1^{\mathrm{T}}\delta\boldsymbol{q}_1 + \boldsymbol{v}_2^{\mathrm{T}}\delta\boldsymbol{q}_2\}\, \mathrm{d}\tau = 0 \tag{4.14}$$

we obtain a set of Lagrange's equations of motion for the overall system as follows:

$$\begin{aligned} &\left\{H_i(\boldsymbol{q}_i)\frac{\mathrm{d}}{\mathrm{d}t} + \frac{1}{2}\dot{H}_i(\boldsymbol{q}_i)\right\}\dot{\boldsymbol{q}}_i + S_i(\boldsymbol{q}_i, \dot{\boldsymbol{q}}_i)\dot{\boldsymbol{q}}_i + \boldsymbol{g}_i(\boldsymbol{q}_i) \\ &\qquad -f_i\left(\frac{\partial Q_i}{\partial \boldsymbol{q}_i}\right) - \lambda_i\left(\frac{\partial R_i}{\partial \boldsymbol{q}_i}\right) = \boldsymbol{v}_i, \quad i = 1, 2 \end{aligned} \tag{4.15}$$

$$\begin{cases} M\ddot{x} - (f_1 - f_2)\cos\theta + (\lambda_1 + \lambda_2)\sin\theta = 0 \\ M\ddot{y} + (f_1 - f_2)\sin\theta + (\lambda_1 + \lambda_2)\cos\theta - Mg = 0 \\ I\ddot{\theta} - f_1 Y_1 + f_2 Y_2 + \lambda_1 l_1 - \lambda_2 l_2 = 0 \end{cases} \tag{4.16}$$

where $S_i(\boldsymbol{q}_i, \dot{\boldsymbol{q}}_i)$ is skew-symmetric. Equation (4.15) expresses the motion of the fingers, and Equation (4.16) that of the object.

Second, the dynamics of a pair of three-DOF finger robots and a 2-D object with non-parallel flat surfaces are derived. The coordinate system is shown in Figure 4.3. The holonomic constraints in this case corresponding to Equations

Table 4.1. Explicit formulae of Q_i and R_i for an object with non-parallel flat surfaces

Point contact constraint:

$$Q_i = -(l_i + r_i) - (-1)^i\{(x - x_{0i})\cos(\theta + (-1)^i\theta_0) - (y - y_{0i})\sin(\theta + (-1)^i\theta_0)\} = 0 \qquad i = 1,2 \quad \text{(T-1)}$$

Rolling constraint:

$$R_i = Y_i - c_{0i} + r_i\left\{\pi - (-1)^i\theta - \theta_0 - \textstyle\sum_{j=1}^{3} q_{ij}\right\} = 0 \qquad i = 1,2 \quad \text{(T-2)}$$

where

$$Y_i = (x_{0i} - x)\sin(\theta + (-1)^i\theta_0) + (y_{0i} - y)\cos(\theta + (-1)^i\theta_0), \; i = 1,2 \qquad \text{(T-3)}$$

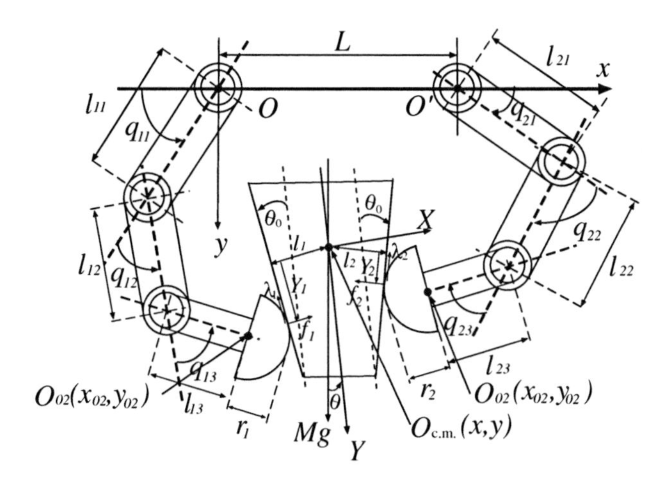

Fig. 4.3. A pair of robot fingers pinching a rigid object with non-parallel surfaces

(4.5) and (4.9) are given in Table 4.1. In a similar way, it is possible to derive the following equations:

$$\left\{H_i(\boldsymbol{q_i})\frac{\mathrm{d}}{\mathrm{d}t} + \frac{1}{2}\dot{H}_i(\boldsymbol{q_i})\right\}\dot{\boldsymbol{q}}_i + S_i(\boldsymbol{q_i}, \dot{\boldsymbol{q}}_i)\dot{\boldsymbol{q}}_i + \boldsymbol{g_i}(\boldsymbol{q_i}) - f_i\left(\frac{\partial Q_i}{\partial \boldsymbol{q_i}}\right) - \lambda_i\left(\frac{\partial R_i}{\partial \boldsymbol{q_i}}\right) = \boldsymbol{v_i}, \quad i = 1,2 \tag{4.17}$$

$$\begin{cases} M\ddot{\boldsymbol{x}} - \sum_{i=1,2}\left(f_i\frac{\partial Q_i}{\partial \boldsymbol{x}} + \lambda_i\frac{\partial R_i}{\partial \boldsymbol{x}}\right) - Mg\begin{pmatrix}0\\1\end{pmatrix} = 0 \\ I\ddot{\theta} - \sum_{i=1,2}\left(f_i\frac{\partial Q_i}{\partial \theta} + \lambda_i\frac{\partial R_i}{\partial \theta}\right) = 0 \end{cases} \tag{4.18}$$

Table 4.2. Explicit formulae of the partial differentials of Q_i and R_i with respect to $\boldsymbol{q}_i$, $\boldsymbol{x}$, θ and ψ for an object with non-parallel flat surfaces

$$Q = f_1 Q_1 + f_2 Q_2 \,!\,\$R = \lambda_1 R_1 + \lambda_2 R_2$$

$$\frac{\partial Q}{\partial \boldsymbol{q}_i} = f_i \left(\frac{\partial Q_i}{\partial \boldsymbol{q}_i} \right) = (-1)^i f_i J_i^{\mathrm{T}}(q_i) \begin{pmatrix} \cos(\theta + (-1)^i \theta_0) \\ -\sin(\theta + (-1)^i \theta_0) \end{pmatrix}, \ i = 1, 2 \tag{T-4}$$

$$\frac{\partial R_0}{\partial \boldsymbol{q}_i} = \lambda_i \left(\frac{\partial R_i}{\partial \boldsymbol{q}_i} \right) = \lambda_i \left\{ J_i^{\mathrm{T}} \begin{pmatrix} \sin(\theta + (-1)^i \theta_0) \\ \cos(\theta + (-1)^i \theta_0) \end{pmatrix} - r_i \boldsymbol{e}_i \right\}, \ i = 1, 2 \tag{T-5}$$

$$\frac{\partial Q}{\partial x} = f_1 \cos(\theta - \theta_0) - f_2 \cos(\theta + \theta_0) \tag{T-6}$$

$$\frac{\partial R}{\partial x} = -\lambda_1 \sin(\theta - \theta_0) - \lambda_2 \sin(\theta + \theta_0) \tag{T-7}$$

$$\frac{\partial Q}{\partial y} = -f_1 \sin(\theta - \theta_0) + f_2 \sin(\theta + \theta_0) \tag{T-8}$$

$$\frac{\partial R}{\partial y} = -\lambda_1 \cos(\theta - \theta_0) - \lambda_2 \cos(\theta + \theta_0) \tag{T-9}$$

$$\frac{\partial Q}{\partial \theta} = f_1 Y_1 - f_2 Y_2 \tag{T-10}$$

$$\frac{\partial R}{\partial \theta} = -\lambda_1 l_1 + \lambda_2 l_2 \tag{T-11}$$

where the partial derivatives of Q_i and R_i ($i = 1, 2$) with respect to $\boldsymbol{q}_i$, $\boldsymbol{x} = (x, y)^{\mathrm{T}}$ and θ are given in Table 4.2.

4.2 Force/Torque Balance

According to the equation of motion of the object expressed by Equation (4.16) or (4.18), the contact constraint forces f_i ($i = 1, 2$) are exerted on the object in the direction normal to the object surfaces, which can be described by means of the following two-dimensional wrench vectors:

$$\boldsymbol{w}_{f1} = \begin{pmatrix} \cos\theta \\ -\sin\theta \\ Y_1 \end{pmatrix}, \quad \boldsymbol{w}_{f2} = - \begin{pmatrix} \cos\theta \\ -\sin\theta \\ Y_2 \end{pmatrix}. \tag{4.19}$$

On the other hand, the rolling contact contraint forces appear in directions tangent to the object surfaces as shown in Figure 4.1, which can be described the same by the wrench vectors:

$$\boldsymbol{w}_{\lambda 1} = \begin{pmatrix} \sin\theta \\ \cos\theta \\ l_1 \end{pmatrix}, \quad \boldsymbol{w}_{\lambda 2} = \begin{pmatrix} \sin\theta \\ \cos\theta \\ -l_2 \end{pmatrix}. \tag{4.20}$$

In order to stop both the translational and rotational movements of the object, the summation of these four wrench vectors should be equal to the external force $-Mg(0,1,0)^{\mathrm{T}}$ on the object due to gravity. This is expressed as

$$f_1\boldsymbol{w}_{f1} + f_2\boldsymbol{w}_{f2} + \lambda_1\boldsymbol{w}_{\lambda 1} + \lambda_2\boldsymbol{w}_{\lambda 2} = -Mg\begin{pmatrix}0\\1\\0\end{pmatrix}. \tag{4.21}$$

Then, it is evident from this expression that, if f_i and λ_i are chosen such that

$$\begin{cases} f_1 = f_0 + \dfrac{Mg}{2}\sin\theta, \ f_2 = f_0 - \dfrac{Mg}{2}\sin\theta, \\ \lambda_1 = \xi + \dfrac{Mg}{2}\cos\theta, \ \lambda_2 = -\xi + \dfrac{Mg}{2}\cos\theta, \end{cases} \tag{4.22}$$

then the first two components of Equation (4.21) are satisfied and the last component implies

$$\begin{aligned} &-\left(f_0 + \frac{Mg}{2}\sin\theta\right)Y_1 + \left(f_0 - \frac{Mg}{2}\sin\theta\right)Y_2 \\ &+\left(\xi + \frac{Mg}{2}\cos\theta\right)l_1 - \left(-\xi + \frac{Mg}{2}\cos\theta\right)l_2 = 0 \end{aligned} \tag{4.23}$$

from which ξ should be equal to

$$\xi = \frac{1}{l_1 + l_2}\left[f_0(Y_1 - Y_2) + \frac{Mg}{2}\left\{(Y_1 + Y_2)\sin\theta - (l_1 - l_2)\cos\theta\right\}\right]. \tag{4.24}$$

Now, denote the second term in the brackets of Equation (4.24) by N, that is,

$$N = \frac{Mg}{2}\left\{(Y_1 + Y_2)\sin\theta - (l_1 - l_2)\cos\theta\right\}. \tag{4.25}$$

This quantity has a physical meaning. In fact, N means the rotational moment around the object centre of mass $O_{\mathrm{c.m.}}$ if the force with the magnitude $Mg/2$ is exerted at both contact points O_1 and O_2 upward in the vertical direction as shown in Figure 4.4. On the other hand the first term $f_0(Y_1 - Y_2)$ in the same brackets of Equation (4.24) is equivalent to the sum of the rotational moments around the object centre of mass $O_{\mathrm{c.m.}}$ when forces with the same magnitude f_0 are exerted at the contact points O_1 and O_2, respectively, in opposite directions normal to the object surfaces (see Figure 4.4). Hence, the force/torque balance about the object is established under the effect of gravity by setting four constraint forces f_i and λ_i ($i = 1, 2$) as in Equation (4.22) with ξ satisfying Equation (4.24) and $f_0 > Mg/2$. However, we must bear in mind that in the construction of the control signals we should not assume knowledge of object kinematics such as the mass M, width $l_1 + l_2$, or even the measurements data on movements of the centre of mass $(x, y)^{\mathrm{T}}$ and rotational angle θ.

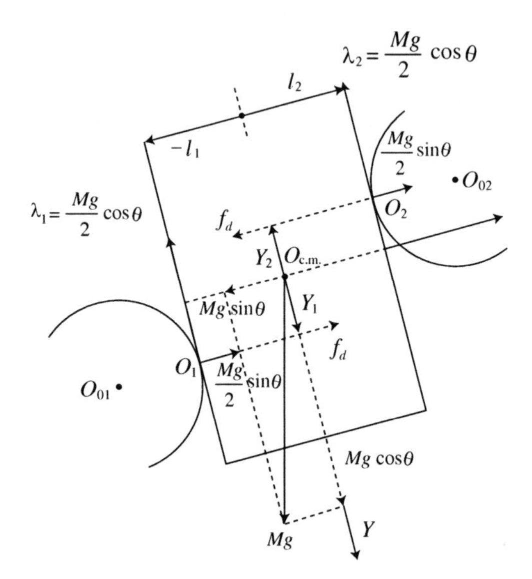

Fig. 4.4. Force/torque balance is established under the effect of gravity by setting four constraint forces f_i and λ_i $(i = 1, 2)$ adequately so that the quantity ξ in Equation (4.24) vanishes

At this stage, let us recall the typical methodology of robot control called PD control with damping shaping, which is based upon the idea of using an artificial potential to compensate for the gravity term or a causal estimator $\hat{M}$ for estimating the unknown mass of a payload. In this problem, the potential function of the object is $-Mgy$. Since we assume that the variable y cannot be measured directly, we must employ some approximate variable that can be accessible or calculated easily and in real time by referring only to the kinematic parameters of the fingers. One possible candidate must be $(y_{01} + y_{02})/2$. Therefore, it is important to evaluate the difference between y and $(y_{01} + y_{02})/2$. To do this, multiply Equation (4.1) by $\cos\theta$ and Equation (4.9) by $\sin\theta$, and take a sum of those two resultant equations, and finally sum these for $i = 1, 2$. Then, we obtain

$$
\begin{aligned}
\Delta y &= y - \frac{y_{01} + y_{02}}{2} \\
&= -\left\{ \frac{Y_1 + Y_2}{2} \cos\theta - \frac{1}{2}(l_1 - l_2 + r_1 - r_2)\sin\theta \right\}.
\end{aligned} \tag{4.26}
$$

This shows that y can be regarded as a function of $\boldsymbol{q}_1$, $\boldsymbol{q}_2$ and θ. Then, it is possible to show that, according to Equation (4.4),

$$
\begin{aligned}
&\frac{\partial y(\boldsymbol{q}_1, \boldsymbol{q}_2, \theta)}{\partial \theta} \\
&= \frac{Y_1 + Y_2}{2} \sin\theta + \frac{1}{2} \cdot \frac{\partial (Y_1 + Y_2)}{\partial \theta} \cos\theta + \frac{1}{2} \{l_1 - l_2 + r_1 - r_2\} \cos\theta \\
&= \frac{Y_1 + Y_2}{2} \sin\theta - \frac{l_1 - l_2}{2} \cos\theta = N. \qquad (4.27)
\end{aligned}
$$

Similarly, it is easy to show that

$$
\frac{\partial y(\boldsymbol{q}_1, \boldsymbol{q}_2, \theta)}{\partial \boldsymbol{q}_i} = \frac{1}{2} \left\{ \frac{\partial y_{0i}}{\partial \boldsymbol{q}_i} - r_i \boldsymbol{e}_i \cos\theta \right\}, \quad i = 1, 2, \qquad (4.28)
$$

where $\boldsymbol{e}_i = (1, 1)^{\mathrm{T}}$. From Equations (4.27) and (4.28) it follows that

$$
\frac{\mathrm{d}}{\mathrm{d}t} Mgy = \dot{\theta} N + \sum_{i=1,2} \frac{Mg}{2} \dot{\boldsymbol{q}}_i^{\mathrm{T}} \left\{ \frac{\partial (y_{01} + y_{02})}{\partial \boldsymbol{q}_i} - r_i \boldsymbol{e}_i \cos\theta \right\}. \qquad (4.29)
$$

In contrast, the opposing forces with magnitude of f_0 are defined to exert at contact points O_1 and O_2 in opposite directions on the line equivalent to the direction of vector $\boldsymbol{x}_1 - \boldsymbol{x}_2$. However, we are unable to access the values of $\boldsymbol{x}_i$, the locations of the contact points O_1 and O_2. Instead of $\boldsymbol{x}_1 - \boldsymbol{x}_2$, it is possible to use the approximate vector $\boldsymbol{x}_{01} - \boldsymbol{x}_{02}$, which is accessible from the finger sides. Then, the approximate opposition forces exerted on the object along the line $\boldsymbol{x}_{01} - \boldsymbol{x}_{02}$ induce reactive rotational moments on the finger joints, which are described as

$$
-(-1)^i \frac{f_d}{r_1 + r_2} J_{0i}^{\mathrm{T}}(\boldsymbol{q}_i) \begin{pmatrix} x_{01} - x_{02} \\ y_{01} - y_{02} \end{pmatrix}, \quad i = 1, 2, \qquad (4.30)
$$

where $f_d > 0$ is a constant specified later. Since it is evident from Equations (4.1) and (4.9) that

$$
\begin{pmatrix} x_{01} - x_{02} \\ y_{01} - y_{02} \end{pmatrix} = \begin{pmatrix} \cos\theta \\ -\sin\theta \end{pmatrix} (-l_w) + \begin{pmatrix} \sin\theta \\ \cos\theta \end{pmatrix} (Y_1 - Y_2), \qquad (4.31)
$$

where

$$
l_w = l_1 + l_2 + r_1 + r_2 \qquad (4.32)
$$

it is possible to obtain the following relation by substituting Equation (4.31) into (4.30):

$$
\begin{aligned}
&-(-1)^i \frac{f_d}{r_1 + r_2} J_{0i}^{\mathrm{T}}(\boldsymbol{q}_i) \begin{pmatrix} x_{01} - x_{02} \\ y_{01} - y_{02} \end{pmatrix} \\
&= (-1)^i f_0 J_{0i}^{\mathrm{T}}(\boldsymbol{q}_i) \begin{pmatrix} \cos\theta \\ -\sin\theta \end{pmatrix} - (-1)^i J_{0i}^{\mathrm{T}}(\boldsymbol{q}_i) \begin{pmatrix} \sin\theta \\ \cos\theta \end{pmatrix} \frac{f_d (Y_1 - Y_2)}{r_1 + r_2} \\
&= f_0 \frac{\partial Q_i}{\partial \boldsymbol{q}_i} - (-1)^i \frac{f_d (Y_1 - Y_2)}{r_1 + r_2} \left(\frac{\partial R_i}{\partial \boldsymbol{q}_i} \right) - (-1)^i \frac{r_i f_d (Y_1 - Y_2)}{r_1 + r_2} \boldsymbol{e}_i, \qquad (4.33)
\end{aligned}
$$

where the magnitude f_0 [N] corresponds to the opposing force $f_0(\partial Q_i/\partial q_i)$, $(i=1,2)$, and hence f_d should be defined as follows:

$$f_0 = \frac{l_w}{r_1+r_2} f_d = \left(1 + \frac{l_1+l_2}{r_1+r_2}\right) f_d. \tag{4.34}$$

Finally, we evaluate the rotational moments at the finger joints evoked in reaction to the constraint forces f_i and λ_i with $\xi = 0$ defined by Equation (4.22), which can be formulated in the following way:

$$\begin{aligned}
&-(-1)^i \frac{Mg}{2}(\sin\theta)\frac{\partial Q_i}{\partial \boldsymbol{q}_i} + \frac{Mg}{2}(\cos\theta)\frac{\partial R_i}{\partial \boldsymbol{q}_i} \\
&\quad = -\frac{Mg}{2} J_{0i}^{\mathrm{T}}(\boldsymbol{q}_i)\begin{pmatrix}\cos\theta \\ -\sin\theta\end{pmatrix}\sin\theta + \frac{Mg}{2}\left\{J_{0i}^{\mathrm{T}}(\boldsymbol{q}_i)\begin{pmatrix}\sin\theta \\ \cos\theta\end{pmatrix}\cos\theta - r_i \boldsymbol{e}_i\right\} \\
&\quad = \frac{Mg}{2}\cdot\frac{\partial y_{0i}}{\partial \boldsymbol{q}_i} - \frac{Mg}{2} r_i \boldsymbol{e}_i \cos\theta, \quad i=1,2.
\end{aligned} \tag{4.35}$$

In order to find a good candidate for control signals that may lead the closed-loop dynamics to the state of force/torque balance, let us assume for the time being that the object mass M is known and the quantity N defined by Equation (4.25) is computationally available based upon the measurement data on Y_i $(i=1,2)$ and θ. Then, let us consider the control signals defined as

$$\begin{aligned}
\boldsymbol{v}_i = \boldsymbol{g}_i(\boldsymbol{q}_i) - c_i\dot{\boldsymbol{q}} + (-1)^i \frac{f_d}{r_1+r_2} J_{0i}^{\mathrm{T}}(\boldsymbol{q}_i)\begin{pmatrix} x_{01}-x_{02} \\ y_{01}-y_{02}\end{pmatrix} \\
-\frac{Mg}{2}\left(\frac{\partial y_{0i}}{\partial \boldsymbol{q}_i}\right) + \frac{Mg}{2} r_i \boldsymbol{e}_i \cos\theta + \frac{(-1)^i N}{l_1+l_2}\left(\frac{\partial R_i}{\partial \boldsymbol{q}_i}\right), \quad i=1,2
\end{aligned} \tag{4.36}$$

and define

$$\left\{\begin{aligned}
f_i &= \Delta f_i + f_0 - (-1)^i \frac{Mg}{2}\sin\theta \\
\lambda_i &= \Delta\lambda_i - (-1)^i \frac{f_d(Y_1-Y_2)}{r_1+r_2} + \frac{Mg}{2}\cos\theta - \frac{(-1)^i N}{l_1+l_2}
\end{aligned}\right. \quad i=1,2. \tag{4.37}$$

Then, substituting Equations (4.36) and (4.37) into Equation (4.15) and referring to Equations (4.33) and (4.35) yields

$$\begin{aligned}
H_i\ddot{\boldsymbol{q}}_i + \left(\frac{1}{2}\dot{H}_i + S_i\right)\dot{\boldsymbol{q}}_i + c_i\dot{\boldsymbol{q}}_i - \Delta f_i\left(\frac{\partial Q_i}{\partial \boldsymbol{q}_i}\right) \\
- \Delta\lambda_i\left(\frac{\partial R_i}{\partial \boldsymbol{q}_i}\right) - (-1)^i \frac{r_i f_d(Y_1-Y_2)}{r_1+r_2}\boldsymbol{e}_i = 0.
\end{aligned} \tag{4.38}$$

At the same time, Equation (4.16) can be rewritten by using Equation (4.37) into the form:

$$M\begin{pmatrix}\ddot{x}\\ \ddot{y}\end{pmatrix} - R_\theta\begin{pmatrix}\Delta f_1 - \Delta f_2\\ -(\Delta\lambda_1 + \Delta\lambda_2)\end{pmatrix} = 0 \tag{4.39}$$

by using the 2-D orthogonal matrix R_θ [see Equation (2.78)], and

$$I\ddot{\theta} - \Delta f_1 Y_1 + \Delta f_2 Y_2 + \Delta\lambda_1 l_1 - \Delta\lambda_2 l_2 - f_d(Y_1 - Y_2) = 0, \tag{4.40}$$

which is reformulated in such a way that

$$\begin{aligned}
0 &= I\ddot{\theta} + \sum_{i=1,2}(-1)^i\{f_i Y_i - \lambda_i l_i\}\\
&= I\ddot{\theta} + \sum_{i=1,2}(-1)^i\left[\left\{\Delta f_i + f_0 - (-1)^i\frac{Mg}{2}\sin\theta\right\}Y_i\right.\\
&\quad\left.-\left\{\Delta\lambda_i - (-1)^i\frac{f_d(Y_1 - Y_2)}{r_1 + r_2} + \frac{Mg}{2}\cos\theta - \frac{(-1)^i N}{l_1 + l_2}\right\}l_i\right]\\
&= I\ddot{\theta} - \Delta f_1 Y_1 + \Delta f_2 Y_2 + \Delta\lambda_1 l_1 - \Delta\lambda_2 l_2 + N\\
&\quad - \left\{\frac{Mg}{2}(Y_1 + Y_2)\sin\theta - (l_1 - l_2)\cos\theta\right\}\\
&\quad -(Y_1 - Y_2)f_0 + \frac{l_1 + l_2}{r_1 + r_2}f_d(Y_1 - Y_2)\\
&= I\ddot{\theta} + \sum_{i=1,2}(-1)^i\{\Delta f_i Y_i - \Delta\lambda_i l_i\} - f_d(Y_1 - Y_2).
\end{aligned} \tag{4.41}$$

Since it is easy to see from Equation (4.4) that

$$\frac{r_1\dot{\boldsymbol{q}}_1^{\mathrm{T}}\boldsymbol{e}_1 - r_2\dot{\boldsymbol{q}}_2^{\mathrm{T}}\boldsymbol{e}_2}{r_1 + r_2} - \dot{\theta} = \dot{Y}_1 - \dot{Y}_2 \tag{4.42}$$

the sum of the inner products between $\dot{\boldsymbol{q}}_i$ and Equation (4.38) for $i = 1, 2$, $(\dot{x}, \dot{y})^{\mathrm{T}}$ and Equation (4.39), and $\dot{\theta}$ and Equation (4.40) yields

$$\frac{\mathrm{d}}{\mathrm{d}t}\left\{K + \frac{f_d}{2(r_1 + r_2)}(Y_1 - Y_2)^2\right\} = -\sum_{i=1,2} c_i\|\dot{\boldsymbol{q}}_i\|^2, \tag{4.43}$$

where K is the kinetic energy given in Equation (4.13). This formula looks like the Lyapunov relation for the establishment of a stability theorem. However note that the quantity

$$E = K + \frac{f_d}{2(r_1 + r_2)}(Y_1 - Y_2)^2 \tag{4.44}$$

is not a Lyapunov function, because E is not positive definite even if all four contact and rolling constraints are taken into account. In fact, E includes only one independent position variable $Y_1 - Y_2$, though the overall fingers–object system has three degrees of freedom. Nevertheless, Equation (5.82) plays an important role in the stability analysis of grasping.

4.3 Control Signals for Grasping in a Blind Manner

The design of such control signals as defined in Equation (4.36) in the previous section is a mathematical fiction, because in our everyday life rigid objects to be grasped are miscellaneous and therefore exact values of their masses are not available. Hence, we replace this term in the definition of the control signals of Equation (4.36) with an estimator $\hat{M}$ for the true but unknown mass M in such a way that

$$-\frac{Mg}{2}\left(\frac{\partial y_{0i}}{\partial \boldsymbol{q}_i}\right) \quad \rightarrow \quad -\frac{\hat{M}g}{2}\left(\frac{\partial y_{0i}}{\partial \boldsymbol{q}_i}\right), \tag{4.45}$$

where $\hat{M}$ should be given by the causal calculation of known parameters and measurement data on finger joint angles. Indeed, one feasible method to construct this estimator is as follows:

$$\begin{aligned}\hat{M}(t) &= \hat{M}(0) + \frac{g\gamma_M^{-1}}{2}\int_0^t \sum_{i=1,2} \dot{\boldsymbol{q}}_i^{\mathrm{T}}\left(\frac{\partial y_{0i}}{\partial \boldsymbol{q}_i}\right)\mathrm{d}\tau \\ &= \hat{M}(0) + \frac{g\gamma_M^{-1}}{2}\sum_{i=1,2}\left\{y_{0i}(t) - y_{0i}(0)\right\},\end{aligned} \tag{4.46}$$

where γ_M is a positive parameter. The term $(Mg/2)r_i\boldsymbol{e}_i\cos\theta$ in Equation (4.36) should be also replaced with another estimator $r_i\hat{N}_i$, where we put

$$N_i = -\frac{Mg}{2}\cos\theta + (-1)^i\frac{f_d}{r_1 + r_2}(Y_1 - Y_2), \quad i = 1, 2 \tag{4.47}$$

and define with positive parameters γ_i $(i = 1, 2)$

$$\begin{aligned}\hat{N}_i(t) &= \hat{N}_i(0) + \gamma_i^{-1}\int_0^t r_i\dot{\boldsymbol{q}}_i^{\mathrm{T}}\boldsymbol{e}_i\mathrm{d}\tau \\ &= \hat{N}_i(0) + \gamma_i^{-1}r_i\sum_{j=1,2}\left\{q_{ij}(t) - q_{ij}(0)\right\}, \quad i = 1, 2.\end{aligned} \tag{4.48}$$

Finally, as to the last term of $\boldsymbol{v}_i$ in Equation (4.36), we had better not employ it in the construction of control signals in practice, because it is assumed that external sensing for the measurement of physical variables Y_i $(i = 1, 2)$ and θ is not available. Thus, we design the following control signals for the pair of robot fingers with two joints depicted in Figure 4.1:

$$\begin{aligned}\boldsymbol{v}_i = \boldsymbol{g}_i(\boldsymbol{q}_i) - c_i\dot{\boldsymbol{q}}_i + (-1)^i\frac{f_d}{r_1 + r_2}J_{0i}^{\mathrm{T}}(\boldsymbol{q}_i)\begin{pmatrix} x_{01} - x_{02} \\ y_{01} - y_{02}\end{pmatrix} \\ -\frac{\hat{M}g}{2}\left(\frac{\partial y_{0i}}{\partial \boldsymbol{q}_i}\right) - r_i\hat{N}_i\boldsymbol{e}_i, \quad i = 1, 2.\end{aligned} \tag{4.49}$$

The first term $\boldsymbol{g}_i(\boldsymbol{q}_i)$ on the right-hand side expresses direct compensation for the effect of gravity for finger i, the second damping injections for finger joints, the third the opposing forces, the fourth plays a role of compensation for the effect of gravity of the object through finger joints, and the fifth plays a role of compensation for the remaining term.

Before explaining how the proposed control signals work in an effective way, we show the closed-loop dynamics of motion of the overall fingers–object system. Differently from the definitions of Δf_i and $\Delta\lambda_i$ $(i = 1, 2)$ in Equation (4.37), we set

$$\begin{cases} \Delta f_i = f_i - f_0 + (-1)^i \dfrac{Mg}{2} \sin\theta \\ \Delta\lambda_i = \lambda_i + (-1)^i \dfrac{f_d(Y_1 - Y_2)}{r_1 + r_2} - \dfrac{Mg}{2} \cos\theta \end{cases} \quad i = 1, 2. \tag{4.50}$$

Note that the last term $(-1)^i N/(l_1 + l_2)$ in Equation (4.37) is excluded in this expression of $\Delta\lambda_i$ corresponding to the fact that the last term of $\boldsymbol{v}_i$ in (4.36) is excluded in the practical design of control signals based on Equation (4.49). Then, by substituting Equation (4.49) into Equation (4.15), we obtain the following:

$$\begin{aligned} H_i \ddot{\boldsymbol{q}}_i + \left(\frac{1}{2}\dot{H}_i + S_i\right)\dot{\boldsymbol{q}}_i + c_i \dot{\boldsymbol{q}}_i - \Delta f_i \left(\frac{\partial Q_i}{\partial \boldsymbol{q}_i}\right) \\ - \Delta\lambda_i \left(\frac{\partial R_i}{\partial \boldsymbol{q}_i}\right) + \frac{\Delta M g}{2}\left(\frac{\partial y_{0i}}{\partial \boldsymbol{q}_i}\right) + r_i \Delta N_i \boldsymbol{e}_i = 0, \quad i = 1, 2 \end{aligned} \tag{4.51}$$

$$\begin{cases} M\ddot{x} - \sum_{i=1,2}\left(\Delta f_i \dfrac{\partial Q_i}{\partial x} + \Delta\lambda_i \dfrac{\partial R_i}{\partial x}\right) = 0 \\ M\ddot{y} - \sum_{i=1,2}\left(\Delta f_i \dfrac{\partial Q_i}{\partial y} + \Delta\lambda_i \dfrac{\partial R_i}{\partial y}\right) = 0 \end{cases} \tag{4.52}$$

$$I\ddot{\theta} - \Delta f_1 Y_1 + \Delta f_2 Y_2 + \Delta\lambda_1 l_1 - \Delta\lambda_2 l_2 + S_N = 0, \tag{4.53}$$

where $\Delta M = \hat{M} - M$ and $\Delta N_i = \hat{N}_i - N_i$ and S_N is defined as

$$S_N = -N - f_d(Y_1 - Y_2). \tag{4.54}$$

Note again that, differently from Equation (4.40), $-N$ appears in Equation (4.53) caused by excluding the last term $(-1)^i N/(l_1 + l_2)$ in Equation (4.37) in the new definition of $\Delta\lambda_i$ given in Equation (4.50).

The major significance of the control signals defined in Equation (4.49) comes from the principle of linear superposition implicitly utilised in the design of the effective control command. Notice again that the right-hand side

Table 4.3. Definitions of N, S_N, N_i, Δf_i and $\Delta\lambda_i$ $(i = 1, 2)$

$$N = \frac{Mg}{2}\{Y_1 \sin(\theta - \theta_0) + Y_2 \sin(\theta + \theta_0) - l_1 \cos(\theta - \theta_0) + l_2 \cos(\theta + \theta_0)\} \qquad \text{(T-12)}$$

$$S_N = -f_d \left\{ d \cos\theta_0 + \frac{r_1 - r_2}{r_1 + r_2} l_w \sin\theta_0 \right\} - N \qquad \text{(T-13)}$$

where

$$\begin{cases} d = Y_1' - Y_2' - (r_1 - r_2)\sin\theta_0 \\ \quad = (x_{01} - x_{02})\sin\theta + (y_{01} - y_{02})\cos\theta \\ l_w = \tilde{l}\cos\theta_0 + (Y_1 + Y_2)\sin\theta_0 \\ \quad = -(x_{01} - x_{02})\cos\theta + (y_{01} - y_{02})\sin\theta \end{cases} \qquad \text{(T-14)}$$

$$\begin{cases} Y_i' = Y_i \cos\theta_0 - l_i \sin\theta_0 \\ \tilde{l} = l_1 + l_2 + r_1 + r_2 \end{cases} \qquad \text{(T-15)}$$

$$N_i = -\frac{Mg}{2}\cos(\theta + (-1)^i\theta_0) - \frac{f_d}{r_1 + r_2}\{l_w \sin\theta_0 - (-1)^i d\cos\theta_0\}, \ i = 1, 2 \qquad \text{(T-16)}$$

$$\Delta f_i = f_i - f_0 + (-1)^i \frac{Mg}{2}\sin(\theta + (-1)^i\theta_0) - \frac{f_d}{r_1 + r_2}\{l_w \cos\theta_0 + (-1)^i d\sin\theta_0\}, \ i = 1, 2 \qquad \text{(T-17)}$$

$$\Delta\lambda_i = \lambda_i - \frac{Mg}{2}\cos(\theta + (-1)^i\theta_0) - \frac{f_d}{r_1 + r_2}\{l_w \sin\theta_0 - (-1)^i d\cos\theta_0\}, \ i = 1, 2 \qquad \text{(T-18)}$$

of Equation (4.49) is composed by superposing each independently and physically significant control signal in a linear sum. If the object is placed horizontally on a desk and the motion of the overall fingers–object system is confined to a horizontal plane, the control signals should be free from the effect of gravity and thereby take the simpler form:

$$\boldsymbol{v}_i = -c_i\dot{\boldsymbol{q}}_i + (-1)^i \frac{f_d}{r_1 + r_2} J_{0i}^{\mathrm{T}}(\boldsymbol{q}) \begin{pmatrix} x_{01} - x_{02} \\ y_{01} - y_{02} \end{pmatrix} - r_i \hat{N}_i \boldsymbol{e}_i, \quad i = 1, 2. \quad (4.55)$$

Further, if the geometric shape of the object is known *apriori*, for example, it were a parallelepiped, then the last term $-r_i\hat{N}_i\boldsymbol{e}_i$ in the right-hand side of Equation (4.55) might be omitted. However, if the object geometry is uncertain, for example, the side surfaces of the object are flat but not in parallel, then the term $-r_i\hat{N}_i\boldsymbol{e}_i$ is indespensible as discussed later. Therefore, in order to realise stable grasping in a blind manner without assuming any prior knowledge of the object geometry, each term of the superposed signals of Equation (4.49) is necessary for robustness against variability in the object geometry.

Table 4.4. Physical parameters of the robot finger (three and three DOFs)

$m_{11} = m_{21}$	link mass	0.045[kg]
$m_{12} = m_{22}$	link mass	0.025[kg]
$m_{13} = m_{23}$	link mass	0.015[kg]
$I_{11} = I_{21}$	inertia moment	1.584×10^{-5}[kg · m^2]
$I_{12} = I_{22}$	inertia moment	3.169×10^{-6}[kg · m^2]
$I_{13} = I_{23}$	inertia moment	8.450×10^{-7}[kg · m^2]
$l_{11} = l_{21}$	link length	0.065[m]
$l_{12} = l_{22}$	link length	0.039[m]
$l_{13} = l_{23}$	link length	0.026[m]
r_1	radius	0.015[m]
r_2	radius	0.020[m]
M	object mass	0.040[kg]
h	object height	0.05000[m]
I	object inertia moment	1.26×10^{-5}[kg · m^2]
θ_0	object inclination angle	−15[deg]
l_1	object length	0.012859[m]
l_2	object length	0.022390[m]

In Section 4.5 we will prove the stability of blind grasping in the case that the same control signals $\boldsymbol{v}_i$ $(i = 1, 2)$ defined by Equation (4.49) is used for a 2-D object with non-parallel but flat surfaces. Further, there is a conjecture that the same control would be valid even for a broader kind of objects with non-flat surfaces, though the mathematical verification seems difficult and has not yet been tackled. Finally, we emphasise that even in the case of objects with non-parallel surfaces the same closed-loop dynamics expressed by Equations (4.51–4.53) are formulated if N, S_N, N_i, Δf_i, $\Delta\lambda_i$ $(i = 1, 2)$ are replaced with those listed in Table 4.3, respectively. Observe that those functions N, S_N, N_i, Δf_i, and $\Delta\lambda_i$ $(i = 1, 2)$ defined in Table 4.3 are reduced to N of Equation (4.25), S_N of Equation (4.54), N_i of Equation (4.47), and Δf_i and $\Delta\lambda_i$ of Equation (4.50), respectively, when θ_0 is zero.

Before presenting theoretical verifications of the effectiveness of the control signals of Equation (4.49) that can be constructed without using the knowledge of object kinematics or any external sensing, we will show some results from computer simulation.

First, we show simulation results for stable grasping of a rigid object with non-parallel flat surfaces by means of a pair of robot fingers with three DOFs as shown in Figure 4.2. In the simulation Lagrange's equations of motion described by Equations (4.17) and (4.18) [not by (4.15) and (4.16)] is used by feeding the same control signals $\boldsymbol{v}_i$ $(i = 1, 2)$ of Equation (4.49). In order to incorporate the holonomic constraints described in Table 4.1, the so-called

Table 4.5. Parameters of the control signals

f_d	internal force	2.0[N]
$c_1 = c_2$	damping coefficient	0.006[msN]
γ_M	regressor gain	0.01
γ_1	regressor gain	0.001
γ_2	regressor gain	0.001

Table 4.6. Initial values of the simulation

$Y_1(0)$	initial value of $Y_1(t)$	0.001[m]
$Y_2(0)$	initial value of $Y_2(t)$	0.000[m]
$\hat{M}(0)$	initial value of $\hat{M}(t)$	0.010[kg]
$\hat{N}_1(0)$	initial value of $\hat{N}_1(t)$	0.000[N]
$\hat{N}_2(0)$	initial value of $\hat{N}_2(t)$	0.000[N]

constraint stabilisation method (CSM) can be used. The details of the construction of a numerical simulator based on the CSM will be presented in the next section. All physical parameters necessary for conducting the simulation are given in Table 4.4, parameters and gains necessary for determing the control signals are summarised in Table 4.5, and the initial values for Y_i' ($i = 1, 2$) and for the estimators $\hat{M}$ and $\hat{N}_i$ ($i = 1, 2$) are presented in Table 4.6. Figure 4.5 shows that all key physical variables $Y_1' - Y_2'$, θ, ΔM, S_N, and Δf_i, $\Delta\lambda_i$ and ΔN_i ($i = 1, 2$) (see Table 4.3) converge to corresponding constant values as $t \to \infty$. Further, observe that in Figure 4.5 all Δf_i, $\Delta\lambda_i$, ΔN_i ($i = 1, 2$), ΔM and S_N asymptotically converge to zero as $t \to \infty$ but θ and $Y_1' - Y_2'$ tend to some constant values as $t \to \infty$. From this figure it can be predicted that the trajectory of the solution to the closed-loop dynamics of Equations (4.51) to (4.53) under the holonomic constraints of Table 4.1 converge asymptotically to some equilibrium state that establishes in the sequel the force/torque balance. Another interesting observation is that the estimate $\hat{M}(t)$ for the unknown object mass converges to its true value through this blind grasping. As discussed later in Section 4.5 this observation comes from redundancy in the total DOFs of the overall fingers–object system.

Another similar simulation was conducted by using a pair of robot fingers, one of which has two joints (two DOFs) and another three joints (three DOFs). Even in this case one degree-of-freedom is redundant and thereby it is observed in the simulation that not only ΔM but also other Δf_i, $\Delta\lambda_i$ and ΔN_i (for $i = 1, 2$) converge to zero as $t \to \infty$.

Next, we conducted one more computer simulation by using a pair of dual robot fingers with the same two DOFs. To show how the control signals of Equation (4.49) are robust against variability in the object shape, we employed a rigid object with non-parallel surfaces and non-uniform mass density. All

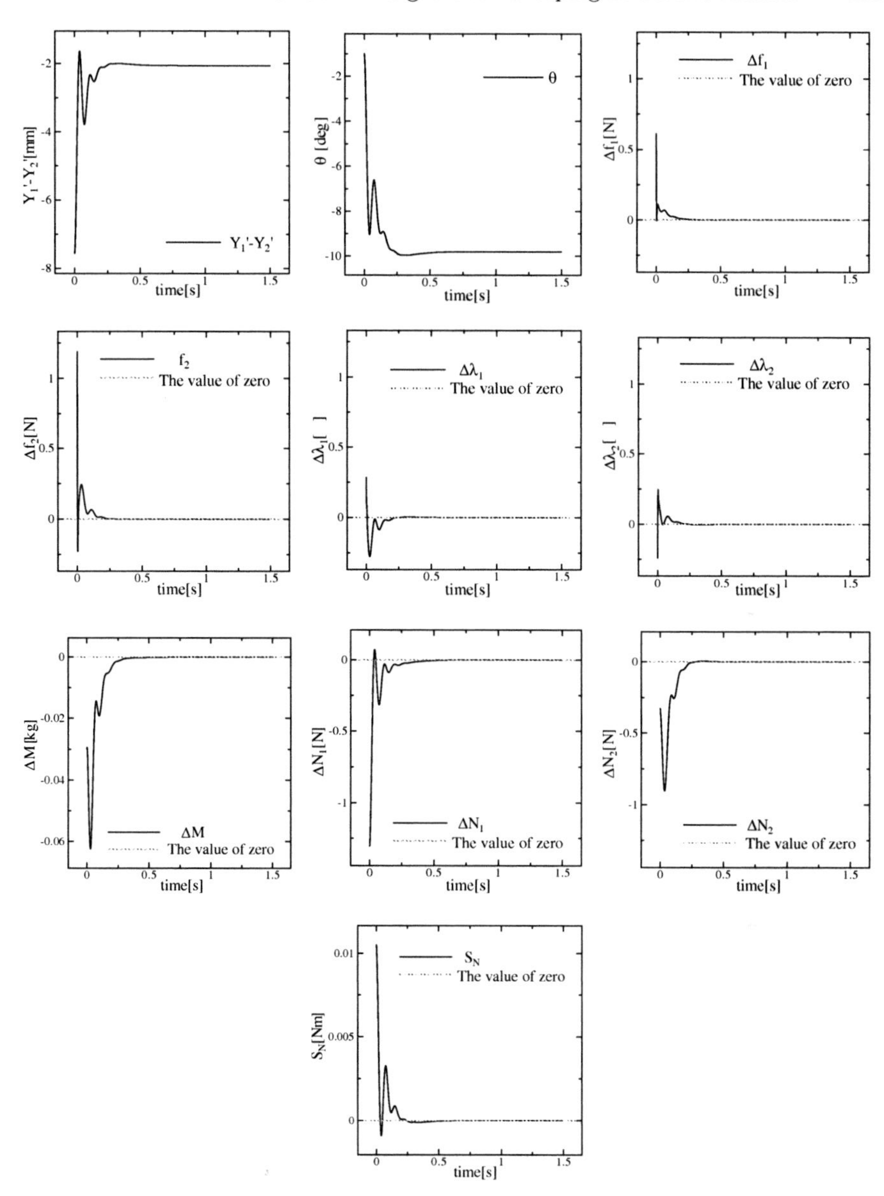

Fig. 4.5. The transient responses of the physical variables in the case of a pair of fingers with three joints

physical parameters, control gains and initial conditions are given in Tables 4.7, 4.8 and 4.9, respectively. It is interesting to see from Figure 4.6 that all physical values $Y_1 - Y_2$, θ, S_N, ΔM, Δf_i, $\Delta \lambda_i$ and ΔN_i $(i = 1, 2)$ converge to

Table 4.7. Physical parameters of the robot fingers (two and two DOFs)

$m_{11} = m_{21}$	link mass	0.0451[kg]
$m_{12} = m_{22}$	link mass	0.0252[kg]
$I_{11} = I_{21}$	inertia moment	2.53×10^{-5}[kg · m²]
$I_{12} = I_{22}$	inertia moment	7.94×10^{-6}[kg · m²]
$l_{11} = l_{21}$	link length	0.082[m]
$l_{12} = l_{22}$	link length	0.0615[m]
r_1	radius	0.015[m]
r_2	radius	0.020[m]
M	object mass	0.040[kg]
h	object height	0.05000[m]
I	object inertia moment	1.26×10^{-5}[kg · m²]
θ_0	object inclination angle	-15[deg]
l_1	object length	0.012859[m]
l_2	object length	0.022390[m]

Table 4.8. Parameters of the control signals

f_d	internal force	2.0[N]
$c_1 = c_2$	damping coefficient	0.006[msN]
γ_M	regressor gain	0.01
γ_1	regressor gain	0.001
γ_2	regressor gain	0.001

Table 4.9. Initial values of the simulation

$Y_1(0)$	initial value of $Y_1(t)$	-0.0035[m]
$Y_2(0)$	initial value of $Y_2(t)$	-0.003[m]
$\hat{M}(0)$	initial value of $\hat{M}(t)$	0.010[kg]
$\hat{N}_1(0)$	initial value of $\hat{N}_1(t)$	0.000[N]
$\hat{N}_2(0)$	initial value of $\hat{N}_2(t)$	0.000[N]

their corresponding constants as $t \to \infty$. Differently from the overall system with redundant DOFs, this configuration of two fingers with the same two joints in planar motion of object grasping is not redundant and thereby the estimation error ΔM for the object mass converges to some constant value that is not necessarily zero. Nevertheless, the force/torque balance was established in a dynamic sense as $t \to \infty$ even in this case, as seen from Figure 4.6.

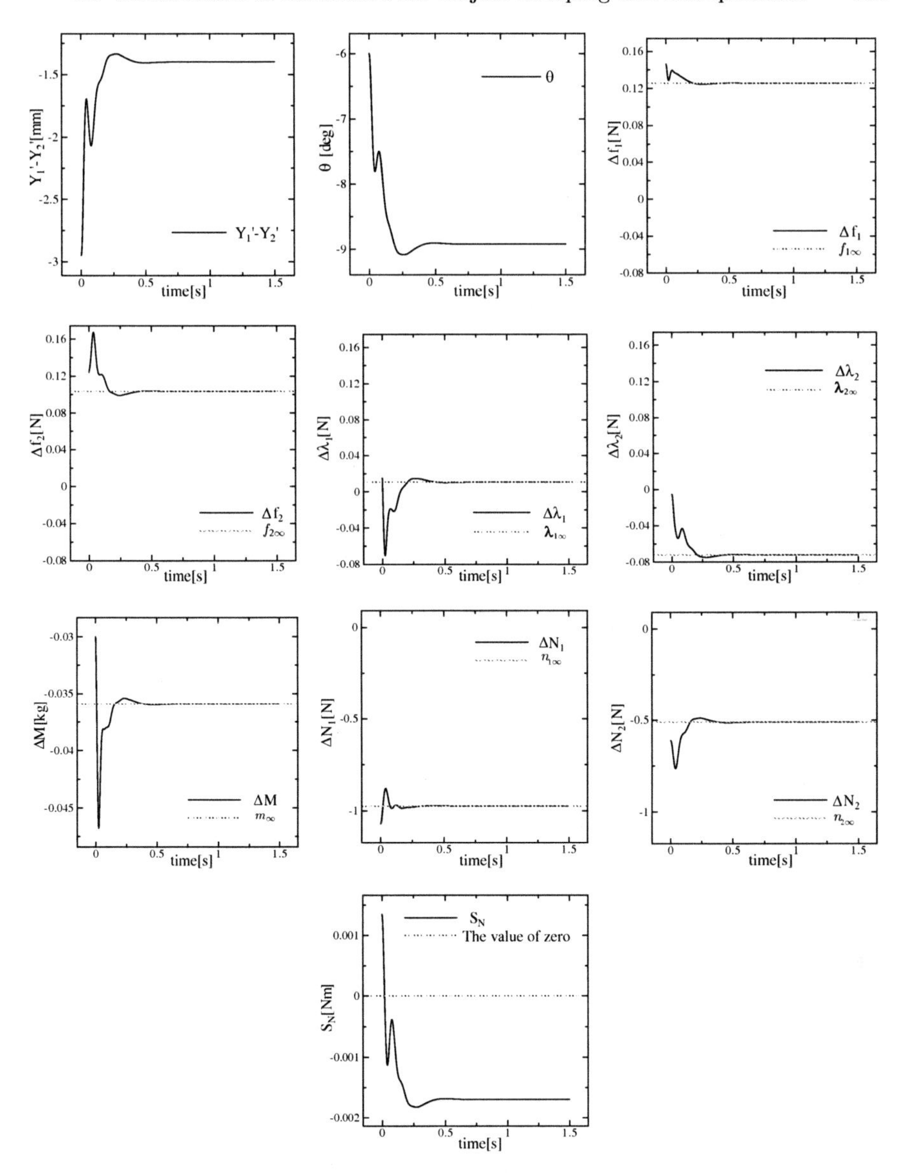

Fig. 4.6. The transient responses of physical variables in the case of fingers with two joints and an object with parallel surfaces

4.4 Construction of Simulators for Object Grasping and Manipulation

Numerical simulation of movements of the whole fingers–object system exhibiting physical interactions between fingertips and a rigid object can be carried out by using Baumgarte's method called the constraint stabilization method together with the Runge–Kutta method for numerically solving a system of differential equations. In order to apply the CSM for a system of differential equations of Equations (4.17–4.18) with control inputs $\boldsymbol{v}_i$ of Equation (4.49) under the holonomic constraints described in Table 4.1, the four constraints $Q_i = 0$ and R_i $(i = 1, 2)$ must be approximated by the following set of over-damped second-order linear differential equations:

$$\begin{cases} \ddot{Q}_i + \gamma_{fi}\dot{Q}_i + \omega_{fi}Q_i = 0 \\ \ddot{R}_i + \gamma_{\lambda i}\dot{R}_i + \omega_{\lambda i}R_i = 0 \end{cases} \quad i = 1, 2 \tag{4.56}$$

where γ_{fi} and $\gamma_{\lambda i}$ are damping coefficients and ω_{fi} and $\omega_{\lambda i}$ stiffness parameters. It is recommended to choose these constants to satisfy

$$\gamma_{fi}^2 = 4\omega_{fi}, \quad \gamma_{\lambda i}^2 = 4\omega_{\lambda i}, \qquad i = 1, 2 \tag{4.57}$$

and in both simulations shown in Figures 4.5 and 4.6 we set $\gamma_{fi} = 3.0 \times 10^3$ and $\omega_{\lambda i} = 2.25 \times 10^6$ for $i = 1, 2$. Next, we must introduce the following symbols for convenience:

$$\begin{cases} J_{Qqi} = \dfrac{\partial Q_i}{\partial \boldsymbol{q}_i}, \quad J_{Rqi} = \dfrac{\partial R_i}{\partial \boldsymbol{q}_i} \\ J_{Qi\boldsymbol{w}} = \dfrac{\partial Q_i}{\partial \boldsymbol{w}}, \quad J_{Ri\boldsymbol{w}} = \dfrac{\partial R_i}{\partial \boldsymbol{w}} \end{cases} \quad i = 1, 2. \tag{4.58}$$

Then, the set of second-order differential equations of Equation (4.56) can be equivalently recast into the form:

$$\begin{cases} J_{Qqi}^{\mathrm{T}}\ddot{\boldsymbol{q}}_i + J_{Qi\boldsymbol{w}}^{\mathrm{T}}\ddot{\boldsymbol{w}} + \left(\dot{J}_{Qqi} + \gamma_{fi}J_{Qqi}\right)^{\mathrm{T}}\dot{\boldsymbol{q}}_i \\ \qquad + \left(\dot{J}_{Qi\boldsymbol{\omega}} + \gamma_{fi}J_{Qi\boldsymbol{\omega}}\right)^{\mathrm{T}}\dot{\boldsymbol{w}} + w_{fi}Q_i = 0, \\ J_{Rqi}^{\mathrm{T}}\ddot{\boldsymbol{q}}_i + J_{Ri\boldsymbol{w}}^{\mathrm{T}}\ddot{\boldsymbol{w}} + \left(\dot{J}_{Rqi} + \gamma_{\lambda i}J_{Rqi}\right)^{\mathrm{T}}\dot{\boldsymbol{q}}_i \\ \qquad + \left(\dot{J}_{Ri\boldsymbol{\omega}} + \gamma_{\lambda i}J_{Ri\boldsymbol{w}}\right)^{\mathrm{T}}\dot{\boldsymbol{w}} + \omega_{\lambda i}R_i = 0, \end{cases} \quad i = 1, 2. \tag{4.59}$$

In order to express the set of all the second-order differential equations of Equations (4.17), (4.18) and (4.59) in a unified vector–matrix form, we define the following:

$$\boldsymbol{\lambda} = \left[\ddot{\boldsymbol{q}}_1^{\mathrm{T}}, \ddot{\boldsymbol{q}}_2^{\mathrm{T}}, \ddot{\boldsymbol{w}}^{\mathrm{T}}, -f_1, -f_2, -\lambda_1, -\lambda_2\right]^{\mathrm{T}}, \tag{4.60}$$

$$H_{[9\times 9]} = \mathrm{diag}\left(H_1(\boldsymbol{q}_i), H_2(\boldsymbol{q}_2), MI_2, I\right), \tag{4.61}$$

$$A_{[9\times4]} = \begin{pmatrix} \frac{\partial Q_1}{\partial \boldsymbol{q}_1} & \mathbf{0}_3 & \frac{\partial R_1}{\partial \boldsymbol{q}_1} & \mathbf{0}_3 \\ \mathbf{0}_3 & \frac{\partial Q_2}{\partial \boldsymbol{q}_2} & \mathbf{0}_3 & \frac{\partial R_2}{\partial \boldsymbol{q}_2} \\ \frac{\partial Q_1}{\partial \boldsymbol{x}} & \frac{\partial Q_2}{\partial \boldsymbol{x}} & \frac{\partial R_1}{\partial \boldsymbol{x}} & \frac{\partial R_2}{\partial \boldsymbol{x}} \\ \frac{\partial Q_1}{\partial \theta} & \frac{\partial Q_2}{\partial \theta} & \frac{\partial R_1}{\partial \theta} & \frac{\partial R_2}{\partial \theta} \end{pmatrix}, \tag{4.62}$$

$$\boldsymbol{h}_i = \boldsymbol{v}_i - \left\{\frac{1}{2}\dot{H}_i(\boldsymbol{q}_i) + S_1(\boldsymbol{q}_i, \dot{\boldsymbol{q}}_i)\right\}\dot{\boldsymbol{q}}_i, \quad i = 1, 2, \tag{4.63}$$

$$\boldsymbol{h}_{\boldsymbol{w}} = \left(\boldsymbol{h}_{\boldsymbol{x}}^{\mathrm{T}}, h_\theta\right)^{\mathrm{T}} = \mathbf{0}_3, \tag{4.64}$$

$$h_{10} = -\left(\dot{J}_{Qq1} + \gamma_{f1} J_{Qq1}\right)^{\mathrm{T}} \dot{\boldsymbol{q}}_1 - \left(\dot{J}_{Q1\boldsymbol{w}} + \gamma_{f1} J_{Q1\boldsymbol{w}}\right)^{\mathrm{T}} \dot{\boldsymbol{w}} - \omega_{f1} Q_1, \tag{4.65}$$

$$h_{11} = -\left(\dot{J}_{Qq2} + \gamma_{f2} J_{Qq2}\right)^{\mathrm{T}} \dot{\boldsymbol{q}}_2 - \left(\dot{J}_{Q2\boldsymbol{w}} + \gamma_{f2} J_{Q2\boldsymbol{w}}\right)^{\mathrm{T}} \dot{\boldsymbol{w}} - \omega_{f2} Q_2, \tag{4.66}$$

$$h_{12} = -\left(\dot{J}_{Rq1} + \gamma_{\lambda1} J_{Rq1}\right)^{\mathrm{T}} \dot{\boldsymbol{q}}_1 - \left(\dot{J}_{R1\boldsymbol{w}} + \gamma_{\lambda2} J_{R1\boldsymbol{w}}\right)^{\mathrm{T}} \dot{\boldsymbol{w}} - \omega_{\lambda1} R_1, \tag{4.67}$$

$$h_{13} = -\left(\dot{J}_{Rq2} + \gamma_{\lambda2} J_{Rq2}\right)^{\mathrm{T}} \dot{\boldsymbol{q}}_2 - \left(\dot{J}_{R2\boldsymbol{w}} + \gamma_{\lambda2} J_{R2\boldsymbol{w}}\right)^{\mathrm{T}} \dot{\boldsymbol{w}} - \omega_{\lambda2} R_2. \tag{4.68}$$

It should be remarked that $J_{Qqi} = \partial Q_i / \partial \boldsymbol{q}_i$, $J_{Qi\boldsymbol{w}} = \partial Q_i / \partial \boldsymbol{w} = ((\partial Q_i / \partial \boldsymbol{x})^{\mathrm{T}}, \partial Q_i / \partial \theta)^{\mathrm{T}}$, and J_{Rqi} and $J_{Ri\boldsymbol{\omega}}$ have analogous meanings. Thus, it is possible to express all the differential equations of Equations (4.17), (4.18) and (4.59) in the following form:

$$\mathrm{T}_{[13\times13]} \times \boldsymbol{\lambda} = \left[\boldsymbol{h}_1^{\mathrm{T}}, \boldsymbol{h}_2^{\mathrm{T}}, \boldsymbol{h}_{\boldsymbol{w}}^{\mathrm{T}}, h_{10}, h_{11}, h_{12}, h_{13}\right]^{\mathrm{T}}, \tag{4.69}$$

where

$$\mathrm{T}_{[13\times13]} = \begin{pmatrix} \mathrm{H}_{[9\times9]} & \mathrm{A}_{[9\times4]} \\ \mathrm{A}_{[4\times9]}^{\mathrm{T}} & 0_{[6\times6]} \end{pmatrix}. \tag{4.70}$$

The final formula of Equation (4.69) can be solved numerically by using the Runge–Kutta method.

4.5 Stability of Blind Grasping

We are now in a position to discuss the problem of the stability of an equilibrium still state satisfying the force/torque balance in a dynamic sense. At the same time, we shall prove a theorem on the convergence of a trajectory of any solution to the closed-loop dynamics of Equations (4.51–4.53) starting from a neighbourhood of the reference equilibrium state with force/torque balance.

Firstly, note that, in the case of a pair of robot fingers depicted in Figure 4.2, the trajectory of any solution to the closed-loop dynamics of Equations (4.51–4.53) should lie on the 10-dimensional manifold in the state space R^{18} defined by

$$CM_{10} = \Big\{ \left(\boldsymbol{q}_1^{\mathrm{T}}, \boldsymbol{q}_2^{\mathrm{T}}, \boldsymbol{w}^{\mathrm{T}}, \dot{\boldsymbol{q}}_1^{\mathrm{T}}, \dot{\boldsymbol{q}}_2^{\mathrm{T}}, \dot{\boldsymbol{w}}^{\mathrm{T}}\right)^{\mathrm{T}} : \\ Q_i = 0, R_i = 0, \dot{Q}_i = 0, \dot{R}_i = 0, \quad i = 1, 2 \Big\}. \tag{4.71}$$

Since the condition of force/torque balance for the 2-D object is described by

$$\sum_{i=1,2} \Delta f_i \boldsymbol{w}_{fi} + \Delta\lambda_i \boldsymbol{w}_{\lambda i} - (0, 0, S_N)^{\mathrm{T}} = 0,$$

where

$$\boldsymbol{w}_{fi} = \left(\frac{\partial Q_i}{\partial x}, \frac{\partial Q_i}{\partial y}, \frac{\partial Q_i}{\partial \theta}\right)^{\mathrm{T}}, \ \boldsymbol{w}_{\lambda i} = \left(\frac{\partial R_i}{\partial x}, \frac{\partial R_i}{\partial y}, \frac{\partial R_i}{\partial \theta}\right)^{\mathrm{T}}, \ i = 1, 2 \tag{4.72}$$

any equilibrium state of force/torque balance and zero-velocity should belong to the two-dimensional manifold defined as

$$EM_2 = \{(\boldsymbol{z}, \dot{\boldsymbol{z}} = 0) : Q_i = 0, R_i = 0, \quad i = 1, 2\}, \tag{4.73}$$

where $\boldsymbol{z} = (\boldsymbol{q}_1^{\mathrm{T}}, \boldsymbol{q}_2^{\mathrm{T}}, \boldsymbol{x}^{\mathrm{T}}, \theta)^{\mathrm{T}}$. This set is called the equilibrium-point manifold (EP-manifold) and can be regarded as the 2-D configuration manifold embedded in the nine-dimensional position state space of $R^9 = \{\boldsymbol{z}\}$.

Second, we show that the closed-loop dynamics of Equations (4.51–4.53) can be regarded as Lagrange's equation of motion concerning the Lagrangian

$$L = K - \tilde{P} + \sum_{i=1,2} (\Delta f_i Q_i + \Delta\lambda_i R_i) \tag{4.74}$$

under the constraints $Q_i = 0$ and $R_i = 0$ $(i = 1, 2)$ described in Table 4.2 and the external finger joint damping force $-c_i \dot{\boldsymbol{q}}_i$ for $i = 1, 2$, where K is the kinetic energy given by Equation (4.13) and $\tilde{P}$ is a scalar function called the artificial potential described as

$$\tilde{P} = \frac{1}{2}\left\{\gamma_M \Delta M^2 + \frac{f_d}{r_1 + r_2}\|\boldsymbol{x}_{01} - \boldsymbol{x}_{02}\|^2 + \gamma_1 \hat{N}_1^2 + \gamma_2 \hat{N}_2^2\right\} - Mg\Delta y, \tag{4.75}$$

where $\Delta y = y - (y_{01} + y_{02})/2$ was defined previously in Equation (4.26) in the case of Figure 4.1 but in a general case of grasping an object with non-parallel surfaces as shown in Figure 4.2 it is given by

$$\Delta y = y - \frac{y_{01} + y_{02}}{2} = -\frac{1}{2}\Bigg[\{Y_1 \cos(\theta - \theta_0) + Y_2 \cos(\theta + \theta_0)\} \\ + \{(l_1 + r_1)\sin(\theta - \theta_0) - (l_2 + r_2)\sin(\theta + \theta_0)\} \Bigg]. \tag{4.76}$$

This can be ascertained by taking the inner products between $\dot{\boldsymbol{q}}_i$ and Equation (4.17) for $i = 1, 2$, $\dot{\boldsymbol{w}}$ and Equation (4.18), and $-\dot{\boldsymbol{q}}_i$ and $\boldsymbol{v}_i$ of Equation (4.49) for $i = 1, 2$ and summing all the resultant products. In fact, the sum of inner products between $\dot{\boldsymbol{q}}_i$ and Equation (4.17) for $i = 1, 2$ and $\dot{\boldsymbol{w}}$ and Equation (4.18) yields

$$\sum_{i=1,2} \dot{\boldsymbol{q}}_i \boldsymbol{v}_i = \frac{\mathrm{d}}{\mathrm{d}t} \{K + P_1(\boldsymbol{q}_1) + P_2(\boldsymbol{q}_2) - Mgy\} \tag{4.77}$$

and the sum of inner products between $-\boldsymbol{v}_i$ of Equation (4.49) and $\dot{\boldsymbol{q}}_i$ for $i = 1, 2$ yields

$$-\sum_{i=1,2} \dot{\boldsymbol{q}}_i^{\mathrm{T}} \boldsymbol{v}_i = \frac{\mathrm{d}}{\mathrm{d}t} \Bigg\{ -P(\boldsymbol{q}_1) - P_2(\boldsymbol{q}_2) + \frac{f_d \|\boldsymbol{x}_{01} - \boldsymbol{x}_{02}\|^2}{2(r_1 + r_2)} \\ + \frac{\gamma_M}{2} \hat{M}^2 + \sum_{i=1,2} \frac{\gamma_i}{2} \hat{N}_i^2 \Bigg\} - \sum_{i=1,2} c_i \|\dot{\boldsymbol{q}}_i\|^2. \tag{4.78}$$

Since

$$\begin{aligned} (\gamma_M/2)\Delta M^2 &= (\gamma_M/2)\left\{\hat{M}^2 - 2\hat{M}M + M^2\right\} \\ &= (\gamma_M/2)\hat{M}^2 - \gamma_M \left\{\hat{M}(0) + (g/2\gamma_M)(y_{01} + y_{02})\right\} M + (\gamma_M/2)M^2 \\ &= (\gamma_M/2)\hat{M}^2 - \frac{Mg}{2}(y_{01} + y_{02}) + (\gamma_M/2)(M^2 - 2\hat{M}(0)) \end{aligned} \tag{4.79}$$

the addition of Equation (4.77) to (4.78) is reduced to

$$\frac{\mathrm{d}}{\mathrm{d}t}(K + \tilde{P}) = -\sum_{i=1,2} c_i \|\dot{\boldsymbol{q}}_i\|^2, \tag{4.80}$$

where $\tilde{P}$ is just given in Equation (4.75).

Third, we shall show that the artificial potential function $\tilde{P}$ has a minimum value at some still state $\boldsymbol{z} = \boldsymbol{z}_\infty$ and $\dot{\boldsymbol{z}} = 0$ under the constraints $Q_i = 0$ and $R_i = 0$ for $i = 1, 2$. To do this, it is important to notice that

$$
\begin{aligned}
&\frac{\partial}{\partial \boldsymbol{q}_i}\left\{\frac{f_d}{2(r_1+r_2)}(l_w^2+d^2)-Mg\Delta y\right\} \\
&\quad=\frac{f_d}{r_1+r_2}\left(l_w\frac{\partial l_w}{\partial \boldsymbol{q}_i}+d\frac{\partial d}{\partial \boldsymbol{q}_i}\right)-Mg\frac{\partial \Delta y}{\partial \boldsymbol{q}} \\
&\quad=\frac{f_d}{r_1+r_2}\left(l_w r_i\sin\theta_0-(-1)^i d r_i\cos\theta_0\right)\boldsymbol{e}_i \\
&\quad\quad -Mg\left(-\frac{r_i}{2}\cos(\theta+(-1)^i\theta_0)\right)\boldsymbol{e}_i \\
&\quad=-r_i N_i \boldsymbol{e}_i, \qquad\qquad i=1,2.
\end{aligned}
\tag{4.81}
$$

Hence, by bearing in mind that $\|\boldsymbol{x}_{01}-\boldsymbol{x}_{02}\|^2=d^2+l_w^2$, it is possible to obtain

$$
\frac{\partial \tilde{P}}{\partial \boldsymbol{q}_i}=\frac{\Delta Mg}{2}\left(\frac{\partial y_{0i}}{\partial \boldsymbol{q}_i}\right)+r_i\Delta N_i\boldsymbol{e}_i,
\tag{4.82}
$$

which is equivalent to the sum of last two terms in the left hand side of Equation (4.51). Further, since Δy can be regarded as a function of Y_1, Y_2, and θ, it follows that

$$
\frac{\partial \tilde{P}}{\partial \boldsymbol{x}}=0.
\tag{4.83}
$$

Finally, it follows that

$$
\begin{aligned}
\frac{\partial \tilde{P}}{\partial \theta}&=\frac{f_d}{r_1+r_2}\left(l_w\frac{\partial l_w}{\partial \theta}+d\frac{\partial d}{\partial \theta}\right)-\frac{\partial Mg\Delta y}{\partial \theta} \\
&=-\frac{f_d}{r_1+r_2}\{(r_1+r_2)d\cos\theta_0+(r_1-r_2)l_w\sin\theta_0\}-N \\
&=S_N,
\end{aligned}
\tag{4.84}
$$

which is equivalent to the last term of the left-hand side of Equation (4.53).

At this stage, it is important to note that if we consider

$$
\begin{aligned}
P_0&=\tilde{P}-(1/2)\gamma_M\Delta M^2 \\
&=\frac{1}{2}\left\{\frac{f_d}{r_1+r_2}(l_w^2+d^2)+\gamma_1\hat{N}_1^2+\gamma_2\hat{N}_2^2\right\}-Mg\Delta y
\end{aligned}
\tag{4.85}
$$

then

$$
\left\{
\begin{aligned}
&\frac{\partial P_0}{\partial \boldsymbol{q}_i}=r_i\Delta N_i\boldsymbol{e}_i \\
&\frac{\partial P_0}{\partial \boldsymbol{x}}=0 \qquad\qquad i=1,2 \\
&\frac{\partial P_0}{\partial \theta}=S_N
\end{aligned}
\right.
\tag{4.86}
$$

Since the gradient vector $\partial P_0/\partial \boldsymbol{q}_i$ has a fixed orientation $\boldsymbol{e}_i$ in R^3, it is possible to reduce the analysis for finding the minimum of P_0 on the configuration space to that in a set of only three variables θ, p_1 and p_2 such that

$$p_1 = \sum_{j=1}^{3} q_{1j}, \qquad p_2 = \sum_{j=1}^{3} q_{2j}. \tag{4.87}$$

In fact, it is possible to confirm from the definitions l_w and d in Tables 4.1 and 4.3 and Δy in Equation (4.26) that P_0 is a function of only p_1, p_2 and θ. Evidently from Equation (4.86) it follows that

$$\begin{cases} \partial P_0/\partial p_i = r_i \Delta N_i, & i = 1, 2 \\ \partial P_0/\partial \theta = S_N. \end{cases} \tag{4.88}$$

Further, the Hessian matrix of P_0 in p_1, p_2 and θ can be obtained in the following way:

$$\frac{\partial^2 P_0}{\partial p_i^2} = r_i^2 \left(\frac{1}{\gamma_i} + \frac{f_d}{r_1 + r_2} \right), \quad i = 1, 2 \tag{4.89a}$$

$$\frac{\partial^2 P_0}{\partial p_2 \partial p_1} = \frac{r_2 r_2 f_d}{r_1 + r_2} \left(\sin^2 \theta_0 - \cos^2 \theta_0 \right) \tag{4.89b}$$

$$\begin{aligned} \frac{\partial^2 P_0}{\partial \theta^2} &= f_d \left\{ (r_1 + r_2) \cos^2 \theta_0 + \frac{(r_1 - r_2)^2}{r_1 + r_2} \sin^2 \theta_0 \right\} - \frac{\partial N}{\partial \theta}, \\ \frac{\partial N}{\partial \theta} &= \frac{Mg}{2} \sum_{i=1,2} \{ Y_i \cos(\theta + (-1)^i \theta_0) \\ &\qquad - (-1)^i (l_i - r_i) \sin(\theta + (-1)^i \theta_0) \} \end{aligned} \tag{4.89c}$$

$$\begin{aligned} \frac{\partial^2 P_0}{\partial \theta \partial p_i} &= \frac{r_i f_d}{r_1 + r_2} \{ (-1)^i (r_1 + r_2) \cos^2 \theta_0 - (r_1 - r_2) \sin^2 \theta_0 \} \\ &\quad - \frac{Mg}{2} r_i \sin(\theta + (-1)^i \theta_0), \quad i = 1, 2. \end{aligned} \tag{4.89d}$$

Since f_d is selected to be considerably larger than Mg and γ_i ($i = 1, 2$) are selected sufficiently small in comparison with $(r_1 + r_2)/f_d$ as seen from Table 4.5, the Hessian matrix is positive definite. Thus, P_0 has a minimum value $P_{0\infty}$ at some point $p_i = p_{i\infty}$ ($i = 1, 2$) and $\theta = \theta_\infty$. We define $\bar{E} = E - P_{0\infty}$ and note that from Equation (4.80) it follows that

$$\frac{\mathrm{d}}{\mathrm{d}t} \bar{E} = \frac{\mathrm{d}}{\mathrm{d}t} \left(K + \tilde{P} - P_{0\infty} \right) = - \sum_{i=1,2} c_i \|\dot{\boldsymbol{q}}_i\|^2. \tag{4.90}$$

This means that

$$\int_0^\infty \|\dot{\boldsymbol{q}}_i(t)\|^2 \mathrm{d}t \leq c_i^{-1} \bar{E}(0), \quad i = 1, 2 \tag{4.91}$$

and $K(t) \leq \bar{E}(0)$ and $\tilde{P}_0 - P_{0\infty} \leq \bar{E}(0)$, where $\bar{E}(0)$ denotes the value of $\bar{E}$ at the initial state $\boldsymbol{z}(0)$. At this stage we reasonably assume that $\bar{E}(0)$ is sufficiently small that (p_1, p_2, θ) remains in the vicinity of $(p_{1\infty}, p_{2\infty}, \theta_\infty)$. Then, all components of velocity vector $\dot{\boldsymbol{z}}(t)$ remain bounded. Further, it is necessary to show that all accelration components of $\ddot{\boldsymbol{z}}(t)$ are bounded. To show this, we rewrite the closed-loop dynamics of Equations (4.51–4.52) in the following vector–matrix form by setting $\bar{\boldsymbol{z}} = (\boldsymbol{q}_1^{\mathrm{T}}, \boldsymbol{q}_2^{\mathrm{T}}, r^{-1}\boldsymbol{x}^{\mathrm{T}}, \theta)^{\mathrm{T}}$ and $\bar{\boldsymbol{x}} = r^{-1}\boldsymbol{x}^{\mathrm{T}}$ with a non-dimensional scale factor r:

$$H\ddot{\bar{\boldsymbol{z}}} + \left(\frac{1}{2}\dot{H} + S\right)\dot{\bar{\boldsymbol{z}}} - A\Delta\boldsymbol{\lambda} - B\Delta\boldsymbol{m} + \boldsymbol{d}_N = 0, \tag{4.92}$$

where

$$\left\{\begin{array}{l}
H = \begin{pmatrix} H_1 & 0_{3\times3} & 0_{3\times2} & 0_{3\times1} \\ 0_{3\times3} & H_2 & 0_{3\times2} & 0_{3\times1} \\ 0_{2\times3} & 0_{2\times3} & r^2 M I_2 & 0_{2\times1} \\ 0_{1\times3} & 0_{1\times3} & 0_{1\times2} & I \end{pmatrix}, \quad S = \begin{pmatrix} S_1 & 0_{3\times3} & 0_{3\times3} \\ 0_{3\times3} & S_2 & 0_{3\times3} \\ 0_{3\times3} & 0_{3\times3} & 0_{3\times3} \end{pmatrix}, \\
A = \begin{pmatrix} \partial Q_1/\partial \boldsymbol{q}_1 & 0_3 & \partial R_1/\partial \boldsymbol{q}_1 & 0_3 \\ 0_3 & \partial Q_2/\partial \boldsymbol{q}_2 & 0_3 & \partial R_2/\partial \boldsymbol{q}_2 \\ \partial Q_1/\partial \bar{\boldsymbol{x}} & \partial Q_2/\partial \bar{\boldsymbol{x}} & \partial R_1/\partial \bar{\boldsymbol{x}} & \partial R_2/\partial \bar{\boldsymbol{x}} \\ \partial Q_1/\partial \theta & \partial Q_2/\partial \theta & \partial R_1/\partial \theta & \partial R_2/\partial \theta \end{pmatrix}, \\
B = \begin{pmatrix} -\partial y_{01}/\partial \boldsymbol{q}_1 & -r_1\boldsymbol{e}_1 & 0_3 \\ -\partial y_{02}/\partial \boldsymbol{q}_2 & 0_3 & -r_2\boldsymbol{e}_2 \\ 0_2 & 0_2 & 0_2 \\ 0 & 0 & 0 \end{pmatrix}, \\
\Delta\boldsymbol{\lambda} = (\Delta f_1, \Delta f_2, \Delta\lambda_1, \Delta\lambda_2)^{\mathrm{T}}, \\
\Delta\boldsymbol{m} = (\Delta M g/2, \Delta N_1, \Delta N_2)^{\mathrm{T}}, \\
\boldsymbol{d}_N = (0, \cdots, 0, S_N)^{\mathrm{T}}.
\end{array}\right. \tag{4.93}$$

Multiplying this equation by $A^{\mathrm{T}}H^{-1}$ from the left yields

$$\begin{aligned} A^{\mathrm{T}}\ddot{\bar{\boldsymbol{z}}} + A^{\mathrm{T}}H^{-1}\left(\frac{1}{2}\dot{H} + S\right)\dot{\bar{\boldsymbol{z}}} - A^{\mathrm{T}}H^{-1}A\Delta\boldsymbol{\lambda} \\ = A^{\mathrm{T}}H^{-1}B\Delta\boldsymbol{m} - A^{\mathrm{T}}H^{-1}\boldsymbol{d}_N. \end{aligned} \tag{4.94}$$

Since

$$A^{\mathrm{T}}\dot{\bar{\boldsymbol{z}}} = 0, \qquad A^{\mathrm{T}}\ddot{\bar{\boldsymbol{z}}} = -\dot{A}^{\mathrm{T}}\dot{\bar{\boldsymbol{z}}} \tag{4.95}$$

Equation (4.94) is reduced to

$$\begin{aligned} \Delta\boldsymbol{\lambda} = (A^{\mathrm{T}}H^{-1}A)^{-1}\left\{-\dot{A}^{\mathrm{T}}\dot{\bar{\boldsymbol{z}}} + A^{\mathrm{T}}H^{-1}\left(\frac{1}{2}\dot{H} + S\right)\dot{\bar{\boldsymbol{z}}}\right. \\ \left. -A^{\mathrm{T}}H^{-1}B\Delta\boldsymbol{m} + A^{\mathrm{T}}H^{-1}\boldsymbol{d}_N\right\} \end{aligned} \tag{4.96}$$

from which it can be concluded that $\Delta\boldsymbol{\lambda}$ is bounded, because the first two terms in the brackets { } are quadratic functions of components of $\dot{\boldsymbol{z}}$ and the last two terms have already been shown to be bounded, provided that A is of full rank for all $t \in [0, \infty)$. This last property about matrix A can be confirmed in a similar manner to in Section 3.2. This shows that boundedness of all $\Delta\boldsymbol{\lambda}$, $\Delta\boldsymbol{m}$, $\boldsymbol{d}_N$ and $\dot{\boldsymbol{z}}$ implies the boundedness of the acceleration vector $\ddot{\boldsymbol{z}}$, which means that $\dot{\boldsymbol{z}}$ is uniformly continuous in t. Then, by virtue of Lemma 2 (see Appendix A), inequality (4.91) implies that $\dot{\boldsymbol{q}}_i(t) \to 0$ as $t \to \infty$ for $i = 1, 2$. Next we show that convergence of $\dot{\boldsymbol{q}}_i(t)$ to zero as $t \to \infty$ for $i = 1, 2$ means convergence of $\dot{\boldsymbol{x}}(t) \to 0$ and $\dot{\theta}(t) \to 0$ as $t \to \infty$ due to the holonomic constraints given in Table 4.1. To do this, we differentiate d and l_w defined by Equations (T-14) and (T-15) in Table 4.3 with respect to t. This results in

$$(\dot{Y}_1 - \dot{Y}_2)\cos\theta_0 = (\dot{x}_{01} - \dot{x}_{02})\sin\theta + (\dot{y}_{01} - \dot{y}_{02})\cos\theta \\ + \dot{\theta}\left\{(x_{01} - x_{02})\cos\theta - (y_{01} - y_{02})\sin\theta\right\}. \quad (4.97)$$

By taking into account the constraint of Equation (T-2) in Table 4.1 and the equations of Equations (T-14) and (T-15) in Table 4.3, we can rewrite Equation (4.97) in the form

$$-(r_1 + r_2)\dot{\theta}\cos\theta_0 + r_1\boldsymbol{e}_1^{\mathrm{T}}\dot{\boldsymbol{q}}_1 - r_2\boldsymbol{e}_2^{\mathrm{T}}\dot{\boldsymbol{q}}_2 \\ = -\left\{(l_1 + l_2 + r_1 + r_2)\cos\theta_0 + (Y_1 + Y_2)\sin\theta_0\right\}\dot{\theta} \\ +(\dot{x}_{01} - \dot{x}_{02})\sin\theta + (\dot{y}_{01} - \dot{y}_{02})\cos\theta, \quad (4.98)$$

which reduces to

$$\left\{(l_1 + l_2)\cos\theta_0 + (Y_1 + Y_2)\sin\theta_0\right\}\dot{\theta} \\ = -r_1\boldsymbol{e}_1^{\mathrm{T}}\dot{\boldsymbol{q}}_1 + r_2\boldsymbol{e}_2^{\mathrm{T}}\dot{\boldsymbol{q}}_2 + (\dot{x}_{01} - \dot{x}_{02})\sin\theta + (\dot{y}_{01} - \dot{y}_{02})\cos\theta. \quad (4.99)$$

At this stage we need to assume that $|\theta_0| \le \pi/6$ and $|Y_1 + Y_2| < l_1 + l_2$ during movements of the overall fingers–object system. The necessity and reasonableness of this assumption will be discussed in detail in the next section. In Equation (4.99) all the $\dot{x}_{0i}$ and $\dot{y}_{0i}$ for $i = 1, 2$ are linear and homogeneous in $\dot{q}_{ij}$ ($i = 1, 2,\ j = 1$–3) and thereby converge to zero as $t \to \infty$. Since $(l_1 + l_2)\cos\theta_0 + (Y_1 + Y_2)\sin\theta_0 > (l_1 + l_2)/3$, Equation (4.99) implies

$$|\dot{\theta}| \le \frac{3}{l_1 + l_2}\left\{\|\dot{\boldsymbol{x}}_{01} - \dot{\boldsymbol{x}}_{02}\| + \sum_{i=1,2} r_i\left|\boldsymbol{e}_i^{\mathrm{T}}\dot{\boldsymbol{q}}_i\right|\right\}, \quad (4.100)$$

which shows that $\dot{\theta}(t) \to 0$ as $t \to \infty$. Next, we rewrite Equations (T-1) and (T-3) in Table 4.1 into the compact form

$$R^{\mathrm{T}}_{(\theta + (-1)^i\theta_0)}\begin{pmatrix} x - x_{0i} \\ y - y_{0i} \end{pmatrix} = -\begin{pmatrix} (-1)^i(l_i + r_i) \\ Y_i \end{pmatrix}, \quad (4.101)$$

which reduces to

$$\boldsymbol{x} - \boldsymbol{x}_{0i} = -R^{\mathrm{T}}_{(\theta+(-1)^i\theta_0)} \begin{pmatrix} (-1)^i(l_i + r_i) \\ Y_i \end{pmatrix}, \tag{4.102}$$

where $R_{(\theta-\theta_0)}$ or $R_{(\theta+\theta_0)}$ is a two-dimensional orthogonal matrix defined in Equation (2.78). By taking the derivative of Equation (4.102) in t, we obtain

$$\dot{\boldsymbol{x}} = \dot{\boldsymbol{x}}_{01} - \dot{\theta}\Omega R^{\mathrm{T}}_{(\theta-\theta_0)} \begin{pmatrix} -l_1 - r_1 \\ Y_1 \end{pmatrix} + r_1 R^{\mathrm{T}}_{(\theta-\theta_0)} \begin{pmatrix} 0 \\ \dot{\theta} - \boldsymbol{e}_1^{\mathrm{T}}\dot{\boldsymbol{q}}_1 \end{pmatrix}, \tag{4.103}$$

where Ω is a 2×2 skew-symmetric matrix defined as

$$\Omega = \begin{pmatrix} 0 & 1 \\ -1 & 0 \end{pmatrix}. \tag{4.104}$$

Since $\dot{\boldsymbol{q}}_1$, $\dot{\theta}$ and $\dot{\boldsymbol{x}}_{01}$ converge to zero as $t \to \infty$, Equation (4.103) implies that $\dot{\boldsymbol{x}}(t) \to 0$ as $t \to \infty$. Thus we conclude that all components of $\dot{\bar{\boldsymbol{z}}}$ tend to zero as $t \to \infty$. Since $\dot{\bar{\boldsymbol{z}}}$ is continuous uniformly in t, $\ddot{\bar{\boldsymbol{z}}}$ tend to zero as $t \to \infty$ according to Lemma 1 (see Appendix A). This concludes that

$$[A, B](\Delta\boldsymbol{\lambda}^{\mathrm{T}}, \Delta\boldsymbol{m})^{\mathrm{T}} - \boldsymbol{d}_N \to 0 \quad \text{as} \quad t \to \infty. \tag{4.105}$$

Since the 9×7 matrix $[A, B]$ is of full rank, Equation (4.104) means that

$$\Delta\boldsymbol{\lambda}(t) \to 0, \quad \Delta\boldsymbol{m}(t) \to 0, \quad S_N(t) \to 0, \quad \text{as} \quad t \to \infty. \tag{4.106}$$

Thus, the proof of the convergence of $\dot{\boldsymbol{z}}(t) \to 0$ and $\boldsymbol{p}(t) \to \boldsymbol{p}_\infty$ together with Equation (4.106) has been completed.

In the case that one of the fingers has two joints and another has three joints, a similar result to the above can be obtained. However, in the case that each robot finger has two joints, as shown in Figure 4.1, the situation of convergence of solution trajectories differs from the case of robot fingers with redundant joints. In this case, the position state vector $\bar{\boldsymbol{z}}$ is seven-dimensional and $[A, B]$ is a 7×7 matrix, that is, it becomes a square matrix. Notwithstanding this fact, Proposition B.1 and B.2 can be applied in an analogous way as discussed in the previous paragraph. Then, Equation (4.105) should be regarded as

$$A \to A_\infty, \; B \to B_\infty, \; \Delta\boldsymbol{\lambda} \to \boldsymbol{\lambda}_\infty, \text{ and } d_N \to d_\infty \tag{4.107}$$

as $t \to \infty$ so that

$$[A_\infty, B_\infty]\boldsymbol{\lambda}_\infty = d_\infty, \tag{4.108}$$

where A_∞ and B_∞ are constant matrices and $\boldsymbol{\lambda}_\infty$ and d_∞ are constant vectors. Furthermore, even in this case, it is possible to show that the speed of asymptotic convergence of $\dot{\boldsymbol{z}} \to 0$, $\boldsymbol{z} \to \boldsymbol{z}_\infty$ and $\Delta\boldsymbol{\lambda} \to \boldsymbol{\lambda}_\infty$ is exponential in time t as discussed in the next section.

Now we are in a position to discuss the problem of the stability of an equilibrium point with a still state satisfying the force/torque balance in a dynamic sense. In the case of a pair of robot fingers with three joints as shown in Figure 4.2, the trajectory of any solution to the closed-loop dynamics of Equations (4.51–4.53) should lie on the 10-dimensional manifold that can be defined as

$$CM_{10} = \left\{(\bar{\boldsymbol{z}}, \dot{\bar{\boldsymbol{z}}}) : Q_i = 0, R_i = 0, \dot{Q}_i = 0, \dot{R}_i = 0, \text{ for } i = 1, 2\right\}. \quad (4.109)$$

Next, we consider the case that once $\hat{N}_i(0)$ $(i = 1, 2)$ and $\hat{M}(0)$ are set as in Table 4.6 and fixed forever. Then, there is an infinite number of equilibrium-point solutions $\bar{\boldsymbol{z}}$ with still state satisfying

$$A\Delta\boldsymbol{\lambda} + B\Delta\boldsymbol{m} - \boldsymbol{d}_N = 0, \quad (4.110)$$

which is equivalent to Equation (4.92) when $\dot{\bar{\boldsymbol{z}}} = 0$. All $\bar{\boldsymbol{z}}$ satisfying Equation (4.110) constitutes a two-dimensional manifold defined as

$$EM_2 = \{(\bar{\boldsymbol{z}}, \dot{\bar{\boldsymbol{z}}} = 0) : Q_i = 0, R_i = 0 \ (i = 1, 2) \text{ and Equation(4.110)}\}. \quad (4.111)$$

This is called in this book the equilibrium-point manifold (EP-manifold) and can be regarded as a 2-D manifold embedded in the nine-dimensional configuration space defined as $C^9\{\bar{\boldsymbol{z}}\} = \{(\boldsymbol{q}_1^{\mathrm{T}}, \boldsymbol{q}_2^{\mathrm{T}}, \bar{\boldsymbol{x}}^{\mathrm{T}}, \theta)\}$. Any point on this EP manifold should minimise the scalar function P_0 defined by Equation (4.85) and satisfies $\Delta M = 0$. That is, any point on EM_2 minimises $\tilde{P}$. Hence, we consider such an equilibrium point $\bar{\boldsymbol{z}}_\infty = \left(\boldsymbol{q}_{1\infty}^{\mathrm{T}}, \boldsymbol{q}_{2\infty}^{\mathrm{T}}, \bar{\boldsymbol{x}}_\infty^{\mathrm{T}}, \theta_\infty\right)^{\mathrm{T}}$ that attains the minimum of $\tilde{P}$ and takes a pose of Figure 4.2, *i.e.*, all finger joint angles q_{ij} $(i = 1, 2, j = 1, 2, 3)$ are in the range $(\pi/12, \pi/2)$. For convenience we further assume that, at $\bar{\boldsymbol{z}} = \bar{\boldsymbol{z}}_\infty$, $|Y_i| \le r_i$ for $i = 1, 2$. Now we introduce the concept of neighbourhoods of the point $\bar{\boldsymbol{z}}_\infty \in EM_2$ on the constraint manifold CM_{10}, which are defined with two positive parameters $\delta > 0$ and $\rho_0 > 0$ as

$$N^{10}(\delta, \rho_0) = \left\{(\bar{\boldsymbol{z}}, \dot{\bar{\boldsymbol{z}}}) : \bar{E}(\bar{\boldsymbol{z}}, \dot{\bar{\boldsymbol{z}}}) \le \delta^2 \text{ and } R(\bar{\boldsymbol{z}}, \bar{\boldsymbol{z}}_\infty) \le \rho_0\right\}, \quad (4.112)$$

where $\bar{E} = K + \tilde{P} - P_{0\infty}$ and $R(\bar{\boldsymbol{z}}, \bar{\boldsymbol{z}}_\infty)$ denotes the Riemannian distance between $\bar{\boldsymbol{z}}$ and $\bar{\boldsymbol{z}}_\infty$ on $CM_5 = \{\bar{\boldsymbol{z}} : Q_i = 0, \ R_i = 0 \text{ for } i = 1, 2\}$, which is defined as (see Section 3.3)

$$R(\bar{\boldsymbol{z}}, \bar{\boldsymbol{z}}_\infty) = \min_{\bar{\boldsymbol{z}}(t)} \int_0^1 \sqrt{\sum_{i,j} \frac{1}{2} h_{ij}(\bar{\boldsymbol{z}}) \dot{\bar{z}}_i(t) \dot{\bar{z}}_j(t)} \, \mathrm{d}t \quad (4.113)$$

and $H(\bar{\boldsymbol{z}}) = (h_{ij}(\bar{\boldsymbol{z}}))$ denotes the 9×9 inertia matrix defined in Equation (4.93). The necessity of imposing the inequality condition $R(\bar{\boldsymbol{z}}, \bar{\boldsymbol{z}}_\infty) \le \rho_0$ comes from avoidance of self-motion that may possibly arise caused by the redundancy of DOF in finger joints. In order to avoid complication of the mathematical argument, we select $\rho_0 > 0$ adequately small so that at any $\bar{\boldsymbol{z}}$

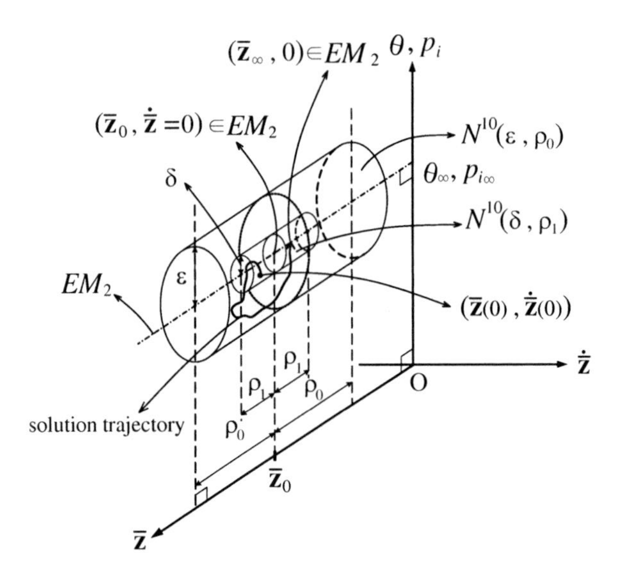

Fig. 4.7. Definitions of stability on a manifold and transferability

satisfying $R(\bar{z}, \bar{z}_\infty) \leq \rho_0$ the 9×4 Jacobian matrix A of the four algebraic constraints with respect to $\bar{z}$ is not degenerate (of full rank). Now we define the stability of such an equilibrium point lying on the manifold EM_2.

Definition 4.1. If for an arbitrarily given $\varepsilon > 0$ there exists a constant $\delta > 0$ depending on $\varepsilon > 0$ and another constant $\rho_1 > 0$ independent of ε and less than ρ_0 such that the trajectory of a solution $(\bar{z}(t), \dot{\bar{z}}(t))$ to the closed-loop dynamics of Equation (4.92) starting from an arbitrary initial state $(\bar{z}(0), \dot{\bar{z}}(0))$ inside $N^{10}(\delta(\varepsilon), \rho_1)$ remains in $N^{10}(\varepsilon, \rho_0)$, then the equilibrium state $(\bar{z}_\infty, \dot{\bar{z}} = 0)$ is said to be stable on a manifold (see Figure 4.7).

Definition 4.2. If for an equilibrium point $\bar{z}_\infty$ there exist constants $\varepsilon_1 > 0$ and $\rho_1 > 0$ such that any solution trajectory to the Equation (4.92) starting from $N^{10}(\varepsilon_1, \rho_1)$ remains in $N^{10}(\varepsilon_1, \rho_0)$ and conveges asymptotically as $t \to \infty$ to some point on $EM_2 \cap N^{10}(\varepsilon_1, \rho_0)$, then this neighbourhood $N^{10}(\varepsilon_1, \rho_1)$ is said to be transferable to a subset of M_2.

It should be remarked that any solution to the closed-loop Equation (4.92) is subject to the eight algebraic constraints $Q_i = 0, R_i = 0, \dot{Q}_i = 0, \dot{R}_i = 0$ $(i = 1, 2)$ and therefore lies on the constraint manifold CM_{10} defined in Equation (4.109). Accordingly, both metrics $\bar{E}(\bar{z}, \dot{\bar{z}})$ and $R(\bar{z}, \bar{z}_\infty)$ should be defined under restrictions on the set of states $(\bar{z}, \dot{\bar{z}})$ that satisfy all such eight algebraic constraints.

4.6 Stability on a Manifold and Transferability to the EP Manifold

As discussed in the paragraph above Equation (4.92), if we choose $\delta > 0$ sufficiently small and $\rho_0 > 0$ adequately then (p_1, p_2, θ) remains in the vicinity of $(p_{1\infty}, p_{2\infty}, \theta_\infty)$. Then, the scalar function $\tilde{P}$ can be written in the form

$$\tilde{P} = \frac{\gamma_M}{2}\Delta M^2 + \frac{1}{2}(\boldsymbol{p} - \boldsymbol{p}_\infty)^{\mathrm{T}} G_\infty (\boldsymbol{p} - \boldsymbol{p}_\infty) + P_{0\infty} + O(\|\Delta \boldsymbol{p}\|^3), \quad (4.114)$$

where $\boldsymbol{p} = (p_1, p_2, p_3)^{\mathrm{T}}$, $p_3 = \theta$ and $\boldsymbol{p}_\infty$ denotes the vector to which $\boldsymbol{p}$ converges as $t \to \infty$, $\Delta\boldsymbol{p} = \boldsymbol{p} - \boldsymbol{p}_\infty$, and G_∞ the Hessian matrix of P_0 in $\boldsymbol{p}$ evaluated at $\boldsymbol{p} = \boldsymbol{p}_\infty$. Similarly, the last two terms of the left-hand side of Equation (4.92) can be expressed as

$$\begin{aligned} -B\Delta\boldsymbol{m} + \boldsymbol{d}_N &= \frac{\Delta M g}{2}\bar{\boldsymbol{e}}_M + \Delta N_2 \bar{\boldsymbol{e}}_1 + \Delta N_2 \bar{\boldsymbol{e}}_2 + r_3^{-1} S_N \bar{\boldsymbol{e}}_3 \\ &= [-B, \bar{\boldsymbol{e}}_3][\Delta\boldsymbol{m}^{\mathrm{T}}, r_3^{-1} S_N]^{\mathrm{T}} \\ &= \frac{\Delta M g}{2}\bar{\boldsymbol{e}}_M + \sum_{i=1}^{3}\sum_{j=1}^{3} g_{ij\infty}\Delta p_j \frac{\bar{\boldsymbol{e}}_j}{r_j} + O(\|\Delta\boldsymbol{p}\|^2), \end{aligned} \quad (4.115)$$

where $g_{ij\infty}$ denotes the ij-th entry of G_∞ and

$$\begin{cases} \bar{\boldsymbol{e}}_M = \begin{pmatrix} \partial y_{01}/\partial \boldsymbol{q}_1 \\ \partial y_{02}/\partial \boldsymbol{q}_2 \\ 0_3 \end{pmatrix}, & \bar{\boldsymbol{e}}_1 = \begin{pmatrix} r_1 \boldsymbol{e}_1 \\ 0_3 \\ 0_3 \end{pmatrix}, \\ \bar{\boldsymbol{e}}_2 = \begin{pmatrix} 0_3 \\ r_2 \boldsymbol{e}_2 \\ 0_3 \end{pmatrix}, & \bar{\boldsymbol{e}}_3 = \begin{pmatrix} 0_3 \\ 0_5 \\ r_3 \end{pmatrix}, \end{cases} \quad (4.116)$$

and $r_3 > 0$ is a constant parameter specified later. Next, we need to introduce the following four 7×9 matrices:

$$\begin{cases} A_M = (A, \bar{\boldsymbol{e}}_1, \bar{\boldsymbol{e}}_2, \bar{\boldsymbol{e}}_3)^{\mathrm{T}}, & A_1 = (A, \bar{\boldsymbol{e}}_M, \bar{\boldsymbol{e}}_2, \bar{\boldsymbol{e}}_3)^{\mathrm{T}}, \\ A_2 = (A, \bar{\boldsymbol{e}}_M, \bar{\boldsymbol{e}}_1, \bar{\boldsymbol{e}}_3)^{\mathrm{T}}, & A_3 = (A, \bar{\boldsymbol{e}}_M, \bar{\boldsymbol{e}}_1, \bar{\boldsymbol{e}}_2)^{\mathrm{T}}, \end{cases} \quad (4.117)$$

and the following four 9×9 projection matrices:

$$\begin{cases} P_M = (I_9 - A_M^+ A_M), & P_1 = (I_9 - A_1^+ A_1), \\ P_2 = (I_9 - A_2^+ A_2), & P_3 = (I_9 - A_3^+ A_3), \end{cases} \quad (4.118)$$

where A_M^+ denotes the pseudo-inverse of 7×9 matrix A_M and A_i^+ that of A_i for $i = 1$–3. Note that

$$P_\xi A = 0, \quad P_\eta \bar{e}_\xi = 0 \quad (\xi \neq \eta), \tag{4.119}$$

where ξ and η denote one of symboles 1, 2, 3 and M.

Before proceeding to prove the exponential convergence of a closed-loop solution to Equation (4.92), we remark on the physical conditions because the proof must be context dependent, that is, dependent on the scales and initial poses of the fingers and object. These are summarised as follows:

$$\begin{aligned}
&1)\quad l_1 + l_2 \geq (r_1 + r_2)/2 \text{ [m]}\\
&2)\quad |Y_i(0)| \leq r_i, \quad i = 1, 2 \quad \text{at} \quad t = 0\\
&3)\quad 5.0 \text{ [N]} \geq f_d \geq 3Mg\\
&4)\quad 0.1 \geq \gamma_M \geq 0.001 \text{ [m}^2/\text{s}^2\text{kg]}\\
&5)\quad 0.001 \geq \gamma_i \geq 0.00001 \text{ [mN}^{-1}\text{]}
\end{aligned}$$

$$6)\quad r_3 = \frac{\max(c_1, c_2)}{f_d(r_1 + r_2)\cos^2\theta_0} = 0.03\text{–}0.3 \text{ [s}^{-1}\text{]} \tag{4.120}$$

$$7)\quad \begin{cases} 1 \geq \bar{e}_M^{\mathrm{T}} P_M \bar{e}_M / \|\bar{e}_M\|^2 \geq \sqrt{\gamma_M}/g \\ 1 \geq \bar{e}_i^{\mathrm{T}} P_i \bar{e}_i / \|\bar{e}_i\|^2 \geq 2r_i, \ i = 1, 2, 3 \end{cases} \tag{4.121}$$

In addition, we implicitly assume that $|\theta_0| < \pi/6$. Conditions 1 and 2 may be relaxed, but they are necessary for rigorous treatment of the convergence proof. Conditions 3–5 together with conditions 1–2 are necessary to assure that the diagonal matrix $G_{D\infty}$ of the Hessian G_∞ defined as $G_{D\infty} = \mathrm{diag}(g_{ii\infty})$ satisfies inequalities

$$\frac{1}{2}G_\infty \leq G_{D\infty} \leq \frac{3}{2}G_\infty. \tag{4.122}$$

Condition 6 suggests the choice of finger joint damping factors c_i $(i = 1, 2)$ based on Hill's model of force/velocity characteristics of muscle contraction (the details of this discussion have been given in the paper [4-9]). The inequalities of Equation (4.121) are reasonably satisfied in most ordinary poses of the fingers and object such as the pose shown in Figure 4.2.

Now we proceed to prove the exponential convergence of a solution trajectory to Equation (4.92) starting from $(\bar{z}(0), \dot{\bar{z}}(0))$ that lies in $N^{10}(\delta, \rho_1)$, where $\delta > 0$ and $\rho_1 > 0$ will be specified later. First, we bear in mind the basic relation of Equation (4.90), from which it follows that $\bar{E}(\bar{z}, \dot{\bar{z}})$ is a non-increasing function of t and therefore each magnitude of ΔM, Δp_1, Δp_2 and $\Delta\theta$ remains small correspondingly to a chosen $\delta > 0$ for an arbitrarily given $\varepsilon > 0$. Second, let us introduce the following nine-dimensional vector:

$$\boldsymbol{w} = \left(\frac{\sqrt{\gamma_M}}{\|\bar{e}_M\|^2} \Delta M \right) P_M \bar{e}_M + \sum_{i=1}^{3} \frac{\Delta p_i}{\|\bar{e}_i\|^2} P_i \bar{e}_i. \tag{4.123}$$

Then, consider a scalar quantity

$$V = \dot{\bar{z}}^{\mathrm{T}} H w \tag{4.124}$$

and notice that every positive eigenvalue of the 9×9 inertia matrix $H(\bar{z})$ is at most of numerical order $O(10^{-5})$ and each diagonal entry g_{ii} of the Hessian matrix G is positive and at least of $O(10^{-1})$. Since it follows that

$$\begin{aligned}
\left| \dot{\bar{z}}^{\mathrm{T}} H P_M \bar{e}_M \frac{\sqrt{\gamma_M}}{\|\bar{e}_M\|^2} \Delta M \right| &\leq \frac{\dot{\bar{z}}^{\mathrm{T}} H H \dot{\bar{z}}}{\|\bar{e}_M\|^2} + \frac{\bar{e}_M^{\mathrm{T}} P_M \bar{e}_M}{4\|\bar{e}_M\|^2} \gamma_M \Delta M^2 \\
&\leq \frac{\dot{\bar{z}}^{\mathrm{T}} H H \dot{\bar{z}}}{\|\bar{e}_M\|^2} + \frac{\gamma_M}{4} \Delta M^2
\end{aligned} \tag{4.125}$$

$$\left| \dot{\bar{z}}^{\mathrm{T}} H P_i \bar{e}_i \frac{\Delta p_i}{\|\bar{e}_i\|^2} \right| \leq \frac{(3/2)}{g_{ii\infty} \|\bar{e}_i\|^2} \dot{\bar{z}}^{\mathrm{T}} H H \dot{\bar{z}} + \frac{g_{ii\infty}}{6} \Delta p_i^2 \tag{4.126}$$

the absolute value of V can be bounded as follows:

$$\begin{aligned}
|V| \leq& \left(\frac{1}{\|\bar{e}_M\|^2} + \sum_{i=1}^{3} \frac{3/2}{g_{ii\infty} \|\bar{e}_i\|^2} \right) \dot{\bar{z}}^{\mathrm{T}} H H \dot{\bar{z}} \\
&+ \frac{1}{3} \left(\frac{3}{4} \gamma_M \Delta M^2 + \sum_{i=1}^{3} \frac{g_{ii\infty}}{2} \Delta p_i^2 \right).
\end{aligned} \tag{4.127}$$

The first term of the right-hand side is less than $(1/2)\dot{\bar{z}}^{\mathrm{T}} H \dot{\bar{z}}$ $(= K/2)$ because the maximum eigenvalue of H is at most of order $O(10^{-5})$. Then, by referring to Equation (4.122) we can rewrite Equation (4.127) in the following way:

$$\begin{aligned}
|V| &\leq \frac{1}{2} K + \frac{1}{3} \left(\frac{3}{4} \gamma_M \Delta M^2 + \frac{3}{4} \Delta \boldsymbol{p}^{\mathrm{T}} G_\infty \Delta \boldsymbol{p} \right) \\
&\leq \frac{1}{2} (K + \tilde{P} - P_{0\infty}) = \frac{1}{2} \bar{E}.
\end{aligned} \tag{4.128}$$

Next, we define a scalar function

$$W(\alpha) = \bar{E} - \alpha V \tag{4.129}$$

with a positive parameter $\alpha > 0$. Then, evidently from Equation (4.128) it follows that

$$\left(1 + \frac{\alpha}{2}\right) \bar{E} \geq W(\alpha) \geq \left(1 - \frac{\alpha}{2}\right) \bar{E}. \tag{4.130}$$

Since we are concerned with the derivative of $W(\alpha)$ in time t, we evaluate $\dot{V}$ in such a manner that

$$\begin{aligned}
\frac{\mathrm{d}}{\mathrm{d}t} V =& \ddot{\bar{z}}^{\mathrm{T}} H w + \dot{\bar{z}} \dot{H} w + \dot{\bar{z}} H \dot{w} \\
=& \left\{ [A, B] \Delta \boldsymbol{\lambda} - \boldsymbol{d}_N - \left(\frac{1}{2} \dot{H} + S \right) \dot{\bar{z}} - C \dot{\bar{z}} \right\}^{\mathrm{T}} w \\
&+ \dot{\bar{z}} \dot{H} w + \dot{\bar{z}} H \dot{w}.
\end{aligned} \tag{4.131}$$

Note that $A^{\mathrm{T}}\boldsymbol{w} = 0$, as shown in Equation (4.119). Hence, Equation (4.131) is reduced to

$$\dot{V} = \boldsymbol{w}^{\mathrm{T}}(B\Delta\boldsymbol{m} - \boldsymbol{d}_N) - \boldsymbol{w}^{\mathrm{T}}C\dot{\boldsymbol{z}} + \dot{\boldsymbol{z}}^{\mathrm{T}}H\dot{\boldsymbol{w}} + h(\dot{\boldsymbol{z}})\boldsymbol{w}, \tag{4.132}$$

where $h(\dot{\boldsymbol{z}})$ is a vector composed of quadratic functions of the components of $\dot{\boldsymbol{z}}$ whose coefficients are at most of order h_M, where we denote by h_M the maximum eigenvalue of H. Substituting Equation (4.115) into Equation (4.132) leads to

$$\begin{aligned}\dot{V} = \boldsymbol{w}^{\mathrm{T}}C\dot{\boldsymbol{z}} + \frac{2\bar{\boldsymbol{e}}_M^{\mathrm{T}}P_M\bar{\boldsymbol{e}}_M}{\sqrt{\gamma_M}\|\bar{\boldsymbol{e}}_M\|^2} \cdot \frac{\gamma_M}{2}\Delta M^2 g \\ + \sum_{i=1}^{3} \frac{\bar{\boldsymbol{e}}_i^{\mathrm{T}}P_i\bar{\boldsymbol{e}}_i}{r_i\|\bar{\boldsymbol{e}}_i\|^2} \cdot g_{ii\infty}\Delta p_i^2 + O(\|\dot{\boldsymbol{z}}\|^2)(\|\boldsymbol{w}\| + 1)h_M,\end{aligned} \tag{4.133}$$

where $O(\|\dot{\boldsymbol{z}}\|^2)$ denotes a vector whose norm is of order of $\|\dot{\boldsymbol{z}}\|^2$. On the other hand, it is possible to show that

$$\begin{aligned}-\boldsymbol{w}^{\mathrm{T}}C\dot{\boldsymbol{z}} &\leq \frac{1}{2}\sum_{i=1,2} c_i\|\dot{\boldsymbol{q}}_i\|^2 + \frac{c_m}{2}\|\boldsymbol{w}\|^2 \\ &\leq \sum_{i=1,2}\frac{c_i}{2}\|\dot{\boldsymbol{q}}_i\|^2 + \frac{c_m}{2}\left\{\frac{\bar{\boldsymbol{e}}_M^{\mathrm{T}}P_M\bar{\boldsymbol{e}}_M}{\|\bar{\boldsymbol{e}}_M\|^4}\gamma_M\Delta M^2 \right. \\ &\qquad \left. + \sum_{i=1}^{3}\frac{\bar{\boldsymbol{e}}_i^{\mathrm{T}}P_i\bar{\boldsymbol{e}}_i}{\|\bar{\boldsymbol{e}}_i\|^4}\Delta p_i^2\right\},\end{aligned} \tag{4.134}$$

where $c_m = \max\{c_1, c_2\}$. Thus, substituting Equations (4.133) and (4.134) into the derivative of $W(\alpha)$ with respect to t yields

$$\begin{aligned}\frac{\mathrm{d}}{\mathrm{d}t}W(\alpha) \leq -\sum_{i=1,2}\frac{c_i}{2}\|\dot{\boldsymbol{q}}_i\|^2 - \alpha\left\{\left(\frac{2g}{\sqrt{\gamma_M}} - \frac{c_m}{\|\bar{\boldsymbol{e}}_M\|^2}\right)\xi_M \cdot \frac{\gamma_M}{2}\Delta M^2 \right. \\ \left. + \sum_{i=1}^{3}\left(\frac{1}{r_i} - \frac{c_m/2}{g_{ii\infty}\|\bar{\boldsymbol{e}}_i\|^2}\right)\xi_i \cdot g_{ii\infty}\Delta p_i^2\right\} \\ +\alpha\left\{O(\|\dot{\boldsymbol{z}}\|^2)(\|\boldsymbol{w}\| + 1)h_M\right\},\end{aligned} \tag{4.135}$$

where we set

$$\xi_M = \frac{\bar{\boldsymbol{e}}_M^{\mathrm{T}}P_M\bar{\boldsymbol{e}}_M}{\|\bar{\boldsymbol{e}}_M\|^2}, \quad \xi_i = \frac{\bar{\boldsymbol{e}}_i^{\mathrm{T}}P_i\bar{\boldsymbol{e}}_i}{\|\bar{\boldsymbol{e}}_i\|^2}, \qquad i = 1, 2, 3. \tag{4.136}$$

From the choice of γ_M, c_i and r_i discussed above, it can be reasonably concluded that

$$\begin{cases}\dfrac{2g}{\sqrt{\gamma_M}} - \dfrac{c_m}{\|\bar{\boldsymbol{e}}_M\|^2} \geq \dfrac{g}{\sqrt{\gamma_M}} \\ \dfrac{1}{r_i} - \dfrac{c_m}{2g_{ii\infty}\|\bar{\boldsymbol{e}}_i\|^2} \geq \dfrac{1}{2r_i}, \quad i = 1, 2, 3.\end{cases} \tag{4.137}$$

Comparing the right-hand sides of this inquality with inequalities of Equation (4.121), we see from Equation (4.135) that

$$\frac{\mathrm{d}}{\mathrm{d}t}W(\alpha) \le -\sum_{i=1,2}\frac{c_i}{2}\|\dot{\boldsymbol{q}}_i\|^2 - \alpha\left\{\frac{\gamma_M}{2}\Delta M^2 + \sum_{i=1}^{3} g_{ii\infty}\Delta p_i^2\right\} + \alpha\left\{O(\|\dot{\boldsymbol{z}}\|^2)(\|\boldsymbol{w}\|+1)h_M\right\}. \tag{4.138}$$

It should be noted again that according to Equation (4.122) we have

$$\begin{aligned}\frac{\gamma_M}{2}\Delta M^2 &+ \sum_{i=1}^{3} g_{ii\infty}\Delta p_i^2 \\ &\ge \frac{\gamma_M}{2}\Delta M^2 + \frac{1}{2}\sum_{i,j} g_{ij\infty}\Delta p_i \Delta p_j + O(\|\Delta\boldsymbol{p}\|^3) \\ &\ge \tilde{P} - P_{0\infty} + O(\|\Delta\boldsymbol{p}\|^3),\end{aligned} \tag{4.139}$$

where $\Delta\boldsymbol{p} = (\Delta\boldsymbol{p}_1^{\mathrm{T}}, \Delta\boldsymbol{p}_2^{\mathrm{T}})^{\mathrm{T}}$. Finally, it should be remarked that from Equations (4.100) and (4.103) derived from the velocity constraints $\dot{Q}_i = 0$ and $\dot{R}_i = 0$ for $i = 1, 2$ it follows that

$$r^{-2}(\dot{x}^2 + \dot{y}^2) + \dot{\theta}^2 \le O\left(\sum_{i=1,2}\|\dot{\boldsymbol{q}}_i\|^2\right). \tag{4.140}$$

Since h_M is far less than c_i ($i = 1, 2$), that is, h_M is of $O(10^{-5})$ and c_i is of $O(10^{-3})$ ($i = 1, 2$), we have

$$\begin{aligned}-\sum_{i=1,2}\frac{c_i}{2}\|\dot{\boldsymbol{q}}_i\|^2 + \alpha\left\{O(\|\dot{\boldsymbol{z}}\|^2)(\|\boldsymbol{w}\|+1)h_M\right\} &\le -\frac{\alpha}{2}\dot{\boldsymbol{z}}^{\mathrm{T}}H\dot{\boldsymbol{z}} \\ &= -\alpha K\end{aligned} \tag{4.141}$$

as long as $0 \le \alpha \le 1.0$. By substituting Equations (4.139) and (4.141) into Equation (4.138), we now obtain

$$\frac{\mathrm{d}}{\mathrm{d}t}W(\alpha) \le -\alpha(K + \tilde{P} - P_{0\infty}) = -\alpha\bar{E}, \tag{4.142}$$

which implies, from Equation (4.130), that

$$\frac{\mathrm{d}}{\mathrm{d}t}W(\alpha) \le -\alpha\bar{E} \le \frac{-2\alpha}{2+\alpha}W(\alpha). \tag{4.143}$$

In particular, set $\alpha = 1$ and denote the value of $W(1)$ at time t by $W(t)$. Then, Equation (4.143) is reduced to

$$\begin{aligned}\bar{E}(t) &\le 2W(t) \le 2\mathrm{e}^{(-2/3)t}W(0) \\ &\le 3\mathrm{e}^{-(2/3)t}\bar{E}(0).\end{aligned} \tag{4.144}$$

This shows exponential convergences of $\dot{\bar{z}}(t)$ to zero as $t \to \infty$, $\Delta \boldsymbol{p}$ to zero and ΔM to zero.

Finally, we prove the stability of such an equilibrium point $\bar{z}_\infty$ on EM_2 on the constraint manifold CM_{10}.

Proposition 1. For a given equilibrium point $\bar{z}_\infty$ lying on EM_2 there exist constants $\varepsilon_1 > 0$ and $\rho_1 > 0$ ($\rho_1 < \rho_0$) such that any solution trajectory starting from an arbitrary initial state $(\bar{z}(0), \dot{\bar{z}}(0))$ belonging to $N^{10}(\varepsilon_1, \rho_1)$ converge asymptotically to some still state on EM_2, that is, $N^{10}(\varepsilon_1, \rho_1)$ is transferable to a subset of EM_2. In particular, the equilibrium state $(\bar{z}_\infty, 0)$ is stable on a manifold.

Proof. We first note that the Riemannian distance between $\bar{z}(0)$ and $\bar{z}(t)$ at any fixed $t > 0$ must be less than the Riemannian metric along a special trajectory, $\tau \in [0, t]$, defined by the position trajectory $\bar{z}(\tau)$ of the solution $(\bar{z}(\tau), \dot{\bar{z}}(\tau))$ to Equation (4.92) starting from $(\bar{z}(0), \dot{\bar{z}}(0))$. This fact is described as

$$R(\bar{z}(0), \bar{z}(t)) \le \int_0^t \sqrt{\frac{1}{2} \sum_{i,j} h_{ij}(\bar{z}(\tau)) \dot{\bar{z}}_i(\tau) \dot{\bar{z}}_j(\tau)}\, \mathrm{d}\tau. \tag{4.145}$$

Since $K \le \bar{E}(\bar{z}, \dot{\bar{z}})$, Equation (4.145) together with (4.144) leads to

$$\begin{aligned} R(\bar{z}(0), \bar{z}(t)) &\le \int_0^t \sqrt{\bar{E}(\bar{z}(\tau), \dot{\bar{z}}(\tau))}\, \mathrm{d}\tau \\ &\le \sqrt{3\bar{E}(0)} \int_0^t \mathrm{e}^{-\tau/3}\, \mathrm{d}\tau \\ &= 3\sqrt{3\bar{E}(0)}(1 - \mathrm{e}^{-t/3}) \le 3\sqrt{3}\varepsilon_1. \end{aligned} \tag{4.146}$$

In particular, we have as $t \to \infty$

$$R(\bar{z}(0), \bar{z}_\infty) \le 3\sqrt{3}\varepsilon_1. \tag{4.147}$$

Hence, if we choose ε_1 and ρ_1 such that

$$\varepsilon_1 = \frac{\rho_0}{6\sqrt{3}}, \quad \rho_1 = \frac{\rho_0}{2} \tag{4.148}$$

then it follows from (4.147) that

$$\begin{aligned} R(\bar{z}(t), \bar{z}_\infty) &\le R(\bar{z}(t), \bar{z}(0)) + R(\bar{z}(0), \bar{z}_\infty) \\ &\le 3\sqrt{3}\varepsilon_1 + \rho_0/2 \le \rho_0 \end{aligned} \tag{4.149}$$

as long as $\bar{E}(t) \le \bar{E}(0) \le \varepsilon_1^2$ and $R(\bar{z}(0), \bar{z}_\infty) \le \rho_1$. This proves the stability of the equilibrium point $\bar{z}_\infty$ on a manifold. At the same time, every component of the velocity vector $\dot{\bar{z}}(t)$ tends to zero exponentially as $t \to \infty$ and $\bar{E}(t) \to 0$ exponentially as $t \to \infty$, too. This means that $\bar{z}(t)$ converges to EM_2 asymptotically and hence $N^8(\varepsilon_1, \rho_1)$ is transferable to a subset of EM_2.

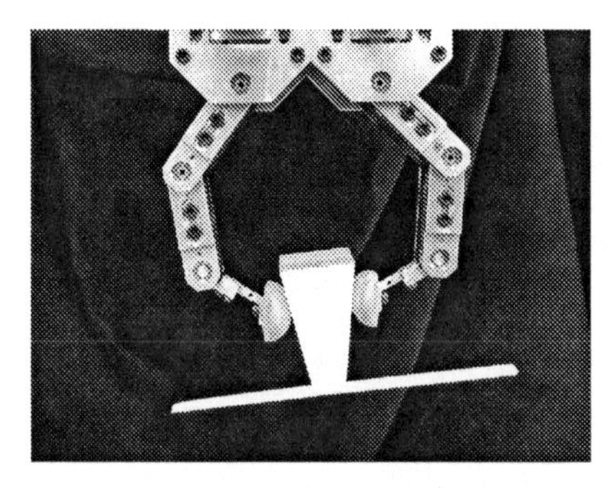
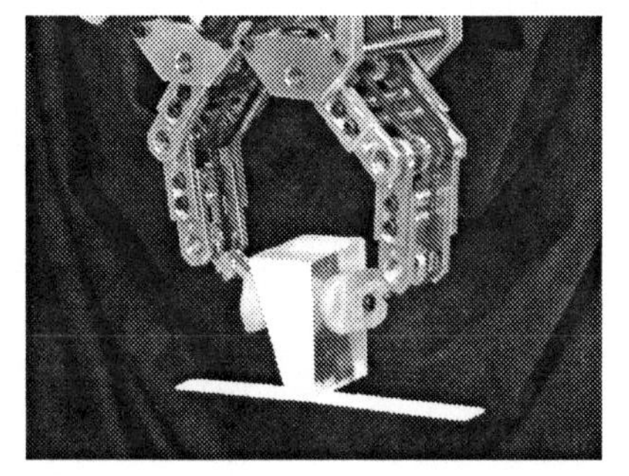

Fig. 4.8. Experimental setup of a pair of robot fingers and a grasped object

Table 4.10. Details of the physical parameters of the experimental setup

Actuator (Maxson DC motor)		power	4.55 [W]
		encoder	512 [p/r]
		gear ratio	23:1
Robot fingers	$l_{11} = l_{21}$	length	0.039 [m]
	$m_{11} = m_{21}$	mass	0.040 [kg]
	$l_{12} = l_\alpha$	length	0.039 [m]
	$m_{12} = m_\alpha$	mass	0.030 [kg]
	$l_{13} = l_{22}$	length	0.033 [m]
	$m_{13} = m_{22}$	mass	0.020 [kg]
	L	base length	0.053 [m]
	α	constant joint angle	50.00 [deg]
	r_i	radius	0.001 [m]
Pinched object	M	object mass	0.410×10^{-1} [kg]
	h	object height	0.050 [m]
	l_i	object length (i=1,2)	0.010 [m]
	θ_0	inclination angle	-10.00 [deg]

4.7 Experiments of Blind Grasping

In this section we show experimental results of stable grasping conducted by using the experimental setup of dual robot fingers shown in Figure 4.8. The physical parameters of robot fingers and the rigid object used in the experiment are given in Table 4.10. In this experiment, all three joints of the left finger were actuated by DC motors with the same characteristics, but only two joints of the right finger were actuated while its second joint was fixed at the angle $q_{22} = \alpha$ (= 50.00 [deg]). Before using the coordinated control signal based upon Equation (4.49), we carried out several experiments on tuning the damping forces ($c_i \dot{q}_i$, $i = 1, 2$) and gravity compensation terms for the

Table 4.11. Physical parameters of the control signals

Control signals	f_d	internal force	0.25 [N]
	γ_M	regressor gain	0.088 [m^2/kgs^2]
	γ_{N1}	regressor gain	0.0002 [s^2/kg]
	γ_{N2}	regressor gain	0.0002 [s^2/kg]
	$\hat{M}(0)$	initial object mass	0.000 [kg]

fingers themselves. Actually, the first two terms of the control signals were adjusted separately from the remaining three terms, which can be generated as output torques from current-controlled DC motors with reduction gears based on tunings of the direct-current amplifications. Therefore, the actual control signals should be expressed as

$$\boldsymbol{v}_i = \hat{\boldsymbol{g}}_i - \hat{c}_i \dot{q}_i + \boldsymbol{u}_i, \quad i = 1, 2 \tag{4.150}$$

$$\boldsymbol{u}_i = \kappa_i \left\{ (-1) \frac{f_d}{r_1 + r_2} J_{0i}^{\mathrm{T}}(\boldsymbol{g}_i) \begin{pmatrix} \boldsymbol{x}_{01} - \boldsymbol{x}_{02} \\ \boldsymbol{y}_{01} - \boldsymbol{y}_{02} \end{pmatrix} - \frac{\hat{M} g}{2} \left(\frac{\partial y_{0i}}{\partial \boldsymbol{q}_i} \right) - r_i \hat{N}_i \boldsymbol{e}_i \right\}, \quad i = 1, 2, \tag{4.151}$$

where κ_i stands for physically appropriate constants related to motor torque constant, reduction gear ratio and current amplication constant. All the net constants $\kappa_i f_d$, $\kappa_i \gamma_M$ and $\kappa_i \gamma_{Ni}$ $(i = 1, 2)$ are evaluated and denoted anew by f_d, γ_M and γ_{Ni} $(i = 1, 2)$ having confirmed that all these constants κ_i are the same for different joints since all the DC motors used in the experiment are the same with the same gear ratios. All such control gains and initial value of $\hat{M}$ are given in Table 4.11. It is diffecult to evaluate the viscosity constants $\hat{c}_i$ $(i = 1, 2)$ for each joint because there exists uncertainty in the frictions inherent to driving mechanisms. The lateral thin plate is attached intentionally to the object with non-parallel flat surfaces to evaluate of the inclination angle $\theta(t)$ through external measurment of both distances from two separate fixed points in the frame coordinates to two corresponding laser spots on the plate. Figure 4.9 shows the transient responses of the physical variables $Y_1' - Y_2'$, θ, $\hat{M}$, $\hat{N}_i$ $(i = 1, 2)$ and S_N'. In this figure, the graphs of θ and $Y_1' - Y_2'$ are based on measurement data obtained by external distance laser sensors and the others are plotted on the basis of computations based on measurement data of the joint angles (measured via internal sensors). It is interesting to note that all $\hat{M}$, $\hat{N}_i$ $(i = 1, 2)$ and S_N' do not converge to their corresponding target constants due to static friction latency existing tangentially between the fingertip and object surfaces. In this experiment, we set $f_d = 0.25$ [N], relatively small in comparison with the numerical value of f_d used in computer simulation specified in Table 4.5. As usual, the regressor gain γ_M is chosen as $\gamma_M = 0.088$ [m^2/s^{2}kg], relatively large in comparison with

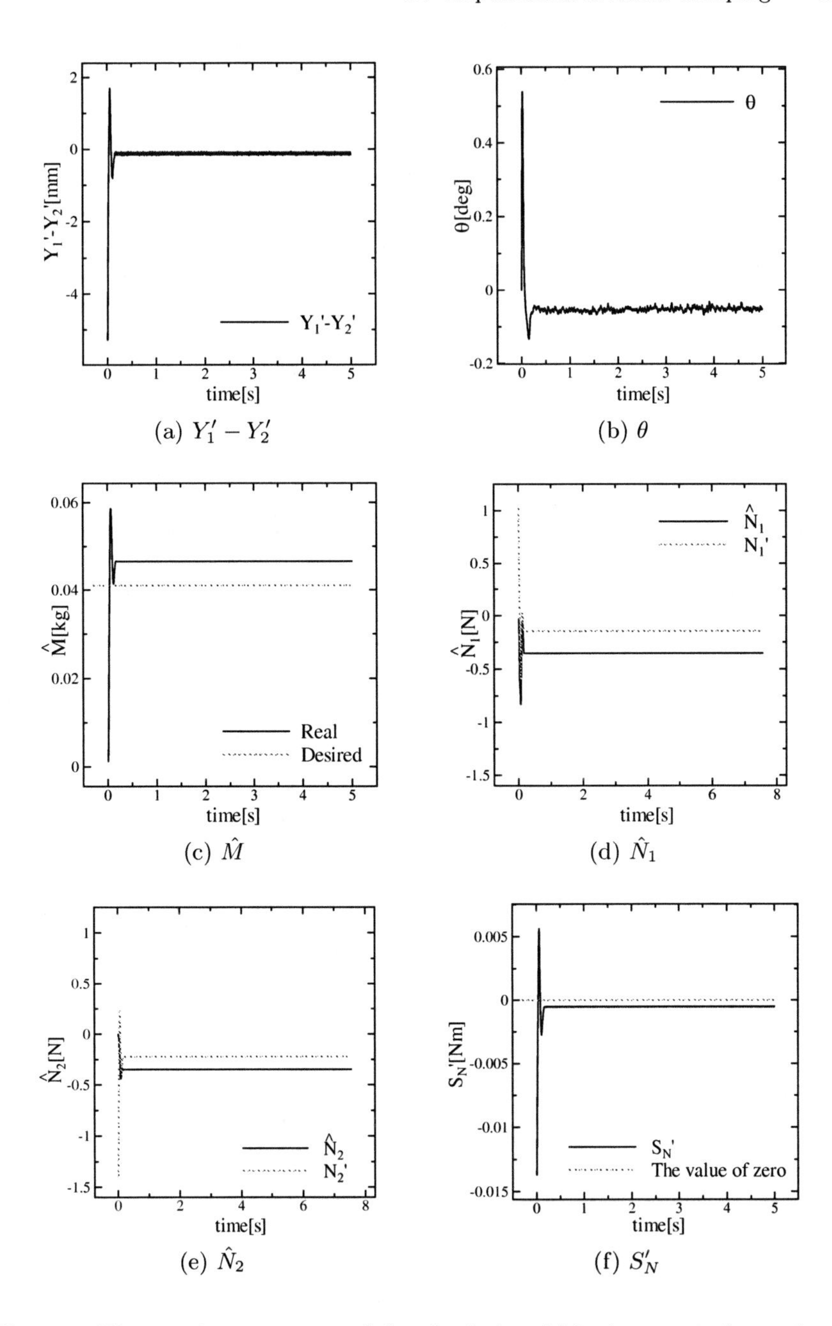

Fig. 4.9. The transient responses of the physical variables in an actual experiment based on the control signal of blind grasping

the $\gamma_M = 0.01$ used in the simulation (see Table 4.5 too). Such careful choices for f_d and γ_M resulted in good performance for the estimator $\hat{M}(t)$, which quickly converged to a constant sufficiently approximating the true value M of the object mass.

According to the experimental observations when we changed the values of f_d and γ_M, grasping of the object would become stabilisable by increasing the gain f_d and decreasing γ_M inverse proportionally, but the estimated value of the object mass varied more from its true value. Large values of f_d and inverse proportionally smaller γ_M were apt to induce slipping between the fingertips and the object rather than rolling. If the object shown in Figure 4.8 was grasped upside down by the robot fingers, stable grasp of control of the object would become hard in practice without using rough surfaces on the fingertip spheres. It should be remarked that throughout this chapter it is assumed that the sum of the pressing force vector $\boldsymbol{f}_i$ and the rolling constraint force vector $\boldsymbol{\lambda}_i$ for $i = 1$ or 2 is always contained in a friction cone at the corresponding contact point, even if in addition the gravity force due to the object weight acted on the contact point.

5

Three-dimensional Grasping by a Pair of Rigid Fingers

This chapter extends the stability theory of 2-D object grasping to cope with three-dimensional (3-D) object grasping by a pair of multi-joint robot fingers with hemispherical ends. It shows that secure grasping of a 3-D object with parallel surfaces in a dynamic sense can be realised in a blind manner like human grasping an object by a thumb and index finger while the eyes closed. Rolling contacts are modelled as Pfaffian constraints that cannot be integrated into holonomic constraints but exert tangential constraint forces on the object surfaces. A noteworthy difference of modelling 3-D object grasping from the 2-D case is that the instantaneous axis of rotation of the object is fixed in the latter case but is time-varying in the former case. Hence, the dynamics of the overall fingers–object system are subject to non-holonomic constraints regarding a 3-D orthogonal matrix consisting of three mutually orthogonal unit vectors fixed at the object. A further difference arises due to the physical assumption that spinning around the opposing axis between the two contact points no long arises, which induces another non-holonomic constraint. Lagrange's equation of motion for the overall system can be derived from the variational principle without violating the causality that governs the non-holonomic constraints. Then, a simple control signal constructed on the basis of finger–thumb opposable forces together with an object-mass estimator is shown to accomplish stable grasping in a dynamic sense without using object information or external sensing. This is called blind grasping if in addition the overall closed-loop dynamics converges to a state of force/torque balance. The closed-loop dynamics can be regarded as Lagrange's equation of motion with an artificial potential function that attains its minimum at some equilibrium state of force/torque balance. A mathematical proof of stability and asymptotic stability on a constraint manifold of the closed-loop dynamics under the non-holonomic constraints is presented. A differential geometric meaning of the exponential convergence of the solution trajectory is also discussed with the aid of Riemannian metrics. In the last section, modelling and control of full dynamics of 3-D grasping admitting spinning around the opposing axis are discussed.

5.1 Introduction

When a human grasps an object securely, the thumb plays a crucial role. Napier [1-1] says that, "The movement of the thumb underlies all the skilled procedures of which the hand is capable," and further, "Without the thumb, the hand is put back 60 million years in evolutionary terms to a stage when the thumb had no independent movement and was just another digit. One cannot emphasize enough the importance of finger–thumb opposition for human emergence from a relatively undistinguished primate background." Finger–thumb opposition is defined by a movement in which the pulp surface of the thumb is placed squarely in contact with the terminal pads of one or all of the remaining digits (Figure 5.1). According to the literature of research works on multi-fingered robotic hands, however, there is a dearth of papers concerned with the dynamics and control of stable precision prehension or grasping of an object through finger-thumb opposition. Rather, most research is concerned with the kinematics and plannings of motions establishing force/torque closure for secure grasping in a static sense by using multi-fingers with frictionless contacts. There are a few exceptional papers [3-1][3-2], however, that treated the problem of rolling contacts and analysed the kinematics and dynamics of multi-fingered hands. However, their proposed control schemes were based on the computed torque method. Furthermore, the previous works in the 1990s overlooked the crucial role of non-holonomic constraints that arise from time-varying change of the instantaneous axis of rotation of the object induced by rolling contacts between the finger-ends and object surfaces. Thus, it had been thought that some difficult problems of coping with such non-holonomic constraints through the overall complicated fingers–object dynamics are veiled

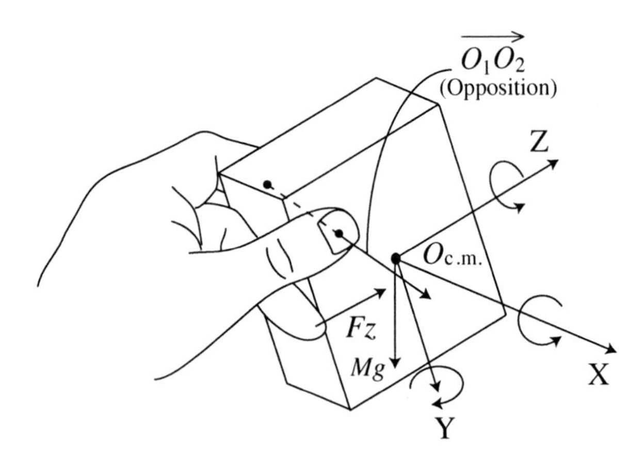

Fig. 5.1. Stable grasping of a 3-D object with parallel surfaces in a dynamic sense. The third finger (middle finger) is used to stop spinning motion about the opposition axis, which may be induced by the gravity

and, regardless of this, finding a simple control signal actualising stable grasping in a dynamic way is indispensable. Thus, the state of the art of control of multi-fingered hands is expressed by Bicchi [1-24] as "a difficult road to simplicity." Very recently in the 2000s, however, it has been shown by the author's group that grasping of a 2-D rigid object by a pair of multi-DOF fingers can be stabilised dynamically under gravity by taking into account the tangential forces induced by rolling contacts [4-5]. Further, a remarkable result has been shown [4-6][4-7] that blind grasping without using knowledge of object kinematics or external sensing can be realised, provided that the overall fingers–object movements are confined to a 2-D plane.

However, modelling of the dynamics of pinching a 3-D rigid object remained unsolved until 2006, because of difficulties coping with non-holonomic constraint problems under redundancy of the total DOFs of fingers and an object and finding a simple control signal that cannot directly control the object but controls it indirectly through constraint forces. In 2006, however, the problem of modelling 3-D object grasping was tackled by the author's group and a mathematical model was derived as a set of Lagrange's equations of motion of the fingers–object system under non-holonomic constraints induced by continuous change of the instantaneous axis of rotation of the object and Pfaffian constraints of rolling contacts.

5.2 Non-holonomic Constraints

Consider the motion of a rigid object with parallel flat surfaces, which is grasped by a pair of robot fingers with three DOFs and four DOFs as shown in Figure 5.2. When the distance from the straight line $\overrightarrow{O_1O_2}$ (the opposition axis) connecting two contact points between the finger-ends and object surfaces to the vertical axis through the object centre of mass in the direction of gravity becomes large, there arises a spinning rotation of the object around that opposition axis. This chapter considers the problem of modelling of pinching in the situation that this spinning motion has ceased after the centre of mass of the object came sufficiently close to a point just beneath the opposing axis and no more such spinning rotation will arise due to dry friction and micro-deformations near the contact points between the finger-ends and object surfaces. The ceasing of spinning, however, induces a non-holonomic constraint among rotational angular velocities ω_x, ω_y and ω_z around the x, y and z axes, respectively, that is, the vector $\boldsymbol{\omega} = (\omega_x, \omega_y, \omega_z)^{\mathrm{T}}$ denotes the vector of rigid-body rotation in terms of the frame coordinates $O - xyz$. At the same time, we introduce the cartesian coordinates $O_{c.m.} - XYZ$ fixed at the object frame and denote three orthogonal unit vectors at the object frame in each corresponding direction X, Y and Z by $\boldsymbol{r}_X$, $\boldsymbol{r}_Y$ and $\boldsymbol{r}_Z$ as shown in Figure 5.3. Since the opposing axis is expressed as $\overrightarrow{O_1O_2} = \boldsymbol{x}_1 - \boldsymbol{x}_2$, where $\boldsymbol{x}_i = (x_i, y_i, z_i)^{\mathrm{T}}$ denotes the Cartesian coordinates of the contact point O_i (see Figure 5.3), the ceasing of spinning motion around the axis $\overrightarrow{O_1O_2}$ implies

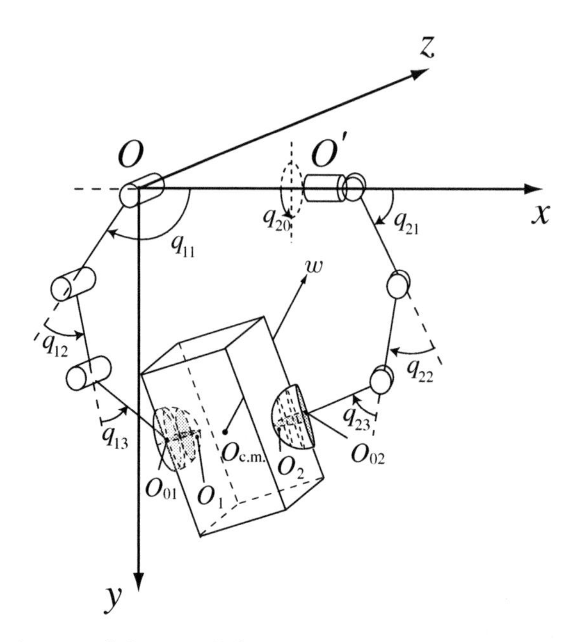

Fig. 5.2. Coordinates of the overall fingers–object system. The coordinates $O-xyz$ are fixed to the frame. The symbol $\boldsymbol{\omega}$ expresses the axis of rotation

that the instantaneous axis of rotation of the object is orthogonal to $\boldsymbol{x}_1 - \boldsymbol{x}_2$, that is,

$$\boldsymbol{\omega}^{\mathrm{T}}(\boldsymbol{x}_1 - \boldsymbol{x}_2) = 0, \tag{5.1}$$

which can be rewritten in the form

$$\omega_x = -\xi_y \omega_y - \xi_z \omega_z \tag{5.2}$$

$$\xi_y = \frac{y_1 - y_2}{x_1 - x_2}, \qquad \xi_z = \frac{z_1 - z_2}{x_1 - x_2}, \tag{5.3}$$

where $\boldsymbol{\omega} = (\omega_x, \omega_y, \omega_z)^{\mathrm{T}}$. On the other hand, denote the Cartesian coordinates of the object mass centre $O_{\mathrm{c.m.}}$ by $\boldsymbol{x} = (x, y, z)^{\mathrm{T}}$ based on the frame coordinates $O - xyz$ and three mutually orthogonal unit vectors fixed at the object frame by $\boldsymbol{r}_X, \boldsymbol{r}_Y$ and $\boldsymbol{r}_Z$, which may rotate dependently on the angular velocity vector $\boldsymbol{\omega}$ of body rotation. Then, the 3×3 rotation matrix

$$R(t) = (\boldsymbol{r}_X, \boldsymbol{r}_Y, \boldsymbol{r}_Z) \tag{5.4}$$

belongs to SO(3) and is subject to the first-order differential equation

$$\frac{\mathrm{d}}{\mathrm{d}t} R(t) = R(t) \Omega(t), \tag{5.5}$$

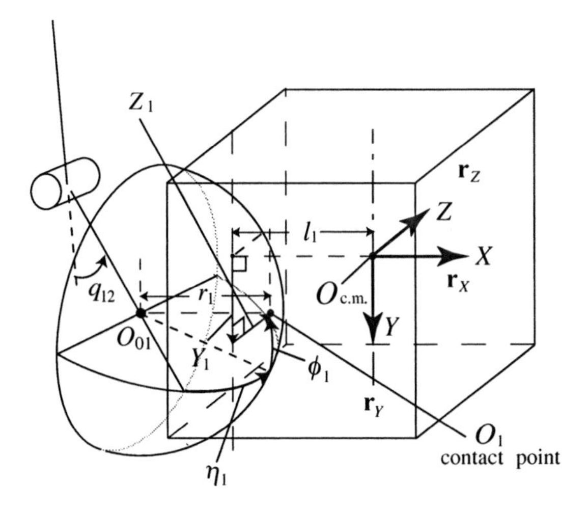

Fig. 5.3. Mutually orthogonal unit vectors $\boldsymbol{r}_X, \boldsymbol{r}_Y$ and $\boldsymbol{r}_Z$ express the rotational motion of the object. The pair (η_i, ϕ_i) for $(i = 1, 2)$ expresses the spherical coordinates of each hemispherical finger end

where

$$\Omega(t) = \begin{pmatrix} 0 & -\omega_z & \omega_y \\ \omega_z & 0 & -\omega_x \\ -\omega_y & \omega_x & 0 \end{pmatrix}. \tag{5.6}$$

Equation (5.5) expresses another non-holonomic constraint on the rotational motion of the object. Next, denote the position of the centre of each hemispherical finger-end by $\boldsymbol{x}_{0i} = (x_{0i}, y_{0i}, z_{0i})^{\mathrm{T}}$. Then, it is possible to notice that (see Figure 5.3)

$$\boldsymbol{x}_i = \boldsymbol{x}_{0i} - (-1)^i r_i \boldsymbol{r}_X, \tag{5.7}$$

$$\boldsymbol{x} = \boldsymbol{x}_{0i} - (-1)^i (r_i + l_i)\boldsymbol{r}_X - Y_i \boldsymbol{r}_Y - Z_i \boldsymbol{r}_Z. \tag{5.8}$$

Since each contact point O_i can be expressed by the coordinates $((-1)^i l_i, Y_i, Z_i)$ based on the object frame $O_{\text{c.m.}} - XYZ$, taking an inner product between Equation (5.8) and $\boldsymbol{r}_X$ gives rise to

$$Q_i = -(r_i + l_i) - (-1)^i (\boldsymbol{x} - \boldsymbol{x}_{0i})^{\mathrm{T}} \boldsymbol{r}_X = 0, \ i = 1, 2, \tag{5.9}$$

which express holonomic constraints of the contacts between finger-ends and the object. A rolling constraint between one finger-end and its contacted object surface can be expressed by equality of the two contact point velocities expressed on either of finger-end spheres and on its corresponding tangent plane (that is, coincident with one of the object's flat surfaces).

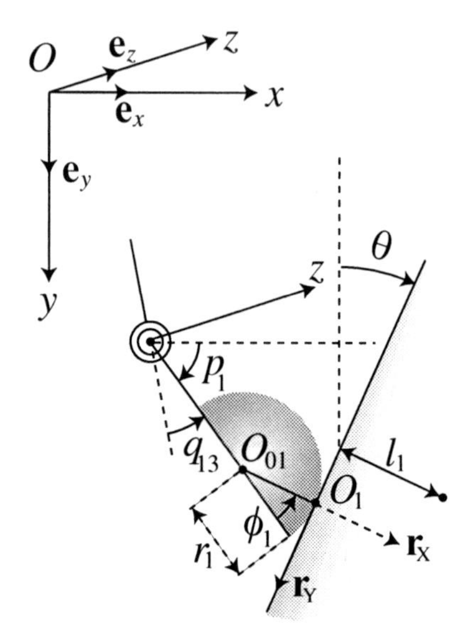

Fig. 5.4. Contact conditions of the rigid object with the left-hand finger-end sphere. The rotational axes of q_{1j} $(j = 1, 2, 3)$, p_1, ϕ_1 and θ are in the direction of the z-axis

5.3 Three-dimensional Rolling Contact Constraints

The derivation of faithful mathematical expressions for the rolling contact constraints between spherical finger-ends and object surfaces was regarded as rather hard to tackle until 2006. Therefore, before constructing mathematical models of 3-D rolling contact constraints, it may be helpful to reconsider the 2-D rolling contact constraints in a physically reasonable way. Thus, it is convenient to see the 2-D setup of rolling contact between a finger-end sphere and a rigid object with a flat surface in the 3-D Euclidean space with orthogonal coordinates $O{-}xyz$ as shown in Figure 5.4. In this figure we assume that the finger has three joints having the same common axis of rotation in the direction of the z-axis and that the axis of rotational movement of the object is also fixed in the z-direction. We also assume that the inertial frame of Figure 5.4 denoted by $O - xyz$ expresses the right-hand orthogonal coordinates and hence all the signs of the angles q_{1i} $(i = 1, 2, 3)$ and θ around the z-axis are taken to be positive in the clockwise direction. At the contact point O_1, define the orthogonal unit vectors $\boldsymbol{r}_X$ and $\boldsymbol{r}_Y$ as shown in Figure 5.4 and also choose a unit vector $\boldsymbol{r}_Z$ in such a way that the frame $(\boldsymbol{r}_X, \boldsymbol{r}_Y, \boldsymbol{r}_Z)$ constitutes the right-hand orthogonal coordinates.

Suppose that the point contact itself between the object and the finger-end sphere is maintained on the xy-plane, that is, the object will not detach from the finger-end. Movement of the contact point O_1 on the circle with radius

r_1 around the centre O_{01} in the xy-plane is induced by both the rotational movement of the finger-end circle around O_{01} with angle p_1 $(= q_{11}+q_{12}+q_{13})$ inside the xy-plane and the rotation of the rigid object in terms of the angle θ around the z-axis. The angular velocity with time rate $\dot{p}_1$ around O_{01} in the xy-plane can be expressed by the vector $\dot{p}_1 \boldsymbol{e}_z$ with $\boldsymbol{e}_z = (0,0,1)^{\mathrm{T}}$ in terms of $O - xyz$. This angular velocity induces the velocity of the contact point O_1 that is expressed as

$$\boldsymbol{v}_{1z} = \dot{p}_1 \boldsymbol{e}_z \times r_1 \boldsymbol{r}_X. \tag{5.10}$$

On the other hand, rotational movement of the xy-plane tangent to the finger-end sphere is expressed by $\dot{\boldsymbol{r}}_X$, the time rate of change of $\boldsymbol{r}_X$. Hence, the difference between these two velocities induces rolling movement of the contact point on the object surface (in this case, the plane perpendicular to the vector $\boldsymbol{r}_X$), which is expressed by

$$r_1 \left\{ -\dot{\boldsymbol{r}}_X + (\boldsymbol{e}_z \times \boldsymbol{r}_X)(\dot{q}_1^{\mathrm{T}} \boldsymbol{e}_1) \right\} = -\dot{Y}_1 \boldsymbol{r}_Y, \tag{5.11}$$

where $\boldsymbol{e}_1 = (1,1,1)^{\mathrm{T}}$. This can be equivalently expressed by

$$r_1 \boldsymbol{r}_Y^{\mathrm{T}} \left\{ -\dot{\boldsymbol{r}}_X + (\boldsymbol{e}_z \times \boldsymbol{r}_X)(\dot{q}_1 \boldsymbol{e}_1) \right\} = -\dot{Y}_1. \tag{5.12}$$

In this planar case, $\boldsymbol{r}_X = (\cos\theta, \sin\theta, 0)^{\mathrm{T}}$ and $\boldsymbol{r}_Y = (-\sin\theta, \cos\theta, 0)^{\mathrm{T}}$ since the clockwise direction of the angle θ in the xy-plane is positive. Hence $\dot{\boldsymbol{r}}_X = \dot{\theta}\boldsymbol{r}_Y$ that corresponds to the equality $\dot{\boldsymbol{r}}_X = \omega_z \boldsymbol{r}_Y - \omega_y \boldsymbol{r}_Z$ with $\omega_z = \dot{\theta}$ and $\omega_y = 0$ as the first column in Equation (5.5). On the other hand, it is easy to see that $\boldsymbol{r}_Y^{\mathrm{T}}(\boldsymbol{e}_z \times \boldsymbol{r}_X) = \boldsymbol{e}_z^{\mathrm{T}}(\boldsymbol{r}_X \times \boldsymbol{r}_Y) = \boldsymbol{e}_z^{\mathrm{T}} \boldsymbol{r}_Z = 1$. Hence, Equation (5.12) is reduced to

$$r_1\{\dot{\theta} - \dot{p}_1\} = \dot{Y}_1, \tag{5.13}$$

which corresponds to the rolling constraint of Equation (4.2) or (4.5) discussed in the case of planar rolling. Unlikely the case of Equation (4.2), $\dot{\theta}$ $(= \omega_z)$ and $\dot{p}_1$ $(= \dot{q}_{11} + \dot{q}_{12} + \dot{q}_{13})$ must be positive when their directions of rotation are clockwise.

Let us now consider the case of the 3-D rolling contact constraint by referring to Figures 5.3 and 5.5. Note that, even if the contact point O_1 slides from the xy-plane and thereby $\boldsymbol{r}_X$ is not inside xy-plane, the velocity vector at O_1 on the left-hand finger-end sphere induced by the net angular velocity $\dot{p}_1 \boldsymbol{e}_z$ is expressed by the form of Equation (5.10). Note that rotational movement of the object surface is also expressed by $\dot{\boldsymbol{r}}_X$, the time rate of change of $\boldsymbol{r}_X$ perpendicular to the object surface. In this 3-D case, note that $\dot{\boldsymbol{r}}_X = \omega_z \boldsymbol{r}_Y - \omega_y \boldsymbol{r}_Z$ according to Equation (5.5). If $r_1 \dot{\boldsymbol{r}}_X$ is coincident with $\boldsymbol{v}_{1z}$ of Equation (5.10), no rolling arises between the left-hand finger-end and object. If $\boldsymbol{v}_{1z}$ minus $r_i \dot{\boldsymbol{r}}_X$ is not zero, rolling of the object on the finger-end sphere arises with the following relative velocity condition:

$$r_1 \left\{ -\dot{\boldsymbol{r}}_X + (\boldsymbol{e}_z \times \boldsymbol{r}_X)\dot{p}_1 \right\} = -\dot{Y}_1 \boldsymbol{r}_Y - \dot{Z}_1 \boldsymbol{r}_Z. \tag{5.14}$$

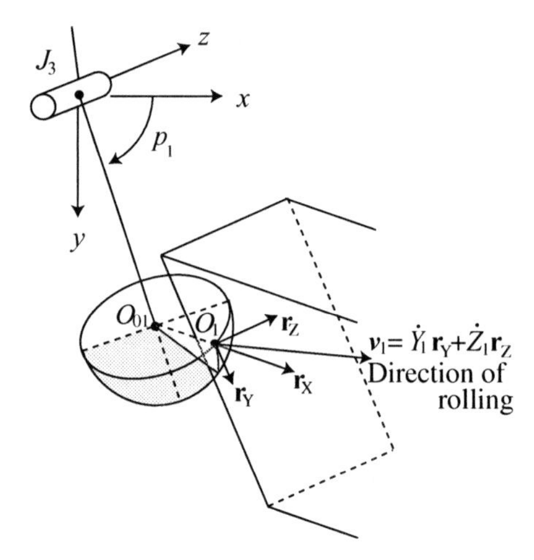

Fig. 5.5. 3-D contact constraint of the rigid object with a flat surface rolling on the left-hand finger-end sphere

This is also expressed componentwise as follows:

$$\begin{cases} r_1 \boldsymbol{r}_Y^{\mathrm{T}} \left\{ -\dot{\boldsymbol{r}}_X + (\boldsymbol{e}_z \times \boldsymbol{r}_X) \dot{q}_1^{\mathrm{T}} \boldsymbol{e}_1 \right\} = -\dot{Y}_1, \\ r_1 \boldsymbol{r}_Z^{\mathrm{T}} \left\{ -\dot{\boldsymbol{r}}_X + (\boldsymbol{e}_z \times \boldsymbol{r}_X) \dot{q}_1^{\mathrm{T}} \boldsymbol{e}_1 \right\} = -\dot{Z}_1, \end{cases} \tag{5.15}$$

which is reduced to

$$\begin{cases} r_1 \left\{ \omega_z - r_{Zz} \dot{p}_1 \right\} = \dot{Y}_1, \\ r_1 \left\{ -\omega_y + r_{Yz} \dot{p}_1 \right\} = \dot{Z}_1. \end{cases} \tag{5.16}$$

Equation (5.14) or equivalently (5.15) or (5.16) can be interpreted as stating that the velocity of the contact point O_1 relative to the finger-end sphere is coincident with the velocity of O_1 on the object surface. We call this the zero-relative-velocity condition. In what follows, we use the symbols $\dot{\theta} = \omega_z$ and $\dot{\psi} = \omega_y$ based on the rule of denoting indefinite integrals of ω_z and ω_y by θ and ψ, respectively.

Next consider the rolling constraint conditions of the object rolling on the right-hand finger-end sphere as shown in Figure 5.6. First, we shall derive the angular velocity vector of O_{02} as the instantaneous axis of rotation of the finger-end sphere originating from the centre O_2. When the first joint rotates around the x-axis with angular velocity $\dot{q}_{20}$, it induces $\dot{q}_{20}\boldsymbol{e}_x$ as one possible component of the axis of rotation of O_{02}. Another component is induced by the net rotational angular velocity $\dot{p}_2$ $(= \dot{q}_{21} + \dot{q}_{22} + \dot{q}_{23} = \dot{q}_2^{\mathrm{T}} \boldsymbol{e}_2)$ around the z'-axis, the z-axis rotated by angle q_{20} around the x-axis, as shown in Figure 5.6, which has the direction $\boldsymbol{e}_{z0} = (0, -\sin q_{20}, \cos q_{20})^{\mathrm{T}}$. This induces the velocity of the contact point on the right-hand finger-end sphere denoted by

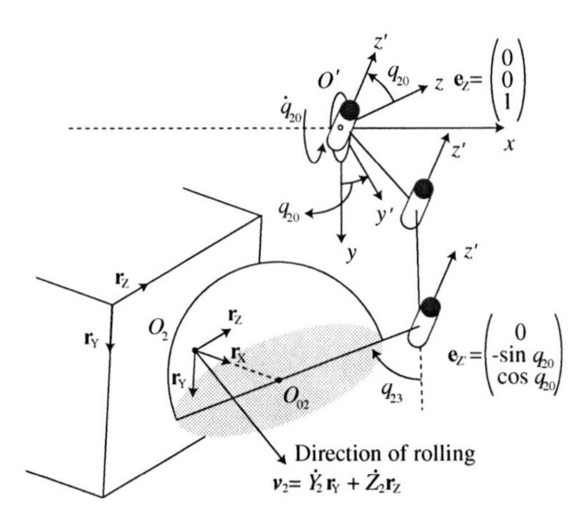

Fig. 5.6. Contact constraint of the object rolling on the right-hand finger-end sphere

$-\dot{p}_2\boldsymbol{e}_{20} \times r_2\boldsymbol{r}_X$. On the other hand, rotation of the right-hand surface of the object can be denoted by $-\dot{\boldsymbol{r}}_X$, whose direction is just reverse to that of $\dot{\boldsymbol{r}}_X$. Thus, rolling of the object on the right-hand finger-end sphere is expressed by the following zero-relative-velocity condition:

$$r_2\left\{\dot{\boldsymbol{r}}_X - (\boldsymbol{e}_{z0} \times \boldsymbol{r}_X)(\dot{q}_2^{\mathrm{T}}\boldsymbol{e}_2) - (\boldsymbol{e}_x \times \boldsymbol{r}_X)\dot{q}_{20}\right\} = -\dot{Y}_2\boldsymbol{r}_Y - \dot{Z}_2\boldsymbol{r}_Z, \quad (5.17)$$

which can be expressed component-wise in the following way:

$$\begin{cases} r_2\boldsymbol{r}_Y^{\mathrm{T}}\left\{\dot{\boldsymbol{r}}_X - (\boldsymbol{e}_{z0} \times \boldsymbol{r}_X)\dot{p}_2 - (\boldsymbol{e}_x \times \boldsymbol{r}_X)\dot{q}_{20}\right\} = -\dot{Y}_2, \\ r_2\boldsymbol{r}_Z^{\mathrm{T}}\left\{\dot{\boldsymbol{r}}_X - (\boldsymbol{e}_{z0} \times \boldsymbol{r}_X)\dot{p}_2 - (\boldsymbol{e}_x \times \boldsymbol{r}_X)\dot{q}_{20}\right\} = -\dot{Z}_2. \end{cases} \quad (5.18)$$

Since $\dot{\boldsymbol{r}}_X = \dot{\theta}\boldsymbol{r}_Y - \dot{\psi}\boldsymbol{r}_Z$ according to Equation (5.5), Equation (5.18) is reduced to

$$\begin{cases} r_2\{-\omega_z + (r_{Zz}\cos q_{20} - r_{Zy}\sin q_{20})\dot{p}_2 + r_{Zx}\dot{q}_{20}\} = \dot{Y}_2, \\ r_2\{\omega_y - (r_{Yz}\cos q_{20} - r_{Yy}\sin q_{20})\dot{p}_2 - r_{Yx}\dot{q}_{20}\} = \dot{Z}_2. \end{cases} \quad (5.19)$$

At this stage, it is important to note that from Equation (5.9) Y_i and Z_i can be expressed as

$$Y_i = (\boldsymbol{x}_{0i} - \boldsymbol{x})^{\mathrm{T}}\boldsymbol{r}_Y, \quad Z_i = (\boldsymbol{x}_{0i} - \boldsymbol{x})^{\mathrm{T}}\boldsymbol{r}_Z, \quad i = 1, 2. \quad (5.20)$$

The rolling constraint conditions expressed through Equations (5.16) and (5.19) are non-holonomic but linear and homogeneous with respect to the velocity variables. Hence, Equations (5.16) and (5.19) can be treated as Pfaffian constraints that can be expressed with accompaning Lagrange multipliers $\{\lambda_{Y1}, \lambda_{Z1}\}$ for Equation (5.16) and $\{\lambda_{Y2}, \lambda_{Z2}\}$ for Equation (5.19) in the forms

$$\begin{cases} \lambda_{Yi}\{\boldsymbol{Y}_{qi}^{\mathrm{T}}\dot{q}_i + \boldsymbol{Y}_{\boldsymbol{x}i}^{\mathrm{T}}\dot{\boldsymbol{x}} + Y_{\theta i}\dot{\theta} + Y_{\psi i}\dot{\psi}\} = 0 \\ \lambda_{Zi}\{\boldsymbol{Z}_{qi}^{\mathrm{T}}\dot{q}_i + \boldsymbol{Z}_{\boldsymbol{x}i}^{\mathrm{T}}\dot{\boldsymbol{x}} + Z_{\theta i}\dot{\theta} + Z_{\psi i}\dot{\psi}\} = 0 \end{cases} \quad i = 1, 2 \tag{5.21}$$

where

$$\begin{cases} \boldsymbol{Y}_{qi} = \dfrac{\partial Y_i}{\partial q_i} - r_i\{(-1)^i(r_{Zz}\cos q_{i0} - r_{Zy}\sin q_{i0})\boldsymbol{e}_i + r_{Zx}\boldsymbol{e}_{0i}\} \\ \boldsymbol{Y}_{\boldsymbol{x}i} = \dfrac{\partial Y_i}{\partial \boldsymbol{x}}, \quad Y_{\theta i} = \dfrac{\partial Y_i}{\partial \theta} + (-1)^i r_i, \quad Y_{\psi i} = \dfrac{\partial Y_i}{\partial \psi} \end{cases} \quad i = 1, 2 \tag{5.22}$$

and

$$\begin{cases} \boldsymbol{Z}_{qi} = \dfrac{\partial Z_i}{\partial q_i} + r_i\{(-1)^i(r_{Yz}\cos q_{i0} - r_{Yy}\sin q_{i0})\boldsymbol{e}_i + r_{Yx}\boldsymbol{e}_{0i}\} \\ \boldsymbol{Z}_{\boldsymbol{x}i} = \dfrac{\partial Z_i}{\partial \boldsymbol{x}}, \quad Z_{\theta i} = \dfrac{\partial Z_i}{\partial \theta}, \quad Z_{\psi i} = \dfrac{\partial Z_i}{\partial \psi} - (-1)^i r_i \end{cases} \quad i = 1, 2 \tag{5.23}$$

and $q_{10} = 0$ and $\boldsymbol{e}_{01}$ denotes the four-dimensional zero vector.

The Lagrangian for the overall fingers–object system can be expressed by the scalar quantity $L = K - P + Q$, where

$$Q = f_1 Q_1 + f_2 Q_2 = 0 \tag{5.24}$$

that corresponds to the holonomic constraints of Equation (5.9) and K denotes the total kinetic energy expressed as

$$\begin{aligned} K = {} & \frac{1}{2}\sum_{i=1,2} \dot{q}_i^{\mathrm{T}} H_i(q_i)\dot{q}_i + \frac{1}{2}M\left(\dot{x}^2 + \dot{y}^2 + \dot{z}^2\right) \\ & + \frac{1}{2}(\omega_z, \omega_y)H_0(\omega_z, \omega_y)^{\mathrm{T}} \end{aligned} \tag{5.25}$$

and P denotes the total potential energy expressed as

$$P = P_1(q_1) + P_2(q_2) - Mgy, \tag{5.26}$$

where $H_i(q_i)$ stands for the inertia matrix for finger i, M the mass of the object, $P_i(q_i)$ the potential energy of finger i, g the gravity constant and H_0 is given by

$$H_0 = \begin{pmatrix} \begin{matrix} I_{ZZ} + \xi_z^2 I_{XX} \\ -2\xi_z I_{ZX} \end{matrix} & \begin{matrix} I_{YZ} + \xi_y\xi_z I_{XX} \\ -\xi_z I_{YX} - \xi_y I_{XZ} \end{matrix} \\ & \\ \begin{matrix} I_{YZ} + \xi_y\xi_z I_{XX} \\ -\xi_z I_{YX} - \xi_y I_{XZ} \end{matrix} & \begin{matrix} I_{YY} + \xi_y^2 I_{XX} \\ -2\xi_y I_{YX} \end{matrix} \end{pmatrix} \tag{5.27}$$

provided that the inertia matrix of the object around its centre of mass $O_{\rm c.m.}$ is expressed as

$$\bar{H} = \begin{pmatrix} I_{XX} & I_{XY} & I_{XZ} \\ I_{XY} & I_{YY} & I_{YZ} \\ I_{XZ} & I_{YZ} & I_{ZZ} \end{pmatrix}. \tag{5.28}$$

It should be noted that the kinetic energy of the object can be expressed as $K_0 = (1/2)\boldsymbol{\omega}_0^{\rm T} H_0 \boldsymbol{\omega}_0 = (1/2)\boldsymbol{\omega}^{\rm T}\bar{H}\boldsymbol{\omega}$ under the non-holonomic constraint of Equation (5.1), where $\boldsymbol{\omega}_0 = (\omega_z, \omega_y)^{\rm T}$.

It should be remarked at this stage that the inertia matrix $\bar{H}$ of the object introduced in Equation (5.28) is not constant but configuration dependent, because the vector $\boldsymbol{\omega}$ of the instantaneous axis of rotation is expressed in terms of the frame coordinates $O-xyz$ (not in terms of the body coordinates $O_{\rm c.m.}-XYZ$). Therefore, if the object inertia matrix is expressed by a constant 3×3 matrix H based on the object coordinates $O_{\rm c.m.}-XYZ$, the kinetic energy of the object should be expressed as $K = (1/2)\bar{\boldsymbol{\omega}}^{\rm T} H \bar{\boldsymbol{\omega}}$, where $\bar{\boldsymbol{\omega}} = (\omega_X, \omega_Y, \omega_Z)^{\rm T}$ expresses the vector of instantaneous axis of rotation in terms of the Cartesian coordinates of the object, that is, $\boldsymbol{\omega} = \omega_X \boldsymbol{r}_X + \omega_Y \boldsymbol{r}_Y + \omega_Z \boldsymbol{r}_Z = R(t)\bar{\boldsymbol{\omega}}$. Hence, $K = (1/2)\boldsymbol{\omega}^{\rm T}(RHR^{\rm T})\boldsymbol{\omega}$ and thereby $\bar{H} = RHR^{\rm T}$. It should be noted, however, that $\bar{H}$ is not dependent on the finger joint angles q_{ij} or the variables x, y, z of the object centre of mass. Partial derivatives of $\bar{H}$ with respect to θ and ψ will be discussed in Section 5.12.

5.4 Lagrange's Equation for the Overall Fingers–Object System

We are now ready to derive Lagrange's equation of motion for the overall fingers–object system depicted in Figure 5.2 under rolling contact constraints. By applying the variational principle for the form

$$\int_{t_0}^{t_1} -\delta L {\rm d}t = \int_{t_0}^{t_1} \left\{ u_1^{\rm T}\delta q_1 + u_2^{\rm T}\delta q_2 + \sum_{i=1,2} \left(\lambda_{Yi} \boldsymbol{Y}_i^{\rm T} + \lambda_{Zi} \boldsymbol{Z}_i^{\rm T} \right) \delta \boldsymbol{X} \right\} {\rm d}t \tag{5.29}$$

where $\boldsymbol{X} = (q_1^{\rm T}, q_2^{\rm T}, \boldsymbol{x}^{\rm T}, \theta, \psi)^{\rm T}$, $\boldsymbol{Y}_1 = (\boldsymbol{Y}_{q1}^{\rm T}, 0_4, \boldsymbol{Y}_{\boldsymbol{x}1}^{\rm T}, Y_{\theta 1}, Y_{\psi 1})$, $\boldsymbol{Y}_2 = (0_3, \boldsymbol{Y}_{q2}^{\rm T}, \boldsymbol{Y}_{\boldsymbol{x}2}^{\rm T}, Y_{\theta 2}, Y_{\psi 2})$, and $\boldsymbol{Z}_1$ and $\boldsymbol{Z}_2$ express similar meanings, we obtain

$$H_i(q_i)\ddot{q}_i + \left\{ \frac{1}{2}\dot{H}_i(q_i) + S_i(q_i, \dot{q}_i) \right\} \dot{q}_i - \frac{\partial}{\partial q_i} K_0 - (-1)^i f_i J_{0i}^{\rm T}(q_i) \boldsymbol{r}_X$$
$$-\lambda_{Yi}\boldsymbol{Y}_{qi} - \lambda_{Zi}\boldsymbol{Z}_{qi} + g_i(q_i) = u_i, \quad i = 1,2 \tag{5.30}$$

$$M\ddot{\boldsymbol{x}} - (f_1 - f_2)\boldsymbol{r}_X - \sum_{i=1,2} \lambda_{Yi} \boldsymbol{Y}_{\boldsymbol{x}i} - \sum_{i=1,2} \lambda_{Zi} \boldsymbol{Z}_{\boldsymbol{x}i} - Mg \begin{pmatrix} 0 \\ 1 \\ 0 \end{pmatrix} = 0, \tag{5.31}$$

Table 5.1. Partial derivatives of holonomic constraints with respect to the position variables

$$\frac{\partial Q}{\partial q_i} = f_i\left(\frac{\partial Q_i}{\partial q_i}\right) = (-1)^i f_i J_i^{\mathrm{T}}(q_i)\boldsymbol{r}_X, \quad i=1,2 \tag{U-1}$$

$$\frac{\partial Q}{\partial \boldsymbol{x}} = f_1\left(\frac{\partial Q_1}{\partial \boldsymbol{x}}\right) + f_2\left(\frac{\partial Q_2}{\partial \boldsymbol{x}}\right) = (f_1 - f_2)\boldsymbol{r}_X \tag{U-2}$$

$$\frac{\partial Q}{\partial \theta} = f_1\frac{\partial Q_1}{\partial \theta} + f_2\frac{\partial Q_2}{\partial \theta} = -f_1 Y_1 + f_2 Y_2 \tag{U-3}$$

$$\frac{\partial Q}{\partial \psi} = f_1\frac{\partial Q_1}{\partial \psi} + f_2\frac{\partial Q_2}{\partial \psi} = f_1 Z_1 - f_2 Z_2 \tag{U-4}$$

$$\boldsymbol{Y}\boldsymbol{q}_i = J_i^{\mathrm{T}}(\boldsymbol{q}_i)\boldsymbol{r}_Y - r_i\left\{(-1)^i(r_{Zz}\cos q_{i0} - r_{Zy}\sin q_{i0})\boldsymbol{e}_i + r_{Zx}\boldsymbol{e}_{0i}\right\} \tag{U-5}$$

$$\boldsymbol{Z}\boldsymbol{q}_i = J_i^{\mathrm{T}}(\boldsymbol{q}_i)\boldsymbol{r}_Z + r_i\left\{(-1)^i(r_{Yz}\cos q_{i0} - r_{Yy}\sin q_{i0})\boldsymbol{e}_i + r_{Yx}\boldsymbol{e}_{0i}\right\} \tag{U-6}$$

$$\boldsymbol{Y}\boldsymbol{x}_i = -\boldsymbol{r}_Y, \quad \boldsymbol{Z}\boldsymbol{x}_i = -\boldsymbol{r}_Z, \quad \text{where} \quad q_{10}=0 \quad \text{and} \quad \boldsymbol{e}_{01}=0 \tag{U-7}$$

$$Y_{\theta i} = \frac{\partial Y_i}{\partial \theta} + (-1)^i r_i = -(-1)^i l_i - \xi_z Z_i \tag{U-8}$$

$$Y_{\psi i} = \frac{\partial Y_i}{\partial \psi} = -\xi_y Z_i, \quad Z_{\theta i} = \frac{\partial Z_i}{\partial \theta} = \xi_z Y_i \tag{U-9}$$

$$Z_{\psi i} = \frac{\partial Z_i}{\partial \psi} - (-1)^i r_i = (-1)^i l_i + \xi_y Y_i \tag{U-10}$$

$$\begin{aligned}\frac{\partial Y_i}{\partial \theta} &= (\boldsymbol{x}_{0i} - \boldsymbol{x})^{\mathrm{T}}\frac{\partial \boldsymbol{r}_Y}{\partial \theta} = (\boldsymbol{x}_{0i} - \boldsymbol{x})^{\mathrm{T}}(-\boldsymbol{r}_X - \xi_z\boldsymbol{r}_Z)\\ &= -(-1)^i(r_i + l_i) - \xi_z Z_i\end{aligned} \tag{U-11}$$

$$\frac{\partial Y_i}{\partial \psi} = (\boldsymbol{x}_{0i} - \boldsymbol{x})^{\mathrm{T}}\frac{\partial \boldsymbol{r}_Y}{\partial \psi} = (\boldsymbol{x}_{0i} - \boldsymbol{x})^{\mathrm{T}}(-\xi_y\boldsymbol{r}_Z) = -\xi_y Z_i \tag{U-12}$$

$$\frac{\partial Z_i}{\partial \theta} = (\boldsymbol{x}_{0i} - \boldsymbol{x})^{\mathrm{T}}\frac{\partial \boldsymbol{r}_Z}{\partial \theta} = (\boldsymbol{x}_{0i} - \boldsymbol{x})^{\mathrm{T}}\xi_z\boldsymbol{r}_Y = \xi_z Y_i \tag{U-13}$$

$$\begin{aligned}\frac{\partial Z_i}{\partial \psi} &= (\boldsymbol{x}_{0i} - \boldsymbol{x})^{\mathrm{T}}\frac{\partial \boldsymbol{r}_Z}{\partial \psi} = (\boldsymbol{x}_{0i} - \boldsymbol{x})^{\mathrm{T}}(\boldsymbol{r}_X + \xi_y\boldsymbol{r}_Y)\\ &= (-1)^i(r_i + l_i) + \xi_y Y_i\end{aligned} \tag{U-14}$$

$$\begin{aligned}&H_0\dot{\boldsymbol{\omega}}_0 + \frac{1}{2}\dot{H}_0\boldsymbol{\omega}_0 + S_0\boldsymbol{\omega}_0 + \sum_{i=1,2}\frac{1}{2}\sum_j\left\{\dot{q}_{ij}\left(\frac{\partial H_0}{\partial q_{ij}}\right)\right\}\boldsymbol{\omega}_0\\ &\quad + \begin{pmatrix} f_1Y_1 - f_2Y_2 \\ -f_1Z_1 + f_2Z_2\end{pmatrix} - \sum_{i=1,2}\begin{pmatrix} Y_{\theta i} & Z_{\theta i} \\ Y_{\psi i} & Z_{\psi i}\end{pmatrix}\begin{pmatrix}\lambda_{Yi} \\ \lambda_{Zi}\end{pmatrix} = 0.\end{aligned} \tag{5.32}$$

The derivation of the partial derivatives of the Lagrangian with respect to $\dot{q}_i$, q_i, $\dot{\boldsymbol{x}}$ and $\boldsymbol{x}$, is straightforward as discussed in Section 4.2 and the results are summarised in Table 5.1. Here, $g_i(q_i) = \{\partial P_i(q_i)/\partial q_i\}$, $i = 1, 2$, and

$$S_0 = \begin{pmatrix} 0 & s_{12} \\ -s_{12} & 0 \end{pmatrix},$$
$$s_{12} = \frac{1}{2}\left(\frac{\partial h_{11}}{\partial \psi} - \frac{\partial h_{12}}{\partial \theta}\right)\omega_z + \frac{1}{2}\left(\frac{\partial h_{12}}{\partial \psi} - \frac{\partial h_{22}}{\partial \theta}\right)\omega_y, \tag{5.33}$$

where we denote the (i, j)-entry of H_0 by h_{ij}. The partial derivatives of h_{11}, h_{12} $(= h_{21})$ and h_{22} in ψ and θ and as well as those of Y_i and Z_i in ψ and θ should be derived in a careful manner so that the resultant formulae do not contradict the non-holonomic constraint of Equation (5.5). The details will be discussed in the last part of this chapter (Section 5.12). Obviously from the variational principle of Equation (5.29), it follows that

$$\int_0^t \left\{ \sum_{i=1,2} \dot{q}_i^{\mathrm{T}} u_i \right\} \mathrm{d}\tau = K(t) + P(t) - K(0) - P(0). \tag{5.34}$$

This relation can be utilised later in the derivation of the passivity for a class of closed-loop dynamics when the control signals u_i are designed in the form of smooth functions of the state variables of robot fingers. Note that the dynamics of the object expressed by Equations (5.31) and (5.32) cannot be controlled directly from the control u_i $(i = 1, 2)$ but must be controlled indirectly through the constraint forces f_i, λ_{Zi} and λ_{Yi} $(i = 1, 2)$. It is also important to note that the position variables θ and ψ do not appear explicitly in Equation (5.32), which expresses the Lagrange equation for rotational motion of the object, because they (θ and ψ) do not appear in the right-hand sides of (U-8) to (U-10) of Table 5.1. Also, note that θ and ψ do not explicitly appear in the partial derivatives of h_{ij} in ψ or θ, as discussed in Section 5.12. All these partial derivatives in ψ and θ can be determined on the basis of non-holonomic constraint expressed by Equation (5.5). Thus, it can be claimed that the overall system dynamics of Equations (5.30), (5.31) and (5.32) together with the non-holonomic constraint of Equation (5.5) do not contradict the causality.

5.5 Physical Meaning of Opposition-based Control Under Rollings

The mechanism of a human-like thumb and index finger illustrated in Figure 5.7 can be roughly represented by the pair of robot fingers shown in Figure 5.2, if fingertips are rigid and their shapes can be approximated by a sphere

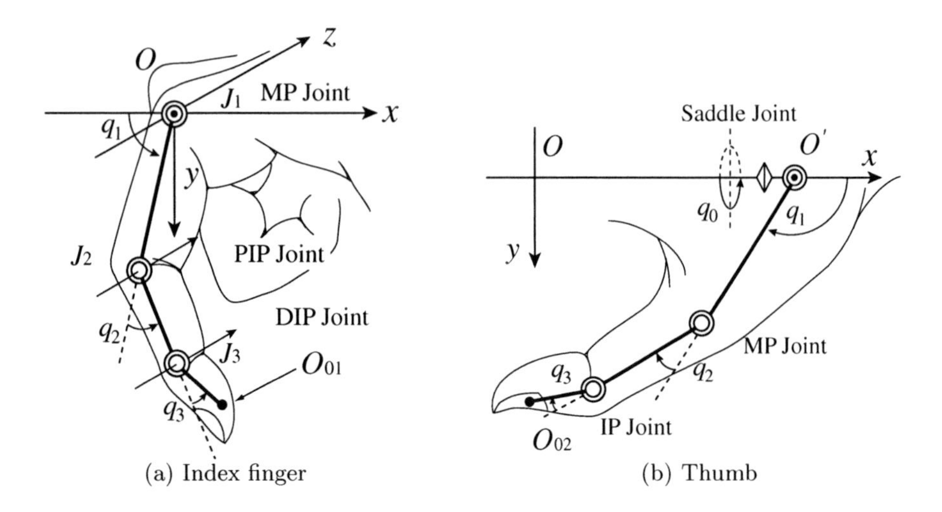

Fig. 5.7. Mechanisms of the human index finger and thumb

around a contact point. The first two joints of the thumb constitute a saddle joint that has two axes of rotation, one axis with joint angle q_0 is fixed and assumed to direct in the x-axis and the other axis with joint angle q_1 is changeable depending on q_0 inside the yz-plane but always orthogonal to the x-axis. More precisely, the second joint with angle q_1 has its axis of rotation in the direction expressed by the vector $\boldsymbol{z}' = (0, -\sin q_2, \cos q_2)^{\mathrm{T}}$ in terms of $O' - xyz$. The other two MP and IP joints, with angles q_2 and q_3, have the same axis of rotation as the second joint of q_1. The index finger is regarded as a planar finger robot because all three joints have a common and fixed axis of rotation in the z-direction but the contact point with a rigid object can move on the finger-end sphere and therefore the contact point can move in a three-dimensional region. Thus, the pinching motion of the robot fingers–object system depicted in Figure 5.2 can be regarded as characterising the essentials of human-like pinching by using a thumb and index finger combination. Then, what control signal can best realise stable precision prehension? It must be designed on the basis of fingers–thumb opposability even in this three-dimensional case.

In order to gain an in-depth insignt into the physical meanings of each term of the motion equations of the fingers–object system, it is necessary to spell out Lagrange's equations of motion of Equations (5.30–5.32) in more explicit forms as in the following:

$$L_1(q_1, \dot{q}_1, \ddot{q}_1, \boldsymbol{\omega}_0) + f_1 J_{01}^{\mathrm{T}}(q_1)\boldsymbol{r}_X - \lambda_{Y1}\{J_{01}^{\mathrm{T}}(q_1)\boldsymbol{r}_Y + r_1 r_{Zz}\boldsymbol{e}_1\} \\ -\lambda_{Z1}\{J_{01}^{\mathrm{T}}(q_1)\boldsymbol{r}_Z - r_1 r_{Yz}\boldsymbol{e}_1\} + g_1(q_1) = u_1 \quad (5.35)$$

$$
\begin{aligned}
&L_2(q_2, \dot{q}_2, \ddot{q}_2, \boldsymbol{\omega}_0) - f_2 J_{02}^{\mathrm{T}}(q_2) \boldsymbol{r}_X \\
&\quad -\lambda_{Y2}\{J_{02}^{\mathrm{T}}(q_2)\boldsymbol{r}_Y - r_2(r_{Zz}\cos q_{20} - r_{Zy}\sin q_{20})\boldsymbol{e}_2 + r_{Zx}\boldsymbol{e}_{02}\} \\
&\quad -\lambda_{Z2}\{J_{02}^{\mathrm{T}}(q_2)\boldsymbol{r}_Z + r_2(r_{Yz}\cos q_{20} - r_{Yy}\sin q_{20})\boldsymbol{e}_2 + r_{Yx}\boldsymbol{e}_{02}\} \\
&\qquad\qquad\qquad\qquad\qquad\qquad\qquad\qquad + g_2(q_2) = u_2, \quad (5.36)
\end{aligned}
$$

where

$$
\begin{aligned}
&L_i(q_i, \dot{q}_i, \ddot{q}_i, \boldsymbol{\omega}_0) \\
&\quad = H_i(q_i)\ddot{q}_i + \left\{\frac{1}{2}\dot{H}_i(q_i) + S_i(q_i, \dot{q}_i)\right\}\dot{q}_i - \frac{\partial}{\partial q_i}\left(\frac{1}{2}\boldsymbol{\omega}_0^{\mathrm{T}} H_0 \boldsymbol{\omega}_0\right), \quad i = 1, 2 \quad (5.37)
\end{aligned}
$$

and

$$
M\ddot{\boldsymbol{x}} - (f_1 - f_2)\boldsymbol{r}_X + (\lambda_{Y1} + \lambda_{Y2})\boldsymbol{r}_Y + (\lambda_{Z1} + \lambda_{Z2})\boldsymbol{r}_Z - Mg\begin{pmatrix} 0 \\ 1 \\ 0 \end{pmatrix} = 0, \quad (5.38)
$$

$$
\begin{aligned}
&L_0(q_1, q_2, \dot{q}_1, \dot{q}_2, \boldsymbol{\omega}_0, \dot{\boldsymbol{\omega}}_0) + \begin{pmatrix} f_1 Y_1 - f_2 Y_2 \\ -f_1 Z_1 + f_2 Z_2 \end{pmatrix} - \begin{pmatrix} l_1 \lambda_{Y1} - l_2 \lambda_{Y2} \\ -l_1 \lambda_{Z1} + l_2 \lambda_{Z2} \end{pmatrix} \\
&\qquad + \begin{pmatrix} \xi_z \\ \xi_y \end{pmatrix} (Z_1 \lambda_{Y1} + Z_2 \lambda_{Y2} - Y_1 \lambda_{Z1} - Y_2 \lambda_{Z2}) = 0. \quad (5.39)
\end{aligned}
$$

The physical meaning of the leading terms L_1, L_2, $M\ddot{\boldsymbol{x}}$ and L_0 becomes apparent if we notice the basic relation:

$$
\begin{aligned}
&\sum_{i=1,2} \dot{q}_i^{\mathrm{T}} L_i(q_i, \dot{q}_i, \ddot{q}_i, \boldsymbol{\omega}_0) + \dot{\boldsymbol{x}}^{\mathrm{T}} M\ddot{\boldsymbol{x}} + \boldsymbol{\omega}_0^{\mathrm{T}} L_0 \\
&\quad = \frac{\mathrm{d}}{\mathrm{d}t}\left\{\sum_{i=1,2} \frac{1}{2}\dot{q}_i^{\mathrm{T}} H_i(q_i)\dot{q}_i + \frac{M}{2}\|\dot{\boldsymbol{x}}\|^2 + \frac{1}{2}\boldsymbol{\omega}_0^{\mathrm{T}} H_0 \boldsymbol{\omega}_0\right\} \\
&\quad = \frac{\mathrm{d}}{\mathrm{d}t} K \qquad\qquad\qquad\qquad\qquad\qquad\qquad\qquad\qquad\qquad (5.40)
\end{aligned}
$$

Equation (5.38) explicitly expresses Newton's second law of motion concerning translational motion of the object. The second and third terms on the left-hand side of Equation (5.39) express rotational moments that are evoked by rolling between the object and finger-ends and affect rotation of the object represented by $\boldsymbol{\omega}_0$. The last term on the left-hand side of Equation (5.39) comes from the implicit effect of rotational moments of spinning around the $\boldsymbol{r}_X$-axis.

Now, for the sake of gaining physical insight into the problem of control for precision prehension, we consider the elementary situation in a state of weightlessness as if the pair of robot fingers is pinching an object in an artificial satellite without the effect of gravity. In this case, the gravity terms $g_i(q_i)$

$(i = 1, 2)$ in Equations (5.35) and (5.36) are missing and the term $Mg(0, 1, 0)^{\mathrm{T}}$ in Equation (5.38) is also missing. Then, let us consider the control signals defined as

$$u_i = -C_i \dot{q}_i + (-1)^i \frac{f_d}{r_1 + r_2} J_{0i}^{\mathrm{T}}(q_i)(\boldsymbol{x}_{01} - \boldsymbol{x}_{02}), \quad i = 1, 2, \tag{5.41}$$

where C_i $(i = 1, 2)$ denote a positive definite matrix for damping and f_d a positive constant appropriately chosen to determine the magnitude of the fingers–thumb opposing force. Note that the right-hand side of Equation (5.41) can be calculated from knowledge of the finger kinematics and measurement data of finger joint angles alone. In other words, the control signals u_i can be constructed without using the kinematics of the object or measured data from external sensors like visual or tactile sensing.

Now, it is important to note that taking inner products between u_i of Equation (5.41) and $\dot{q}_i$ for $i = 1, 2$ and summing these two products yields

$$\sum_{i=1,2} \dot{q}_i^{\mathrm{T}} u_i = -\sum_{i=1,2} \frac{\mathrm{d}}{\mathrm{d}t} \left\{ \frac{f_d}{2(r_1 + r_2)} \|\boldsymbol{x}_{01} - \boldsymbol{x}_{02}\|^2 \right\} - \sum_{i=1,2} \dot{q}_i^{\mathrm{T}} C_i \dot{q}_i. \tag{5.42}$$

Since from Equation (5.20) it follows that

$$\begin{cases} Y_1 - Y_2 = (\boldsymbol{x}_{01} - \boldsymbol{x}_{02})^{\mathrm{T}} \boldsymbol{r}_Y \\ Z_1 - Z_2 = (\boldsymbol{x}_{01} - \boldsymbol{x}_{02})^{\mathrm{T}} \boldsymbol{r}_Z \end{cases} \tag{5.43}$$

it is easy to see that

$$f_0 \boldsymbol{r}_X + \frac{f_d}{r_1 + r_2}(\boldsymbol{x}_{01} - \boldsymbol{x}_{02}) = \frac{f_d(Y_1 - Y_2)}{r_1 + r_2} \boldsymbol{r}_Y + \frac{f_d(Z_1 - Z_2)}{r_1 + r_2} \boldsymbol{r}_Z \tag{5.44}$$

$$\|\boldsymbol{x}_{01} - \boldsymbol{x}_{02}\|^2 = l_w^2 + (Y_1 - Y_2)^2 + (Z_1 - Z_2)^2, \tag{5.45}$$

where l_w and f_0 are positive constants defined as

$$l_w = l_1 + l_2 + r_1 + r_2, \quad f_0 = \left(1 + \frac{l_1 + l_2}{r_1 + r_2}\right) f_d. \tag{5.46}$$

Hence, by substituting Equation (5.42) into Equation (5.34) and referring to Equations (5.44) and (5.45), we obtain

$$\frac{\mathrm{d}}{\mathrm{d}t} E = -\sum_{i=1,2} \dot{q}_i^{\mathrm{T}} C_i \dot{q}_i, \tag{5.47}$$

where

$$E = K + \frac{f_d}{2(r_1 + r_2)} \left\{ (Y_1 - Y_2)^2 + (Z_1 - Z_2)^2 \right\} \tag{5.48}$$

and K denotes the kinetic energy defined by Equation (5.25).

5.6 Stability of Blind Grasping under the Circumstances of Weightlessness

Under the circumstances of weightlessness explained in the previous section, the closed-loop dynamics of the fingers–object system when the control signals of Equation (5.41) are applied can be expressed as

$$
\begin{aligned}
&L_i(q_i, \dot{q}_i, \ddot{q}_i, \boldsymbol{\omega}_O) - (-1)^i J_{0i}^{\mathrm{T}}(q_i)\Delta f_i \boldsymbol{r}_X + C_i\dot{q}_i \\
&\quad - (-1)^i J_{0i}^{\mathrm{T}}(q_i)\frac{f_d}{r_1+r_2}\{(Y_1-Y_2)\boldsymbol{r}_Y + (Z_1-Z_2)\boldsymbol{r}_Z\} \\
&\quad - \lambda_{Yi}\left\{J_{0i}^{\mathrm{T}}(q_i)\boldsymbol{r}_Y - r_i\left((-1)^i r_Z(q_{i0})\boldsymbol{e}_i + r_{Zx}\boldsymbol{e}_{0i}\right)\right\} \\
&\quad - \lambda_{Zi}\left\{J_{0i}^{\mathrm{T}}(q_i)\boldsymbol{r}_Z + r_i\left((-1)^i r_Y(q_{i0})\boldsymbol{e}_i + r_{Yx}\boldsymbol{e}_{0i}\right)\right\} = 0, \quad i = 1, 2, \qquad (5.49)
\end{aligned}
$$

where $\boldsymbol{e}_{01} = 0$, $q_{10} = 0$, $\Delta f_i = f_i - f_0$, and

$$
\begin{cases}
r_Z(q_{i0}) = r_{Zz}\cos q_{i0} - r_{Zy}\sin q_{i0}, \\
r_Y(q_{i0}) = r_{Yz}\cos q_{i0} - r_{Yy}\sin q_{i0},
\end{cases} \qquad (5.50)
$$

$$
M\ddot{\boldsymbol{x}} - (\Delta f_1 - \Delta f_2)\boldsymbol{r}_X + (\lambda_{Y1} + \lambda_{Y2})\boldsymbol{r}_Y + (\lambda_{Z1} + \lambda_{Z2})\boldsymbol{r}_Z = 0 \qquad (5.51)
$$

and

$$
\begin{aligned}
L_O &+ \begin{pmatrix} \Delta f_1 Y_1 - \Delta f_2 Y_2 \\ -\Delta f_1 Z_1 + \Delta f_2 Z_2 \end{pmatrix} + f_0 \begin{pmatrix} Y_1 - Y_2 \\ -Z_1 + Z_2 \end{pmatrix} - \begin{pmatrix} l_1\lambda_{Y1} - l_2\lambda_{Y2} \\ -l_1\lambda_{Z1} + l_2\lambda_{Z2} \end{pmatrix} \\
&+ \begin{pmatrix} \xi_z \\ \xi_y \end{pmatrix}(Z_1\lambda_{Y1} + Z_2\lambda_{Y2} - Y_1\lambda_{Z1} - Y_2\lambda_{Z2}) = 0. \qquad (5.52)
\end{aligned}
$$

We now show a computer simulation result based upon the mathematical model of the closed-loop dynamics described above with physical parameters given in Table 5.2 and the parameters of the control signals in Table 5.3, where $C_1 = \mathrm{diag}(c_1, c_1, c_1)$ and $C_2 = \mathrm{diag}(c_{20}, c_2, c_2, c_2)$. The transient responses of the key variables of the dynamics are shown in Figure 5.8. Apparently from the figure we see that $Y_1 - Y_2 \to 0$, $Z_1 - Z_2 \to 0$, $f_i \to f_0$, $\lambda_{Yi} \to 0$ and $\lambda_{Zi} \to 0$ $(i = 1, 2)$ as $t \to \infty$. It is also noticeable that the set of closed-loop equations expressed through Equations (5.49–5.52) has a special equilibrium solution expressed as

$$
\begin{cases}
Y_1 = Y_2, \quad Z_1 = Z_2, \\
f_1 = f_2 = f_0, \quad \lambda_{Y1} = \lambda_{Y2} = 0, \quad \lambda_{Z1} = \lambda_{Z2} = 0, \\
\dot{q}_1 = 0, \quad \dot{q}_2 = 0, \quad \dot{\boldsymbol{x}} = 0, \quad \omega_y = 0, \quad \omega_z = 0.
\end{cases} \qquad (5.53)
$$

Evidently this satisfies the closed-loop dynamics of Equations (5.49–5.50). It should be noted that there is an infinite number of equilibrium states satisfying Equation (5.53), which constitutes a four-dimensional manifold in the

Table 5.2. Physical parameters of the fingers and object

$l_{11} = l_{21}$	length	0.040 [m]
$l_{12} = l_{22}$	length	0.040 [m]
$l_{13} = l_{23}$	length	0.030 [m]
m_{11}	weight	0.043 [kg]
m_{12}	weight	0.031 [kg]
m_{13}	weight	0.020 [kg]
l_{20}	length	0.000 [m]
m_{20}	weight	0.000 [kg]
m_{21}	weight	0.060 [kg]
m_{22}	weight	0.031 [kg]
m_{23}	weight	0.020 [kg]
I_{XX11}	inertia moment	5.375×10^{-7}[kgm^2]
$I_{YY11} = I_{ZZ11}$	inertia moment	6.002×10^{-6}[kgm^2]
I_{XX12}	inertia moment	3.875×10^{-7}[kgm^2]
$I_{YY12} = I_{ZZ12}$	inertia moment	4.327×10^{-6}[kgm^2]
I_{XX13}	inertia moment	2.500×10^{-7}[kgm^2]
$I_{YY13} = I_{ZZ13}$	inertia moment	1.625×10^{-6}[kgm^2]
I_{XX21}	inertia moment	7.500×10^{-7}[kgm^2]
$I_{YY21} = I_{ZZ21}$	inertia moment	8.375×10^{-6}[kgm^2]
I_{XX22}	inertia moment	3.875×10^{-7}[kgm^2]
$I_{YY22} = I_{ZZ22}$	inertia moment	4.327×10^{-6}[kgm^2]
I_{XX23}	inertia moment	2.500×10^{-7}[kgm^2]
$I_{YY23} = I_{ZZ23}$	inertia moment	1.625×10^{-6}[kgm^2]
$I_{XX} = I_{ZZ}$	inertia moment(object)	1.133×10^{-5}[kgm^2]
I_{YY}	inertia moment(object)	6.000×10^{-6}[kgm^2]
r_0	link radius	0.005 [m]
$r_i(i = 1, 2)$	radius	0.010 [m]
L	base length	0.063 [m]
M	object weight	0.040 [kg]
$l_i(i = 1, 2)$	object width	0.015 [m]
h	object height	0.050 [m]

Table 5.3. Parameters of the control signals

f_d	internal force	1.000 [N]
$c_1 = c_2$	damping coefficient	0.001 [Nms]
c_{20}	damping coefficient	0.006 [Nms]

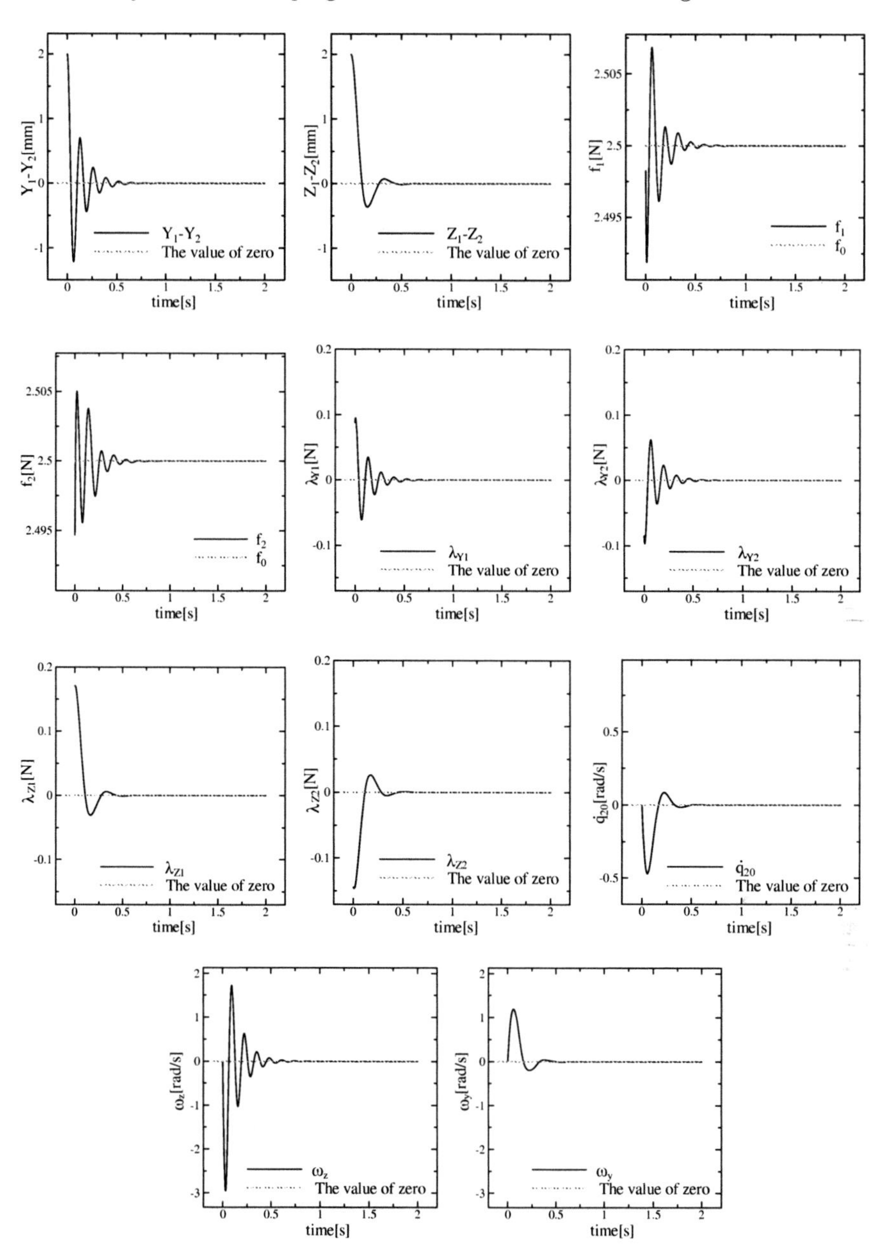

Fig. 5.8. The transient responses of the physical variables including all contact constraint forces

six-dimensional configuration manifold CM^6, because there are 12 position variables but two holonomic constraints and four constraints among infinitesimal displacements due to the four Pfaffian constraints. Note that $E = 0$ on this equilibrium manifold. Since E includes only two positive quadratic terms of position variables, $(Y_1 - Y_2)^2$ and $(Z_1 - Z_2)^2$, the scalar quantity E is not a Lyapunov function for the closed-loop dynamics of Equations (5.49–5.52). Nevertheless, E is non-negative definite with respect to the state variables even under these constraints, but is positive definite with respect to the velocity variables. Further, the time rate of E is only non-positive definite, but it is possible to show that $\dot{q}_i(t) \to 0$ as $t \to \infty$ for $i = 1, 2$ implies, by virtue of the rolling constraints, that $\dot{\boldsymbol{x}}(t) \to 0$, $\omega_z(t) \to 0$, $\omega_y(t) \to 0$ as $t \to \infty$. We see from Figure 5.8 that the transient responses of $Z_1 - Z_2$, λ_{Zi} $(i = 1, 2)$ and ω_y are well behaved and comparable with those of the corresponding physical variables $Y_1 - Y_2$, λ_{Yi} $(i = 1, 2)$, and ω_z. Note that such mild responses of λ_{Zi} $(i = 1, 2)$ and ω_y can be obtained by setting a larger damping factor $c_{20} = 0.006$ [Nms] for the single joint q_{20} rotating around the x-axis than other damping factors $c_i = 0.001$ [Nms] $(i = 1, 2)$ for other joints q_{ij} $(i = 1, 2,\ j = 1, 2, 3)$. This simulation was started from a still state $(\dot{q}_{ij}(0) = 0 (i = 1, 2,\ j = 1, 2, 3),\ \omega_z(0) = 0,\ \omega_y(0) = 0,\ \dot{\boldsymbol{x}}(0) = 0)$ but two variables $Y_1 - Y_2$ and $Z_1 - Z_2$ are set at the initial time.

We now prove the convergence of a solution of Equations (5.49–5.52) to some equilibrium state satisfying Equation (5.53) at $t \to \infty$ under the assumption that the object width $l_1 + l_2$ is of the same order as the radius r_i of each finger-end sphere. First note that from Equation (5.47) the non-negative function E is non-increasing with t and hence both angular velocities $\dot{q}_i(t)$ $(i = 1, 2)$ belong to $L^2(0, \infty)$. Then, by applying a similar argument developed to that in the case of planar pinching in the previous chapter, it is possible to show that $\ddot{q}_i(t)$ for $i = 1, 2$ become bounded and therefore $\dot{q}_i(t)$ for $i = 1, 2$ are continuous uniformly in t provided that the following 6×12 Jacobian matrix A is of full rank:

$$A^{\mathrm{T}} = \begin{pmatrix} -J_{01}^{\mathrm{T}}\boldsymbol{r}_X & 0_3 & J_{01}^{\mathrm{T}}\boldsymbol{r}_Y + r_1 r_{Zz}\boldsymbol{e}_1 & 0_3 & J_{01}^{\mathrm{T}}\boldsymbol{r}_Z - r_1 r_{Yz}\boldsymbol{e}_1 & 0_3 \\ 0_4 & J_{02}^{\mathrm{T}}\boldsymbol{r}_X & 0_4 & J_{02}^{\mathrm{T}}\boldsymbol{r}_Y - r_2 r_Z(q_{20})\boldsymbol{e}_2 - r_2 r_{Zx}\boldsymbol{e}_{02} & 0_4 & J_{02}^{\mathrm{T}}\boldsymbol{r}_Z + r_2 r_Y(q_{20})\boldsymbol{e}_2 + r_2 r_{Yx}\boldsymbol{e}_{02} \\ \boldsymbol{r}_X & -\boldsymbol{r}_X & -\boldsymbol{r}_Y & -\boldsymbol{r}_Y & -\boldsymbol{r}_Z & -\boldsymbol{r}_Z \\ -Y_1 & Y_2 & l_1 - \xi_z Z_1 & -l_2 - \xi_z Z_2 & \xi_z Y_1 & \xi_z Y_2 \\ Z_1 & -Z_2 & -\xi_y Z_1 & -\xi_y Z_2 & -l_1 + \xi_y Y_1 & l_2 + \xi_y Y_2 \end{pmatrix}. \quad (5.54)$$

Next, it is important to note that differentiation of Equation (5.43) with respect to time t leads to

$$\begin{aligned}\dot{Y}_1 - \dot{Y}_2 &= (\dot{\boldsymbol{x}}_{01} - \dot{\boldsymbol{x}}_{02})^{\mathrm{T}} \boldsymbol{r}_Y + (\boldsymbol{x}_{01} - \boldsymbol{x}_{02})^{\mathrm{T}} \dot{\boldsymbol{r}}_Y \\ &= (\dot{\boldsymbol{x}}_{01} - \dot{\boldsymbol{x}}_{02})^{\mathrm{T}} \boldsymbol{r}_Y + \{l_w - \xi_z (Z_1 - Z_2)\}\omega_z - \xi_y (Z_1 - Z_2)\omega_y, \end{aligned} \tag{5.55}$$

where the second equality is derived by referring to Equations (5.2), (5.5), (5.9) and (5.43). Analogously, it follows from Equation (5.43) that

$$\dot{Z}_1 - \dot{Z}_2 = (\dot{\boldsymbol{x}}_{01} - \dot{\boldsymbol{x}}_{02}) \boldsymbol{r}_Z + \{-l_w + \xi_y (Y_1 - Y_2)\omega_y\} + \xi_y (Y_1 - Y_2)\omega_z. \tag{5.56}$$

Note that $\dot{\boldsymbol{x}}_{0i}(t) \to 0$ for $i = 1, 2$ as $t \to \infty$ because $\dot{q}_i(t) \to 0$ for $i = 1, 2$ as $t \to \infty$. On the other hand, the rolling contact constraints expressed by Equations (5.16) and (5.19) mean that

$$\begin{aligned}\dot{Y}_1 - \dot{Y}_2 &= r_1\{\omega_z - r_{Zz}\dot{p}_1\} - r_2\{-\omega_z + r_Z(q_{20})\dot{p}_2 + r_{Zx}\dot{q}_{20}\} \\ &= (r_1 + r_2)\omega_z - h_z(\dot{p}_1, \dot{p}_2, \dot{q}_{20}), \end{aligned} \tag{5.57}$$

$$\dot{Z}_1 - \dot{Z}_2 = -(r_1 + r_2)\omega_y + h_y(\dot{p}_1, \dot{p}_2, \dot{q}_{20}), \tag{5.58}$$

where

$$\begin{cases} h_z = r_1 r_{Zz}\dot{p}_1 + r_2\{r_Z(q_{20})\dot{p}_2 + r_{Zx}\dot{q}_{20}\}, \\ h_y = r_1 r_{Yz}\dot{p}_1 + r_2\{r_Y(q_{20})\dot{p}_2 + r_{Yx}\dot{q}_{20}\}. \end{cases} \tag{5.59}$$

Comparing these equations with Equations (5.55) and (5.56), we obtain the following relation:

$$\begin{pmatrix} l_1 + l_2 - \xi_z(Z_1 - Z_2) & -\xi_y(Z_1 - Z_2) \\ \xi_z(Y_1 - Y_2) & -(l_1 + l_2) + \xi_y(Y_1 - Y_2) \end{pmatrix} \begin{pmatrix} \omega_z \\ \omega_y \end{pmatrix} = \begin{pmatrix} -h_z + (\dot{\boldsymbol{x}}_{01} - \dot{\boldsymbol{x}}_{02})^{\mathrm{T}} \boldsymbol{r}_Y \\ h_y + (\dot{\boldsymbol{x}}_{01} - \dot{\boldsymbol{x}}_{02})^{\mathrm{T}} \boldsymbol{r}_Z \end{pmatrix}. \tag{5.60}$$

Since the magnitudes of $Y_1 - Y_2$ and $Z_1 - Z_2$ can remain arbitrarily small by selecting an arbitrary small value $E(0)$ for the value of the total energy function E in Equation (5.48) at the initial time $t = 0$, the absolute values of both ξ_y and ξ_z remain of $O(1)$, the diagonal entries of the 2×2 coefiicient matrix of the left-hand side of Equation (5.60) are dominantly approximated by the matrix $\mathrm{diag}(l_1 + l_2, -l_1 - l_2)$. Further, since all velocities $\dot{\boldsymbol{x}}_{0i}$, $\dot{p}_i$ for $i = 1, 2$ and $\dot{q}_{20}$, and therefore both h_z and h_y, tend to vanish as $t \to \infty$, it is evidently seen from Equation (5.60) that

$$\omega_y \to 0 \quad \text{and} \quad \omega_z \to 0 \quad \text{as} \quad t \to \infty. \tag{5.61}$$

Then, it is easy to check from the rolling constraints expressed by Equations (5.16) and (5.19) that $\dot{Y}_i \to 0$ and $\dot{Z}_i \to 0$ as $t \to \infty$ for $i = 1, 2$. Finally, it is possible to see from differentiation of Equation (5.20) with respect to t that $\dot{\boldsymbol{x}} \to 0$ as $t \to \infty$. Since all the velocity variables are uniformly continuous in t and tend to zero as $t \to \infty$, all the acceleration variables $\ddot{\boldsymbol{x}}(t)$, $\dot{\omega}_y(t)$ and $\dot{\omega}_z(t)$ tend to zero as $t \to \infty$ according to Lemma 1 of Appendix A. Thus, it is possible to conclude from Equations (5.49–5.52) that

$$-A^{\mathrm{T}}\Delta\boldsymbol{\lambda} + B^{\mathrm{T}}\Delta\boldsymbol{m} \to 0_{12} \quad \text{as} \quad t \to \infty, \tag{5.62}$$

where

$$\Delta\boldsymbol{\lambda} = (\Delta f_1, \Delta f_2, \lambda_{Y1}, \lambda_{Y2}, \lambda_{Z1}, \lambda_{Z2})^{\mathrm{T}}, \tag{5.63}$$

$$\Delta\boldsymbol{m} = (Y_1 - Y_2, Z_1 - Z_2)^{\mathrm{T}}, \tag{5.64}$$

$$B = \begin{pmatrix} \dfrac{f_d}{r_1+r_2}\boldsymbol{r}_Y^{\mathrm{T}} J_{01}, & -\dfrac{f_d}{r_1+r_2}\boldsymbol{r}_Y^{\mathrm{T}} J_{02}, & 0_3, & f_0, & 0 \\ \dfrac{f_d}{r_1+r_2}\boldsymbol{r}_Z^{\mathrm{T}} J_{01}, & -\dfrac{f_d}{r_1+r_2}\boldsymbol{r}_Z^{\mathrm{T}} J_{02}, & 0_3, & 0, & -f_0 \end{pmatrix}. \tag{5.65}$$

Since the 12×8 matrix $[A^{\mathrm{T}}, B^{\mathrm{T}}]$ is of full rank, Equation (5.62) implies that

$$Y_1 - Y_2 \to 0, \quad Z_1 - Z_2 \to 0, \quad \Delta\boldsymbol{\lambda} \to 0 \tag{5.66}$$

as $t \to \infty$. This concludes that the solution trajectory of the closed-loop system consisting of Equations (5.49–5.52) together with the constraints expressed by Equations (5.9), (5.16) and (5.19) converges asymptotically to the four-dimensional equilibrium manifold on which the condition of Equation (5.53) is satisfied. Unfortunately, asymptotic convergence of each individual finger joint trajectory q_{ij} ($i = 1, 2,\ j = 1, 2, 3$) and q_{20} to each corresponding constant has not yet been proved by the argument developed above. This will be treated in a more general situation in the next two sections.

5.7 Control for Stable Blind Grasping

Even if gravity affects the motion of the object and fingers as shown in Figures 5.1 and 5.2, it is possible to construct a control signal for stable grasping in a dynamic sense without knowing the object kinematics or using external sensing (visual or tactile). The signal is defined as

$$u_i = g_i(q_i) - C_i\dot{q}_i + \frac{(-1)^i f_d}{r_1+r_2} J_i^{\mathrm{T}}(q_i)(\boldsymbol{x}_{01} - \boldsymbol{x}_{02})$$
$$-\frac{\hat{M}g}{2}\frac{\partial y_{0i}}{\partial q_i} - r_i\hat{N}_i\boldsymbol{e}_i - r_i\hat{N}_{0i}\boldsymbol{e}_{0i}, \quad i = 1, 2, \tag{5.67}$$

where $\boldsymbol{e}_{01} = 0$ and $\boldsymbol{e}_{02} = (1, 0, 0, 0)^{\mathrm{T}}$, and

$$\hat{M}(t) = \hat{M}(0) + (g/2\gamma_M)\sum_{i=1,2}(y_{0i}(t) - y_{0i}(0)), \tag{5.68}$$

$$\hat{N}_i(t) = \hat{N}_i(0) + (r_i/\gamma_i)\sum_{j=1}^{3}\{q_{ij}(t) - q_{ij}(0)\},\ i = 1, 2, \tag{5.69}$$

$$\hat{N}_{02}(t) = \hat{N}_{02}(0) + (r_2/\gamma_{02})\{q_{20}(t) - q_{20}(0)\}. \tag{5.70}$$

Note that $\hat{M}(t)$ plays the role of an estimator for the unknown object mass M. The signals $\hat{N}_i(t)$ and $\hat{N}_{0i}(t)$ are not estimators but play an important role in suppressing excess rotation of the finger joints together with the damping terms $-C_i\dot{q}_i$. Next, note that from Equation (5.9) it follows that

$$\boldsymbol{x}_{01} - \boldsymbol{x}_{02} = -l_w \boldsymbol{r}_X + (Y_1 - Y_2)\boldsymbol{r}_Y + (Z_1 - Z_2)\boldsymbol{r}_Z \tag{5.71}$$

$$\boldsymbol{x} - \frac{1}{2}(\boldsymbol{x}_{01} + \boldsymbol{x}_{02}) = \frac{l_0}{2}\boldsymbol{r}_X - \frac{Y_1 + Y_2}{2}\boldsymbol{r}_Y - \frac{Z_1 + Z_2}{2}\boldsymbol{r}_Z, \tag{5.72}$$

where

$$\begin{cases} l_w = r_1 + r_2 + l_1 + l_2, \\ l_0 = (r_1 - r_2) + (l_1 - l_2). \end{cases} \tag{5.73}$$

It is also important to note that

$$MgR(t)\begin{pmatrix} r_{Xy} \\ r_{Yy} \\ r_{Zy} \end{pmatrix} - Mg\begin{pmatrix} 0 \\ 1 \\ 0 \end{pmatrix} = 0, \tag{5.74}$$

where r_{Xy}, r_{Yy} or r_{Zy} denotes the y-component of the vectors $\boldsymbol{r}_X$, $\boldsymbol{r}_Y$ or $\boldsymbol{r}_Z$ respectively. Finally, it is necessary to define

$$\begin{cases} \Delta\boldsymbol{\lambda} = [\Delta f_1, \Delta f_2, \Delta\lambda_{Y1}, \Delta\lambda_{Y2}, \Delta\lambda_{Z1}, \Delta\lambda_{Z2}]^{\mathrm{T}}, \\ \Delta\boldsymbol{m} = [\Delta M, \Delta N_1, \Delta N_2, \Delta N_{02}]^{\mathrm{T}}. \end{cases} \tag{5.75}$$

$$\begin{cases} \Delta f_i = f_i - f_0 - (-1)^i \dfrac{Mg}{2} r_{Xy}, \quad f_0 = \left(1 + \dfrac{l_1 + l_2}{r_1 + r_2}\right) f_d \\ \Delta\lambda_{Yi} = \lambda_{Yi} - \dfrac{Mg}{2} r_{Yy} + (-1)^i \dfrac{f_d(Y_1 - Y_2)}{r_1 + r_2} \\ \Delta\lambda_{Zi} = \lambda_{Zi} - \dfrac{Mg}{2} r_{Zy} + (-1)^i \dfrac{f_d(Z_1 - Z_2)}{r_1 + r_2} \\ \Delta M = \hat{M} - M \\ \Delta N_i = \hat{N}_i - N_i, \quad \Delta N_{02} = \hat{N}_{02} - N_{02} \end{cases} \quad i = 1,2 \tag{5.76}$$

$$\begin{cases} N_i = -\dfrac{Mg}{2}(-1)^i \{r_{Yy} r_Z(q_{i0}) - r_{Zy} r_Y(q_{i0})\} \\ \qquad + \dfrac{f_d}{r_1 + r_2}\{(Y_1 - Y_2) r_Z(q_{i0}) - (Z_1 - Z_2) r_Y(q_{i0})\}, \quad i = 1,2 \\ N_{02} = -\dfrac{Mg}{2}(r_{Yy} r_{Zx} - r_{Zy} r_{Yx}) + \dfrac{f_d}{r_1 + r_2}\{(Y_1 - Y_2) r_{Zx} - (Z_1 - Z_2) r_{Yx}\}. \end{cases} \tag{5.77}$$

Thus, by substituting Equation (5.67) into Equation (5.30) and rewriting Equations (5.31) and (5.32) with reference to Equations (5.74–5.77), we obtain the following closed-loop dynamics of the overall fingers–object system:

$$H\ddot{\boldsymbol{X}} + \left(\frac{1}{2}\dot{H} + S\right)\dot{\boldsymbol{X}} + C\dot{\boldsymbol{X}} - A^{\mathrm{T}}\Delta\boldsymbol{\lambda} + B^{\mathrm{T}}\Delta\boldsymbol{m} - \boldsymbol{d} = 0, \tag{5.78}$$

where

$$\boldsymbol{X} = \begin{pmatrix} q_1 \\ q_2 \\ r^{-1}\boldsymbol{x} \\ \theta \\ \psi \end{pmatrix}, \quad H = \begin{pmatrix} H_1 & 0_{3\times4} & 0_{3\times3} & 0_{3\times2} \\ 0_{4\times3} & H_2 & 0_{4\times3} & 0_{4\times2} \\ 0_{3\times3} & 0_{3\times4} & Mr^2I_3 & 0_{3\times2} \\ 0_{2\times3} & 0_{2\times4} & 0_{2\times3} & H_0 \end{pmatrix}, \tag{5.79}$$

$$\begin{cases} S = \begin{pmatrix} S_1 & 0_{3\times4} & 0_{3\times3} & S_{14} \\ 0_{4\times3} & S_2 & 0_{4\times3} & S_{24} \\ 0_{3\times3} & 0_{3\times4} & 0_{3\times3} & 0_{3\times2} \\ -S_{14}^{\mathrm{T}} & -S_{24}^{\mathrm{T}} & 0_{2\times3} & S_0 \end{pmatrix}, \\ C = \mathrm{diag}(C_1, C_2, 0_3, 0_2), \\ \boldsymbol{d} = (0, \cdots, 0, S_Z, S_Y)^{\mathrm{T}}. \end{cases} \tag{5.80}$$

Here, r denotes a scale factor introduced to balance the magnitude r^2M with the other magnitudes of eigenvalues of H_1, H_2 and H_0. Note that $K + K_0 = (1/2)\dot{\boldsymbol{X}}^{\mathrm{T}}H\dot{\boldsymbol{X}}$, $S_{i4} = -(1/2)\left\{\partial(\boldsymbol{\omega}_0^{\mathrm{T}}H_0)/\partial q_i\right\}$ $(i = 1, 2)$, $S_0^{\mathrm{T}} = -S_0$, and

$$S_Z = S_{ZM} + S_{Zf}, \quad S_Y = S_{YM} + S_{Yf}, \tag{5.81}$$

$$S_{ZM} = \frac{Mg}{2}\Big\{(Y_1 + Y_2)(r_{Xy} + \xi_z r_{Zy}) - (Z_1 + Z_2)r_{Yy}\xi_z + (l_1 - l_2)r_{Yy}\Big\}, \tag{5.82}$$

$$S_{YM} = -\frac{Mg}{2}\Big\{(Z_1 + Z_2)(r_{Xy} + r_{Yy}\xi_y) - (Y_1 + Y_2)r_{Zy}\xi_y + (l_1 - l_2)r_{Zy}\Big\}, \tag{5.83}$$

$$S_{Zf} = -f_d(Y_1 - Y_2), \tag{5.84}$$

$$S_{Yf} = f_d(Z_1 - Z_2) \tag{5.85}$$

and A is defined in Equation (5.54), and

$$B^{\mathrm{T}} = \begin{pmatrix} \frac{g}{2}\left(\frac{\partial y_{01}}{\partial q_1}\right) & r_1\boldsymbol{e}_1 & 0 & 0 \\ \frac{g}{2}\left(\frac{\partial y_{02}}{\partial q_2}\right) & 0 & r_2\boldsymbol{e}_2 & r_2\boldsymbol{e}_{02} \\ & 0_{5\times5} & & \end{pmatrix}, \tag{5.86}$$

where $\bar{\boldsymbol{x}} = r^{-1}\boldsymbol{x}$. It is interesting to note that the matrix S in Equation (5.80) is again skew-symmetric. Then, note that taking the inner product between Equation (5.78) and $\dot{\boldsymbol{X}}$ yields the following energy relation

$$\frac{\mathrm{d}}{\mathrm{d}t}(K+W) = \sum_{i=1,2} -\dot{q}_i^{\mathrm{T}} C_i \dot{q}_i, \tag{5.87}$$

where

$$\begin{aligned} W = & \frac{f_d}{2(r_1+r_2)} \left\{(Y_1-Y_2)^2 + (Z_1-Z_2)^2\right\} + \frac{\gamma_M}{2}\Delta M^2 \\ & + \sum_{i=1,2} \left\{\frac{\gamma_i}{2}\hat{N}_i^2 + \frac{\gamma_{0i}}{2}\hat{N}_{0i}^2\right\} \\ & + \frac{Mg}{2}\left\{(Y_1+Y_2)r_{Yy} - l_0 r_{Xy} + (Z_1+Z_2)r_{Zy}\right\}. \end{aligned} \tag{5.88}$$

In the derivation of this relation the skew-symmetry of the matrix S is used in such a way that $\dot{\boldsymbol{X}}^{\mathrm{T}} S \dot{\boldsymbol{X}} = 0$ and the relation of the holonomic contact point and non-holonomic rolling constraints with the velocity vector $\dot{\boldsymbol{X}}$ is referenced in such a way that $A\dot{\boldsymbol{X}} = 0$. Further, it is interesting to note that the last term on the right-hand side of Equation (5.88) plus $-Mg(y_{01}+y_{02})/2$ is equivalent to cancellation of the potential energy of the object caused by the gravity (that is, $P = -Mgy$), as seen from Equation (5.72). Then, it is easy to check that the scalar quantity W is quadratic in terms of the position state variables θ, ψ, q_{i0} $(i=1,2), p_i (= q_{i1}+q_{i2})$ $(i=1,2)$, and $y_{01}+y_{02}$. Further, by choosing constant gains γ_M, γ_i $(i=1,2)$, γ_{0i} $(i=1,2)$ and f_d (which must be bigger than Mg) appropriately, it is possible to verify that W has a minimum W_m in a neighbourhood of a given initial position state $\boldsymbol{X}(0)$. Then, define

$$E_m = K + W - W_m. \tag{5.89}$$

This quantity becomes non-negative definite in $(\boldsymbol{X}, \dot{\boldsymbol{X}})$ even if the value of W_m is dependent on the unit vector (r_{Xy}, r_{Yy}, r_{Zy}) at initial time $t=0$. Nevertheless, it is possible to assume that W is always bounded from below by some constant value W_0.

Finally, it is crucial to remark that the closed-loop dynamics of Equation (5.78) can be re-derived by applying the variational principle to the Lagrangian $L(= K - W)$ by using partial derivatives of W with respect to θ and ψ that do not explicitly appear in the original equation of motion described as Equations (5.30–5.32). Partial derivatives of W in θ and ψ can be obtained by using the partial derivatives of unit vectors $\boldsymbol{r}_X$, $\boldsymbol{r}_Y$ and $\boldsymbol{r}_Z$ in θ and ψ, which can be calculated from the concept of infinitesimal rotation discussed fully in Section 5.12.

5.8 Numerical Simulation Results

In this section, we will give only a sketch of the proof of trajectory convergence. Since $E_m = K + W - W_m \geq 0$ for any (r_{Xy}, r_{Yy}, r_{Zy}), Equation (5.87) implies

$$\int_0^\infty \sum_{i=1,2} \dot{q}_i^{\mathrm{T}}(t) C_i \dot{q}_i(t) \mathrm{d}t \leq E_m(0) - E_m(t) \leq E_m(0). \tag{5.90}$$

This shows that $\dot{q}_i(t) \in L^2(0, \infty)$ $(i = 1, 2)$. Since there are six independent constraints of Equations (5.10), (5.11) and $\dot{Q}_i = 0$ $(i = 1, 2)$, it is possible to verify that the other velocity variables $\dot{\boldsymbol{x}}(= (\dot{x}, \dot{y}, \dot{z})^{\mathrm{T}})$, ω_z and ω_y also become square-integrable, that is, $\boldsymbol{\omega}_0(t) \in L^2(0, \infty)$. Then, it is also possible to verify that, according to the boundedness of the overall kinetic energy $K(t)$ and the artificial potential energy $W(t)$, every component of the velocity vector $\dot{\boldsymbol{X}}(t)$ is uniformly continuous in t. Hence, by virtue of Lemma 2 in Appendix A, it can be concluded that $\dot{\boldsymbol{X}}(t) \to 0$ as $t \to \infty$. This also implies that $\ddot{\boldsymbol{X}}(t) \to 0$ as $t \to \infty$. Thus, it follows from Equation (5.78) that

$$A^{\mathrm{T}} \Delta \boldsymbol{\lambda} - B^{\mathrm{T}} \Delta \boldsymbol{m} + \boldsymbol{d} \to 0 \tag{5.91}$$

as $t \to \infty$. Since the 12×10 matrix $[A^{\mathrm{T}}, -B^{\mathrm{T}}]$ is non-degenerate at any ordinary posture of the fingers–object system as depicted in Figure 5.2, it is reasonable to expect that $\Delta\boldsymbol{\lambda}(t) \to \boldsymbol{\lambda}_\infty$ and $\Delta\boldsymbol{m} \to \boldsymbol{m}_\infty$ as $t \to \infty$, where $\boldsymbol{\lambda}_\infty$ and $\boldsymbol{m}_\infty$ are constant vectors. At the same time, the convergence of $\Delta\boldsymbol{m}$ to $\boldsymbol{m}_\infty$ as $t \to \infty$ intuitively implies that all six variables θ, ψ, q_{20} and p_i $(i = 1, 2)$ converge to their corresponding constant values θ_∞, ψ_∞, $q_{20\infty}$ and $p_{i\infty}$ $(i = 1, 2)$, and these convergences together with contact constraints of Equations (5.9–5.11) may imply that all q_{ij} (for $i = 1, 2$ and $j = 1, 2, 3$) converge respectively to constant values $q_{ij\infty}$ $(i = 1, 2$ and $j = 1, 2, 3)$ as $t \to \infty$. Then, the matrices A and B, and vector $\boldsymbol{d}$ in Equation (5.78) are also convergent to constant matrices A_∞ and B_∞, and constant vector $\boldsymbol{d}_\infty$, respectively. However, all these convergences should be treated and proved in a more rigorous way, which will be presented in the next section.

Nevertheless, in order to ascertain this intuitive argument, we have conducted a numerical simulation based on a physical model of such fingers–object grasping with the physical parameters shown in Table 5.4. Then, a control signal defined by Equation (5.67) with the constant gains in Table 5.5 and $C_i = c_i I_3$ for $i = 1, 2$ is fed into the finger dynamics of Equation (5.30) with initial conditions that are also specified in Table 5.5. As shown in Figure 5.9, all the key variables $Y_1 - Y_2$, $Z_1 - Z_2$, S_Y, S_Z, and all components of $\Delta\boldsymbol{\lambda}$ and $\Delta\boldsymbol{m}$ converge asymptotically to their corresponding constant values. It is also confirmed from Figure 5.9 that $|\Delta f_i| < 0.05 (<< 1.0)$ for all $t \geq 0$. This means that both contacts between finger-ends and object surfaces are maintained throughout the pinching motion. Further, it is possible to evaluate convergent values of $\Delta\boldsymbol{\lambda}$ and $\Delta\boldsymbol{m}$ based on the relation

$$-\left[A^{\mathrm{T}}, -B^{\mathrm{T}}\right] \begin{pmatrix} \Delta\boldsymbol{\lambda} \\ \Delta\boldsymbol{m} \end{pmatrix} - \boldsymbol{d} \to 0 \quad \text{as} \quad t \to \infty \tag{5.92}$$

from which

$$\begin{pmatrix} \boldsymbol{\lambda}_\infty \\ \boldsymbol{m}_\infty \end{pmatrix} = \lim_{t \to \infty} \begin{pmatrix} AA^{\mathrm{T}} & -AA^{\mathrm{T}} \\ -BA^{\mathrm{T}} & BB^{\mathrm{T}} \end{pmatrix}^{-1} \begin{pmatrix} -A \\ B \end{pmatrix} \boldsymbol{d}. \tag{5.93}$$

Table 5.4. Physical parameters for three and four DOFs

$l_{11} = l_{21}$	length	0.040 [m]
$l_{12} = l_{22}$	length	0.040 [m]
$l_{13} = l_{23}$	length	0.030 [m]
m_{11}	weight	0.043 [kg]
m_{12}	weight	0.031 [kg]
m_{13}	weight	0.020 [kg]
l_{20}	length	0.000 [m]
m_{20}	weight	0.000 [kg]
m_{21}	weight	0.060 [kg]
m_{22}	weight	0.031 [kg]
m_{23}	weight	0.020 [kg]
I_{XX11}	inertia moment	5.375×10^{-7}[kgm^2]
$I_{YY11} = I_{ZZ11}$	inertia moment	6.002×10^{-6}[kgm^2]
I_{XX12}	inertia moment	3.875×10^{-7}[kgm^2]
$I_{YY12} = I_{ZZ12}$	inertia moment	4.327×10^{-6}[kgm^2]
I_{XX13}	inertia moment	2.500×10^{-7}[kgm^2]
$I_{YY13} = I_{ZZ13}$	inertia moment	1.625×10^{-6}[kgm^2]
I_{XX21}	inertia moment	7.500×10^{-7}[kgm^2]
$I_{YY21} = I_{ZZ21}$	inertia moment	8.375×10^{-6}[kgm^2]
I_{XX22}	inertia moment	3.875×10^{-7}[kgm^2]
$I_{YY22} = I_{ZZ22}$	inertia moment	4.327×10^{-6}[kgm^2]
I_{XX23}	inertia moment	2.500×10^{-7}[kgm^2]
$I_{YY23} = I_{ZZ23}$	inertia moment	1.625×10^{-6}[kgm^2]
$I_{XX} = I_{ZZ}$	inertia moment (object)	1.133×10^{-5}[kgm^2]
I_{YY}	inertia moment (object)	6.000×10^{-6}[kgm^2]
r_0	link radius	0.005 [m]
$r_i (i = 1, 2)$	radius	0.010 [m]
L	base length	0.063 [m]
M	object weight	0.040 [kg]
$l_i (i = 1, 2)$	object width	0.015 [m]
h	object height	0.050 [m]

Table 5.5. Parameters of the control signals

f_d	internal force	1.000 [N]
$c_1 = c_2$	damping coefficient	0.001 [Nms]
c_{20}	damping coefficient	0.006 [Nms]
γ_M	regressor gain	0.05 [m^2/kgs^2]
γ_0	regressor gain	0.0005 [s^2/kg]
$\gamma_i \ (i = 1, 2)$	regressor gain	0.0005 [s^2/kg]

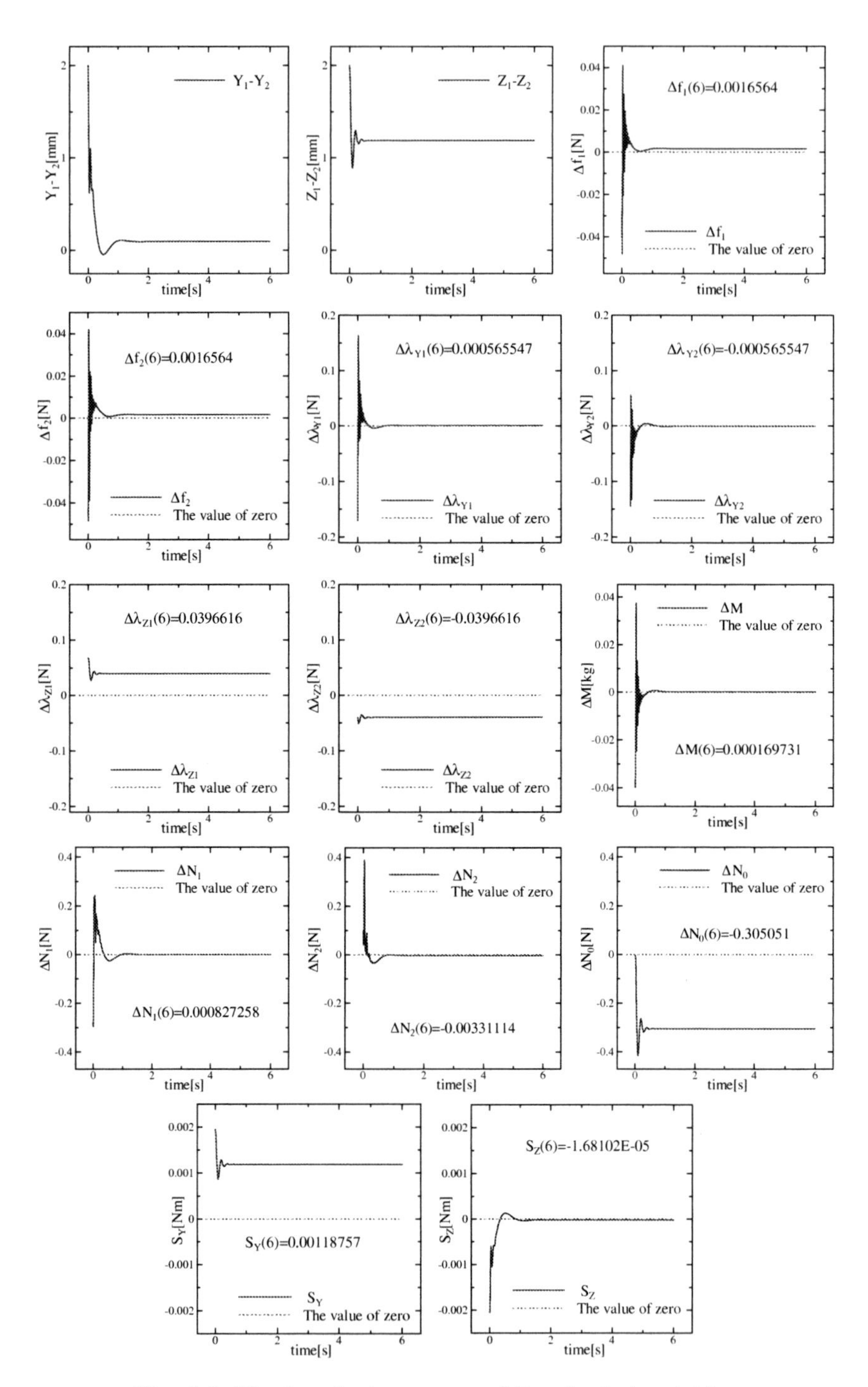

Fig. 5.9. The transient responses of the physical variables

When we construct Figure 5.9 from the numerical simulation, we calculate the vector value of the limit in Equation (5.93) at transient time t and have certified that around $t = 1.5$ [s] its value becomes steady and almost converges to $(\boldsymbol{\lambda}_\infty^{\mathrm{T}}, \boldsymbol{m}_\infty^{\mathrm{T}})^{\mathrm{T}}$.

Finally, we remark that it is possible to prove that the speed of convergences of $\Delta\boldsymbol{m}(t)$ and $\Delta\boldsymbol{\lambda}(t)$ to $\boldsymbol{m}_\infty$ and $\boldsymbol{\lambda}_\infty$ is exponential in t as $t \to \infty$. Thus, it is possible to conclude that the Riemannian metric

$$\int_0^t \sqrt{\frac{1}{2}\dot{\boldsymbol{X}}^{\mathrm{T}}(\tau)H(\boldsymbol{X}(\tau))\dot{\boldsymbol{X}}(\tau)}\ \mathrm{d}\tau \tag{5.94}$$

remains finite when t tends to infinity, that is,

$$\int_0^\infty \sqrt{\frac{1}{2}\dot{\boldsymbol{X}}^{\mathrm{T}}H(\boldsymbol{X})\dot{\boldsymbol{X}}}\ \mathrm{d}t < +\infty. \tag{5.95}$$

Further, in a similar argument, it is possible to show that

$$\int_0^\infty \|\Omega(t)\|\mathrm{d}t < +\infty. \tag{5.96}$$

From these results it can be concluded that, at the equilibrium state ($\boldsymbol{X} = \boldsymbol{X}_\infty, \dot{\boldsymbol{X}} = 0$), the force/torque balance is established except for rotation around the opposition axis. It should be claimed that the equilibrium state ($\boldsymbol{X}_\infty, \dot{\boldsymbol{X}} = 0$) is stable on a 12-dimensional constraint manifold M_{12} that is defined as

$$M_{12} = \Big\{(\boldsymbol{X}, \dot{\boldsymbol{X}}) : Q_i = 0,\ \dot{Q}_i = 0,\ \boldsymbol{Y}_i^{\mathrm{T}}\delta\boldsymbol{X} = 0,\ \boldsymbol{Z}_i^{\mathrm{T}}\delta\boldsymbol{X} = 0,$$
$$\text{Equations (5.16) and (5.19)}\quad (i = 1, 2)\Big\}. \tag{5.97}$$

On the other hand, the set of all vectors $\boldsymbol{X}$ constitutes superficially a 12-dimensional space C^{12}, which can be called the configuration space in the terminology of differential geometry. Then, the constraint manifold of six-dimension can be defined as follows:

$$CM_6 = \{\boldsymbol{X} : Q_i = 0,\ \boldsymbol{Y}_i^{\mathrm{T}}\delta\boldsymbol{X} = 0,\ \boldsymbol{Z}_i^{\mathrm{T}}\delta\boldsymbol{X} = 0,\quad i = 1, 2\}, \tag{5.98}$$

which is regarded as a Riemannian manifold embedded in the configuration space C^{12}. Further, at each point $\boldsymbol{X}$ on CM_6 it is possible to define a tangent space TM_6 in such a way that

$$TM_6(\boldsymbol{X}) = \Big\{\dot{\boldsymbol{X}} : A\dot{\boldsymbol{X}} = 0\Big\}. \tag{5.99}$$

Thus, the 12-dimensional tangent bundle is constituted in the following way:

$$T_{12} = \Big\{\dot{\boldsymbol{X}}(\boldsymbol{X}) : \boldsymbol{X} \in CM_6\Big\}, \tag{5.100}$$

which is equivalent to M_{12} defined in Equation (5.97). By using such terminology from differential geometry, it is possible to deal with the stability problem mentioned above in a rigorous mathematical way. The details are presented in the next section.

5.9 A Mathematical Proof of the Stability of Blind Grasping

In this section, we present a rigorous mathematical proof of asymptotic convergence of the whole solution trajectory $(\boldsymbol{X}(t), \dot{\boldsymbol{X}}(t))$ of the closed loop of Equation (5.78) to an equilibrium state of force/torque balance. First we show that the gradient vector of W of Equation (5.88) with respect to $\boldsymbol{X}$ defined by Equation (5.79) is deduced as follows:

$$\frac{\partial W}{\partial \boldsymbol{X}} = B^{\mathrm{T}} \Delta \boldsymbol{m} - \boldsymbol{d}. \tag{5.101}$$

In fact, it follows from Equation (5.88) that

$$\begin{aligned}\frac{\partial W}{\partial \theta} = &\frac{f_d}{r_1 + r_2}\left\{(Y_1 - Y_2)\frac{\partial(Y_1 - Y_2)}{\partial \theta} + (Z_1 - Z_2)\frac{\partial(Z_1 - Z_2)}{\partial \theta}\right\} \\ &+\frac{Mg}{2}\left\{(Y_1 + Y_2)\frac{\partial r_{Yy}}{\partial \theta} - l_0\frac{\partial r_{Xy}}{\partial \theta} + (Z_1 + Z_2)\frac{\partial r_{Zy}}{\partial \theta}\right\} \\ &+\frac{Mg}{2}\left\{\frac{\partial(Y_1 + Y_2)}{\partial \theta} r_{Yy} + \frac{\partial(Z_1 + Z_2)}{\partial \theta} r_{Zy}\right\}.\end{aligned} \tag{5.102}$$

From Section 5.12 we see that

$$\frac{\partial r_{Xy}}{\partial \theta} = r_{Yy}, \quad \frac{\partial r_{Yy}}{\partial \theta} = -r_{Xy} - \xi_z r_{Zy}, \quad \frac{\partial r_{Zy}}{\partial \theta} = \xi_z r_{Yy}. \tag{5.103}$$

On the other hand, we see from the rolling constraints of Equations (5.16) and (5.19), referring to the last paragraph of Section 5.12,

$$\frac{\partial Y_1}{\partial \theta} = r_1, \quad \frac{\partial Y_2}{\partial \theta} = -r_2, \quad \frac{\partial Z_1}{\partial \theta} = \frac{\partial Z_2}{\partial \theta} = 0. \tag{5.104}$$

Substituting Equations (5.103) and (5.104) into Equation (5.102) yields

$$\begin{aligned}\frac{\partial W}{\partial \theta} &= f_d(Y_1 - Y_2) + \frac{Mg}{2}\Big\{(Y_1 + Y_2)(-r_{Xy} - \xi_z r_{Zy}) - l_0 r_{Yy} \\ &\quad +(Z_1 + Z_2)\xi_z r_{Yy}\Big\} + \frac{Mg}{2}\left\{(r_1 - r_2) r_{Yy}\right\} \\ &= f_d(Y_1 - Y_2) - \frac{Mg}{2}\Big\{(Y_1 + Y_2) r_{Xy} - (l_1 - l_2) r_{Yy} \\ &\quad +\xi_z(Y_1 + Y_2) r_{Zy} - \xi_z(Z_1 + Z_2) r_{Yy}\Big\} \\ &= -S_Z.\end{aligned} \tag{5.105}$$

Similarly, we see from Section 5.12 that

$$\frac{\partial r_{Xy}}{\partial \psi} = -r_{Zy}, \quad \frac{\partial r_{Yy}}{\partial \psi} = -\xi_y r_{Zy}, \quad \frac{\partial r_{Zy}}{\partial \psi} = r_{Xy} + \xi_y r_{Yy}, \tag{5.106}$$

$$\frac{\partial Z_1}{\partial \psi} = -r_1, \quad \frac{\partial Z_2}{\partial \psi} = r_2, \quad \frac{\partial Y_1}{\partial \psi} = \frac{\partial Y_2}{\partial \psi} = 0. \tag{5.107}$$

By substituting these relations into the mathematical formula for $\partial W/\partial\psi$, we obtain

$$\frac{\partial W}{\partial \psi} = -S_Y. \tag{5.108}$$

Since W of Equation (5.88) is not dependent on $\boldsymbol{x}$ in an explicit form, it is reasonable to see that

$$\frac{\partial W}{\partial \boldsymbol{x}} = 0. \tag{5.109}$$

Finally, it is necessary to derive the following partial derivatives:

$$\begin{aligned}\frac{\partial W}{\partial q_1} &= \frac{f_d}{r_1+r_2}\left\{(Y_1-Y_2)\frac{\partial Y_1}{\partial q_1} + (Z_1-Z_2)\frac{\partial Z_1}{\partial q_1}\right\}\\ &\quad +\gamma_M \Delta M \frac{\partial \Delta M}{\partial q_1} + \gamma_1 \hat{N}_1 \frac{\partial \hat{N}_1}{\partial q_1}\\ &\quad +\frac{Mg}{2}\left\{\frac{\partial Y_1}{\partial q_1} r_{Yy} + \frac{\partial Z_1}{\partial q_1} r_{Zy}\right\}\end{aligned} \tag{5.110}$$

$$\begin{aligned}\frac{\partial W}{\partial q_2} &= -\frac{f_d}{r_1+r_2}\left\{(Y_1-Y_2)\frac{\partial Y_2}{\partial q_2} + (Z_1-Z_2)\frac{\partial Z_2}{\partial q_2}\right\}\\ &\quad +\gamma_M \Delta M \frac{\partial \Delta M}{\partial q_2} + \gamma_2 \hat{N}_2 \frac{\partial \hat{N}_2}{\partial q_2} + \gamma_{02}\hat{N}_{02}\frac{\partial \hat{N}_{02}}{\partial q_2}\\ &\quad +\frac{Mg}{2}\left\{\frac{\partial Y_2}{\partial q_2} r_{Yy} + \frac{\partial Z_2}{\partial q_2} r_{Zy}\right\}.\end{aligned} \tag{5.111}$$

In a similar manner to the derivation of Equations (5.102) and (5.105), it is possible to deduce from the rolling constraints of Equations (5.16) and (5.19) with the aid of the variational principle (see Section 5.12)

$$\frac{\partial Y_1}{\partial q_1} = -r_1 r_{Zz} e_1, \qquad \frac{\partial Z_1}{\partial q_1} = r_1 r_{Yz} e_1, \tag{5.112}$$

$$\left\{\begin{aligned}\frac{\partial Y_2}{\partial q_2} &= r_2(r_{Zz}\cos q_{20} - r_{Zy}\sin q_{20})e_2 + r_2 r_{Zx} e_{02},\\ \frac{\partial Z_2}{\partial q_2} &= -r_2(r_{Yz}\cos q_{20} - r_{Yy}\sin q_{20})e_2 - r_2 r_{Yx} e_{02}.\end{aligned}\right. \tag{5.113}$$

Substituting Equations (5.112) into (5.110) and calculating $\partial\Delta M/\partial q_1$ and $\partial\Delta\hat{N}_1/\partial q_1$, directly we obtain

Table 5.6. Gradient vector of $\bar{W}$

$\dfrac{\partial \bar{W}}{\partial p_i} = r_i \Delta N_i, \qquad i = 1, 2$
$\dfrac{\partial \bar{W}}{\partial q_{20}} = r_2 \Delta N_{02}$
$\dfrac{\partial \bar{W}}{\partial \theta} = -S_Z, \quad \dfrac{\partial \bar{W}}{\partial \psi} = -S_Y$

$$\begin{aligned}\frac{\partial W}{\partial q_1} &= -\frac{r_1 f_d}{r_1 + r_2}\left\{(Y_1 - Y_2) r_{Zz} - (Z_1 - Z_2) r_{Yz}\right\} \boldsymbol{e}_1 \\ &\quad -\frac{\Delta Mg}{2}\frac{\partial y_{01}}{\partial q_1} + r_1 \hat{N}_1 \boldsymbol{e}_1 \\ &\quad -\frac{Mg}{2}\left\{r_{Yy} r_{Zz} - r_{Zy} r_{Yz}\right\} r_1 \boldsymbol{e}_1 \\ &= r_1 \Delta N_1 \boldsymbol{e}_1 - \frac{\Delta Mg}{2}\left(\frac{\partial y_{01}}{\partial q_1}\right). \end{aligned} \tag{5.114}$$

Analogously, it follows from Equation (5.111) together with Equation (5.113) that

$$\frac{\partial W}{\partial q_2} = r_1 \Delta N_2 \boldsymbol{e}_2 + r_2 \Delta N_{02} \boldsymbol{e}_{02} - \frac{\Delta Mg}{2}\left(\frac{\partial y_{02}}{\partial q_2}\right). \tag{5.115}$$

In the proof we implicitly assume that the numerical values of the physical parameters of the robot fingers and the object have similar orders correspondingly to those given in Table 5.4. We set damping coefficient matrices as $C_i = c_i I_3$ with c_i given in Table 5.5 and other gains f_d, γ_M, γ_i $(i = 1, 2)$ and γ_{0i} $(i = 1, 2)$ as in Table 5.5.

Now consider the following scalar quantity

$$\begin{aligned}\bar{W} &= W - \frac{\gamma_M}{2}\Delta M^2 \\ &= \frac{f_d}{2(r_1 + r_2)}\left\{(Y_1 - Y_2)^2 + (Z_1 - Z_2)^2\right\} \\ &\quad + \sum_{i=1,2}\left\{\frac{\gamma_i}{2}\hat{N}_i^2 + \frac{\gamma_{0i}}{2}\hat{N}_{0i}^2\right\} + \frac{Mg}{2}\tilde{y}, \end{aligned} \tag{5.116}$$

where

$$\tilde{y} = \frac{1}{2}\left\{(Y_1 + Y_2) r_{Yy} - l_0 r_{Xy} + (Z_1 + Z_2) r_{Zy}\right\}. \tag{5.117}$$

Evidently this scalar function depends on p_1, p_2, q_{20}, θ and ψ. Partial derivatives of $\bar{W}$ with respect p_i, q_{20}, θ and ψ can be obtained as in Table 5.6. In this derivation we used the following relations:

Table 5.7. The hessian matrix of $\bar{W}$

$$\frac{\partial^2 \bar{W}}{\partial p_1^2} = r_1^2 \left\{ \frac{1}{\gamma_1} + \frac{f_d}{r_1 + r_2}(r_{Zz}^2 + r_{Yz}^2) \right\} \quad \text{(V-1)}$$

$$\frac{\partial^2 \bar{W}}{\partial p_2^2} = r_2^2 \left\{ \frac{1}{\gamma_2} + \frac{f_d}{r_1 + r_2}\left(|r_Z(q_{20})|^2 + |r_Y(q_{20})|^2\right) \right\} \quad \text{(V-2)}$$

$$\frac{\partial^2 \bar{W}}{\partial q_{20}^2} = r_2^2 \left\{ \frac{1}{\gamma_{02}} + \frac{f_d}{r_1 + r_2}(r_{Zx}^2 + r_{Yx}^2) \right\} \quad \text{(V-3)}$$

$$\frac{\partial^2 \bar{W}}{\partial p_2 \partial p_1} = \frac{r_1 r_2 f_d}{r_1 + r_2} \{r_{Zz} r_Z(q_{20}) + r_{Yz} r_Y(q_{20})\} \quad \text{(V-4)}$$

$$\frac{\partial^2 \bar{W}}{\partial q_{20} \partial p_1} = \frac{r_1 r_2 f_d}{r_1 + r_2}(r_{Zz} r_{Zx} + r_{Yz} r_{Yx}) \quad \text{(V-5)}$$

$$\begin{aligned}\frac{\partial^2 \bar{W}}{\partial q_{20} \partial p_2} &= \frac{f_d}{r_1 + r_2}\{(Y_1 - Y_2)\bar{r}_Z(q_{20}) - (Z_1 - Z_2)\bar{r}_Y(q_{20})\} \\ &\quad + \frac{r_2 f_d}{r_1 + r_2}\{r_{Zx} r_Z(q_{20}) - r_{Yx} r_Y(q_{20})\} \\ &\quad - \frac{Mg}{2}\{r_{Yy}\bar{r}_Z(q_{20}) - r_{Zy}\bar{r}_Y(q_{20})\} \quad \text{(V-6)}\end{aligned}$$

where

$$\bar{r}_Z(q_{20}) = r_{Zz}\sin q_{20} + r_{Zy}\cos q_{20}$$
$$\bar{r}_Y(q_{20}) = r_{Yz}\sin q_{20} + r_{Yy}\cos q_{20}$$

$$\begin{aligned}\frac{\partial^2 \bar{W}}{\partial \theta^2} &= (r_1 + r_2) f_d - \frac{Mg}{2}(Y_1 + Y_2) r_{Yy} \\ &\quad - \frac{Mg}{2}\left\{ l_0 r_{Xy} + (l_1 - l_2)\xi_z r_{Zy} + \frac{\partial}{\partial \theta}(\xi_z \eta) \right\} \quad \text{(V-7)}\end{aligned}$$

$$\eta = (Y_1 + Y_2) r_{Zy} - (Z_1 + Z_2) r_{Yy}, \quad \bar{l}_0 = (r_1 - r_2) - (l_1 - l_2)$$

$$\begin{aligned}\frac{\partial^2 \bar{W}}{\partial \psi^2} &= (r_1 + r_2) f_d \\ &\quad - \frac{Mg}{2}\left\{ \bar{l}_0 r_{Xy} + (Z_1 + Z_2) r_{Zy} - (l_1 - l_2)\xi_y r_{Yy} + \frac{\partial}{\partial \psi}(\xi_y \eta) \right\} \quad \text{(V-8)}\end{aligned}$$

$$\frac{\partial^2 \bar{W}}{\partial \psi \partial \theta} = \frac{Mg}{2}\left\{ (Y_1 + Y_2) r_{Zy} - (l_1 - l_2)\xi_y r_{Zy} - \frac{\partial}{\partial \psi}(\xi_z \eta) \right\} \quad \text{(V-9)}$$

$$\left\{ \begin{aligned} &\frac{\partial Y_1}{\partial p_1} = -r_1 r_{Zz}, \quad \frac{\partial Z_1}{\partial p_1} = r_1 r_{Yz}, \\ &\frac{\partial Y_2}{\partial p_2} = r_2(r_{Zz}\cos q_{20} - r_{Zy}\sin q_{20}), \quad \frac{\partial Y_2}{\partial q_{20}} = r_2 r_{Zx}, \\ &\frac{\partial Z_2}{\partial p_2} = -r_2(r_{Yz}\cos q_{20} - r_{Yy}\sin q_{20}), \quad \frac{\partial Z_2}{\partial q_{20}} = r_2 r_{Yx}. \end{aligned} \right. \quad (5.118)$$

Table 5.7. The hessian matrix of $\bar{W}$ (continued)

$$\frac{\partial^2 \bar{W}}{\partial \theta \partial p_1} = \frac{-r_1 f_d}{r_1 + r_2} \{(r_1 + r_2) r_{Zz} + (Y_1 - Y_2)\xi_z r_{Yz} + (Z_1 - Z_2)(r_{Xz} + \xi_z r_{Zz})\} + \frac{r_1 Mg}{2} \{r_{Xy} r_{Zz} - r_{Zy} r_{Xz}\} \tag{V-10}$$

$$\frac{\partial^2 \bar{W}}{\partial \psi \partial p_1} = -\frac{r_1 f_d}{r_1 + r_2} \{(r_1 + r_2) r_{Yz} + (Y_1 - Y_2)(r_{Xz} + \xi_y r_{Yz}) + (Z_1 - Z_2)\xi_y r_{Zz}\} - \frac{r_1 Mg}{2} \{r_{Xz} r_{Yy} - r_{Yz} r_{Xy}\} \tag{V-11}$$

$$\frac{\partial^2 \bar{W}}{\partial \theta \partial q_{20}} = \frac{-r_2 f_d}{r_1 + r_2} \{(r_1 + r_2) r_{Zx} + (Y_1 - Y_2)\xi_z r_{Yx} + (Z_1 - Z_2)(r_{Xx} + \xi_x r_{Zx})\} - \frac{r_2 Mg}{2} (r_{Xy} r_{Zx} - r_{Zy} r_{Xx}) \tag{V-12}$$

$$\frac{\partial^2 \bar{W}}{\partial \psi \partial q_{20}} = -\frac{r_2 f_d}{r_1 + r_2} \{(r_1 + r_2) r_{Yx} + (Y_1 - Y_2)(r_{Xx} + \xi_y r_{Yx}) + (Z_1 - Z_2)\xi_y r_{Zx}\} + \frac{r_2 Mg}{2} (r_{Yy} r_{Xx} - r_{Xy} r_{Yx}) \tag{V-13}$$

$$\frac{\partial^2 \bar{W}}{\partial \theta \partial p_2} = \frac{-r_2 f_d}{r_1 + r_2} \{(r_1 + r_2) r_Z(q_{20}) + (Y_1 - Y_2)\xi_z r_Y(q_{20}) + (Z_1 - Z_2)(r_X(q_{20}) + \xi_x r_Z(q_{20}))\} - \frac{r_2 Mg}{2} \{r_{Xy} r_Z(q_{20}) - r_{Zy} r_X(q_{20})\} \tag{V-14}$$

where $r_X(q_{20}) = r_{Xz} \cos q_{20} - r_{Xy} \sin q_{20}$

$$\frac{\partial^2 \bar{W}}{\partial \psi \partial p_2} = \frac{-r_2 f_d}{r_1 + r_2} \{(r_1 + r_2) r_Y(q_{20}) + (Y_1 - Y_2)(r_X(q_{20}) + \xi_y r_Y(q_{20})) + (Z_1 - Z_2)\xi_y r_Z(q_{20})\} - \frac{r_2 Mg}{2} \{r_{Xy} r_Y(q_{20}) - r_{Yy} r_X(q_{20})\} \tag{V-15}$$

Next, the Hessian matrix of $\bar{W}$ with respect to those five variables can be derived as shown in Tables 5.7 by using Equation (5.118) together with the relations presented in Section 5.12. Then, it is possible to verify that the diagonal entries of this Hessian matrix are dominant relatively to their corresponding off-diagonal entries if γ_1, γ_2, and γ_{02} are properly chosen as in Table 5.5, $f_d \gg Mg$, and $Y_1 + Y_2 < 0$. In other words, if the 5×5 Hessian matrix is denoted by $G(\boldsymbol{X})$ and its diagonal part by the form

$$G_D(\boldsymbol{X}) = \mathrm{diag}\left(\frac{\partial^2 \bar{W}}{\partial p_1^2}, \frac{\partial^2 \bar{W}}{\partial p_2^2}, \frac{\partial^2 \bar{W}}{\partial q_{20}^2}, \frac{\partial^2 \bar{W}}{\partial \theta^2}, \frac{\partial^2 \bar{W}}{\partial \psi^2}\right) \tag{5.119}$$

then it is possible to expect that

$$\frac{1}{2}G(\boldsymbol{X}) \leq G_D(\boldsymbol{X}) \leq \frac{3}{2}G(\boldsymbol{X}) \tag{5.120}$$

for any $\boldsymbol{X}$ in a neighbourhood of $\boldsymbol{X}(0)$ in the configuration space and lying on CM_6 defined by Equation (5.98).

Dominance of diagonal entries of the Hessian matrix shown in Equation (5.120) implies that the Morse function is no-negative. Hence, it is possible to construct a modified scalar function like V of Equation (4.124) and $W(\alpha)$ of Equation (4.129) with a similar property to that of Equation (4.130) by using a similar argument that given in Section 4.6 for the proof of exponential convergence in the case of 2-D grasping. The details of the proof, however, must be left to the readers.

Finally, we should remark that the necessary combination of numbers of finger joints can be reduced to $(2, 3)$ as far as stable prehension is concerned, two DOFs for the left planar finger and three DOFs for the right 3-D finger with a saddle joint. Under weightless circumstances, it can be reduced to $(1, 2)$.

5.10 Stable Manipulation of a 3-D Rigid Object

Once a rigid object is grasped in a blind manner stably so that force/torque balance is attained, it is possible to manoeuvre the object by specifying an ideal 3-D trajectory as a time function described by the form $\boldsymbol{x}_d(t) = (x_d(t), y_d(t), z_d(t))$ so that a superficial object centre $\hat{\boldsymbol{x}}(t)$ $(= (\boldsymbol{x}_{01}(t) + \boldsymbol{x}_{02}(t))/2)$ should track the trajectory $\boldsymbol{x}_d(t)$. Since $\boldsymbol{x}_{0i}(t)$ can be easily calculated in real time from measurement data of finger joint angles and knowledge of the finger kinematic parameters, it is possible to devise the feedback signal

$$u_{pi} = -J_{0i}^{\mathrm{T}}(q_i)(\gamma_x \Delta x, \gamma_y \Delta y, \gamma_z \Delta z)^{\mathrm{T}}, \quad i = 1, 2 \tag{5.121}$$

where γ_x, γ_y and γ_z are positive constants that express feedback gains, and $\Delta x = \hat{x} - x_d$, $\Delta y = \hat{y} - y_d$ and $\Delta z = \hat{z} - z_d$. This control signal can be superimposed linearly with the control signal u_i defined by Equation (5.67) in such a way that

$$\begin{aligned} u_i = {} & g_i(q_i) - c_i \dot{q}_i + \frac{(-1)^i f_d}{r_1 + r_2} J_i^{\mathrm{T}}(q_i)(\boldsymbol{x}_{01} - \boldsymbol{x}_{02}) \\ & - \frac{\hat{M} g}{2} \frac{\partial y_{0i}}{\partial q_i} - r_i \hat{N}_i \boldsymbol{e}_i - r_i \hat{N}_{0i} \boldsymbol{e}_{0i} - J_i^{\mathrm{T}}(q_i) \Gamma \Delta \boldsymbol{x}, \quad i = 1, 2, \end{aligned} \tag{5.122}$$

where $\Gamma = \mathrm{diag}(\gamma_x, \gamma_y, \gamma_z)$ and $\Delta \boldsymbol{x} = (\Delta x, \Delta y, \Delta z)^{\mathrm{T}}$.

We show one numerical simulation result of manipulation of an object through imposing such a trajectory tracking task for the pair of finger robots shown in Figure 5.10, where the left finger is planar with three joints in a common z-axis and the right finger is 3-dimensional with one joint in the x-axis and the other three joints in a common z-axis. All the physical parameters

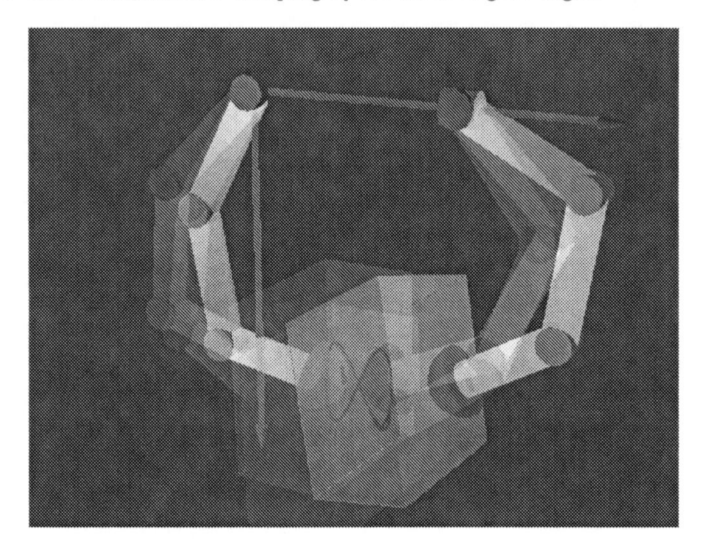

Fig. 5.10. Motions of manipulating an object in 3-D space

Table 5.8. Initial values

q_{11}	$6.000 \times \pi/18$ [rad]	q_{12}	$2.678 \times \pi/18$ [rad]
q_{13}	$8.474 \times \pi/18$ [rad]	q_{20}	$-1.975 \times \pi/18$ [rad]
q_{21}	$4.311 \times \pi/18$ [rad]	q_{22}	$5.687 \times \pi/18$ [rad]
q_{23}	$6.000 \times \pi/18$ [rad]	$Y_1 - Y_2$	0.002 [m]
$Z_1 - Z_2$	0.002 [m]	$\hat{M}(0)$	0.04 [kg]
	$\hat{N}_1(0),\hat{N}_2(0),\hat{N}_0(0)$	0.000,0.000,0.000 [N]	

Table 5.9. Parameters of the control signals

f_d	internal force	1.000 [N]
$c_1 = c_2$	damping coefficient	0.001 [Nms]
c_{20}	damping coefficient	0.006 [Nms]
γ_M	regressor gain	10.00 [m^2/kgs^2]
$\gamma_i(i = 1, 2)$	regressor gain	0.010 [s^2/kg]
γ_0	regressor gain	0.010 [s^2/kg]
γ_x	regressor gain	100.0 [N/m]
γ_y	regressor gain	100.0 [N/m]
γ_z	regressor gain	50.0 [N/m]

of the fingers and the object used in the simulations are given in Table 5.8 and the initial conditions are shown in Table 5.9. It should be remarked that both robot fingers are a little bigger than parameters given in Table 5.4. The ideal trajectory is given as

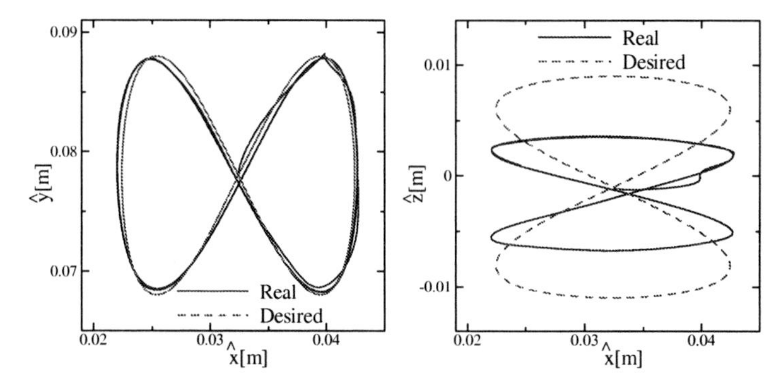

Fig. 5.11. Loci of $(\hat{x},\hat{y})$

Fig. 5.12. Loci of $(\hat{x},\hat{z})$

$$\begin{cases} x_d(t) = \hat{x}(0) + 0.01\sin(6t), \\ y_d(t) = \hat{y}(0) + 0.01\sin(12t), \\ z_d(t) = \hat{z}(0) + 0.01\sin(3t), \end{cases} \tag{5.123}$$

which is a Lissajous curve and control gains are set as in Table 5.9. Note that in this simulation the initial estimate $\hat{M}(0)$ for the object mass is set as $\hat{M}(0) = 0.04$ [kg], which is the true value of M. We show numerically obtained trajectories of $\hat{x}(t)$, $\hat{y}(t)$ and $\hat{z}(t)$ in Figures 5.11 and 5.12. As predicted from physical reasoning, the performance of trajectory tracking in the xy-plane is quite satisfactory owing to the excess joints that have the common z-axis of rotation as seen from Figure 5.11. On the contrary, both trajectory trackings in the xz-plane and yz-plane are considerably deteriorated due to shortage of freedoms of rotational motion in the x- and z-axes. It is interesting to remark that during trajectory tracking of the object centre through movements of the overall system, the object mass estimator $\hat{M}(t)$ fluctuates within $\pm$ 20% of the true value $M = 0.04$ [kg], even if the starting value $\hat{M}(0)$ is adjusted to the true value $M = 0.04$ [kg]. However, it is important to remark that this object mass estimator is indispensable even for trajectory tracking. Another important change is made in re-adjustment of the choice for the gain γ_M of the object mass estimator and other gains γ_i $(i = 1, 2)$. All these gains are chosen to be 3×10^2 times or more larger than the values given for blind grasping (compare Table 5.9 with Table 5.5).

5.11 Full-DOF Model of 3-D Grasping

Throughout the previous sections of this chapter, the physical and mathematical analysis has been developed based upon the assumption that spinning around the opposing axis connecting the two contact points O_1 and O_2 (see Figure 5.2) will no longer arise after the spinning motion has ceased due to

Table 5.10. Partial derivatives of constraints in (φ, ψ, θ)

$$\frac{\partial Q_1}{\partial \varphi} = \frac{\partial Q_2}{\partial \varphi} = 0, \qquad \frac{\partial Q}{\partial \varphi} = 0$$

$$\frac{\partial Y_i}{\partial \varphi} = (\boldsymbol{x}_{0i} - \boldsymbol{x})^{\mathrm{T}} \frac{\partial \boldsymbol{r}_Y}{\partial \varphi} = (\boldsymbol{x}_{0i} - \boldsymbol{x})^{\mathrm{T}} \boldsymbol{r}_Z = Z_i$$

$$\frac{\partial Y_i}{\partial \psi} = (\boldsymbol{x}_{0i} - \boldsymbol{x})^{\mathrm{T}} \frac{\partial \boldsymbol{r}_Y}{\partial \psi} = 0$$

$$\frac{\partial Y_i}{\partial \theta} = (\boldsymbol{x}_{0i} - \boldsymbol{x})^{\mathrm{T}} \frac{\partial \boldsymbol{r}_Y}{\partial \theta} = -(\boldsymbol{x}_{0i} - \boldsymbol{x})^{\mathrm{T}} \boldsymbol{r}_X = -(-1)^i (r_i + l_i)$$

$$\frac{\partial Z_i}{\partial \varphi} = (\boldsymbol{x}_{0i} - \boldsymbol{x})^{\mathrm{T}} \frac{\partial \boldsymbol{r}_Z}{\partial \varphi} = -(\boldsymbol{x}_{0i} - \boldsymbol{x})^{\mathrm{T}} \boldsymbol{r}_Y = -Y_i$$

$$\frac{\partial Z_i}{\partial \psi} = (\boldsymbol{x}_{0i} - \boldsymbol{x})^{\mathrm{T}} \frac{\partial \boldsymbol{r}_Z}{\partial \psi} = (\boldsymbol{x}_{0i} - \boldsymbol{x})^{\mathrm{T}} \boldsymbol{r}_X = (-1)^i (r_i + l_i)$$

$$\frac{\partial Z_i}{\partial \theta} = (\boldsymbol{x}_{0i} - \boldsymbol{x})^{\mathrm{T}} \frac{\partial \boldsymbol{r}_Z}{\partial \theta} = 0$$

$$Y_{\varphi i} = \frac{\partial Y_i}{\partial \varphi} = Z_i, \qquad Z_{\varphi i} = \frac{\partial Z_i}{\partial \varphi} = -Y_i$$

$$Y_{\psi i} = \frac{\partial Y_i}{\partial \psi} = 0, \qquad Z_{\psi i} = \frac{\partial Z_i}{\partial \psi} - (-1)^i r_i = (-1)^i l_i$$

$$Y_{\theta i} = \frac{\partial Y_i}{\partial \theta} + (-1)^i r_i = -(-1)^i l_i, \qquad Z_{\theta i} = 0$$

static friction when the object centre of mass approaches nearly to a point beneath the opposing axis. This assumption induces another non-holonomic constraint expressed as Equation (5.1), which implies that one of the three angular velocities $\boldsymbol{\omega} = (\omega_x, \omega_y, \omega_z)^{\mathrm{T}}$ is not independent. Therefore, motion of the grasped object has been expressed by Lagrange's equation in terms of five variables $(\dot{x}, \dot{y}, \dot{z}, \omega_y, \omega_z)$.

Instead of the assumption of Equation (5.1) based on static friction, we assume that there arises an external force of viscous friction $c_\varphi \omega_x$ effective for damping the spinning around the x-axis in relation to rotational motion of the grasped rigid object. Even in this case, the derivation of rolling contact contraints at both the contact points O_1 and O_2 developed in Section 5.3 is valid as far as Equations (5.14–5.20) are concerned. However, the Pfaffian constraints of Equation (5.21) should be expressed in this case by introducing the state vector $\boldsymbol{X} = (q_1^{\mathrm{T}}, q_2^{\mathrm{T}}, \boldsymbol{x}^{\mathrm{T}}, \varphi, \psi, \theta)^{\mathrm{T}}$ as follows:

$$\begin{cases} \lambda_{Yi} \left\{ \boldsymbol{Y}_{qi}^{\mathrm{T}} \dot{q}_i + \boldsymbol{Y}_{\boldsymbol{x}i}^{\mathrm{T}} \dot{\boldsymbol{x}} + Y_{\varphi i} \omega_x + Y_{\psi i} \omega_y + Y_{\theta i} \omega_z \right\} = 0 \\ \lambda_{Zi} \left\{ \boldsymbol{Z}_{qi}^{\mathrm{T}} \dot{q}_i + \boldsymbol{Z}_{\boldsymbol{x}i}^{\mathrm{T}} \dot{\boldsymbol{x}} + Z_{\varphi i} \omega_x + Z_{\psi i} \omega_y + Z_{\theta i} \omega_z \right\} = 0 \end{cases} \quad i = 1, 2 \qquad (5.124)$$

where we add

$$Y_{\varphi i} = \frac{\partial Y_i}{\partial \varphi}, \qquad Z_{\varphi i} = \frac{\partial Z_i}{\partial \varphi} \tag{5.125}$$

to Equations (5.22) and (5.23), respectively. Here, we use the symbol φ to express an indefinite integral of ω_x, that is, $\dot{\varphi} = \omega_x$. In relation to this modification, it is necessary to modify some formulae of Table 5.1 and add the partial derivatives of Y_i and Z_i in φ. All these can be recast into Table 5.10. The overall kinetic energy of the fingers–object system should be described, instead of Equation (5.25), as follows:

$$K = \frac{1}{2} \sum_{i=1,2} \dot{q}_i^{\mathrm{T}} H_i(q_i) \dot{q}_i + \frac{1}{2} M \|\dot{\boldsymbol{x}}\|^2 + \frac{1}{2} \boldsymbol{\omega}^{\mathrm{T}} R H R^{\mathrm{T}} \boldsymbol{\omega}, \tag{5.126}$$

where $\boldsymbol{\omega} = (\omega_x, \omega_y, \omega_z)^{\mathrm{T}}$, R denotes the orthogonal matrix defined by Equation (5.4) and H stands for the constant inertia matrix of the object evaluated on the basis of object coordinates $O_{\mathrm{c.m.}} - XYZ$ (see Figure 5.1 or 5.2). The total potential energy can be expressed as in Equation (5.26). Thus, owing to the variational principle applied in the form

$$\int_{t_0}^{t_1} \delta L \,\mathrm{d}t = \int_{t_0}^{t_1} \left\{ c_\varphi \omega_x \delta\varphi - \sum_{i=1,2} \left\{ u_i \delta q_i + \left(\lambda_{Yi} \boldsymbol{Y}_i^{\mathrm{T}} + \lambda_{Zi} \boldsymbol{Z}_i^{\mathrm{T}} \right) \delta \boldsymbol{X} \right\} \right\} \mathrm{d}t, \tag{5.127}$$

where $L = K - P + Q$, $\boldsymbol{Y}_1 = (\boldsymbol{Y}_{q1}^{\mathrm{T}}, 0_4, \boldsymbol{Y}_{\boldsymbol{x}1}^{\mathrm{T}}, Y_{\varphi 1}, Y_{\psi 1}, Y_{\theta 1})^{\mathrm{T}}$, $\boldsymbol{Y}_2 = (0_3, \boldsymbol{Y}_{q2}^{\mathrm{T}}, \boldsymbol{Y}_{\boldsymbol{x}2}^{\mathrm{T}}, Y_{\varphi 2}, Y_{\psi 2}, Y_{\theta 2})^{\mathrm{T}}$ and $\boldsymbol{Z}_1$ and $\boldsymbol{Z}_2$ express similar meanings, we obtain

$$\begin{aligned} H_i(q_i)\ddot{q}_i + \left\{ \frac{1}{2} \dot{H}_i(q_i) + S_i(q_i, \dot{q}_i) \right\} \dot{q}_i - (-1)^i f_i J_{0i}^{\mathrm{T}}(q_i) \boldsymbol{r}_X \\ - \lambda_{Yi} \boldsymbol{Y}_{qi} - \lambda_{Zi} \boldsymbol{Z}_{qi} + g_i(q_i) = u_i, \; i = 1, 2, \end{aligned} \tag{5.128}$$

$$\begin{aligned} M\ddot{\boldsymbol{x}} - (f_1 - f_2)\boldsymbol{r}_X + (\lambda_{Y1} + \lambda_{Y2})\boldsymbol{r}_Y \\ + (\lambda_{Z1} + \lambda_{Z2})\boldsymbol{r}_Z - Mg \begin{pmatrix} 0 \\ 1 \\ 0 \end{pmatrix} = 0, \end{aligned} \tag{5.129}$$

$$\begin{aligned} \bar{H}\dot{\boldsymbol{\omega}} + \left(\frac{1}{2} \dot{\bar{H}} + S \right) \boldsymbol{\omega} + c_\varphi \begin{pmatrix} \omega_x \\ 0 \\ 0 \end{pmatrix} - f_1 \begin{pmatrix} 0 \\ Z_1 \\ -Y_1 \end{pmatrix} - f_2 \begin{pmatrix} 0 \\ -Z_2 \\ Y_2 \end{pmatrix} \\ -\lambda_{Y1} \begin{pmatrix} Z_1 \\ 0 \\ l_1 \end{pmatrix} - \lambda_{Y2} \begin{pmatrix} Z_2 \\ 0 \\ -l_2 \end{pmatrix} - \lambda_{Z1} \begin{pmatrix} -Y_1 \\ -l_1 \\ 0 \end{pmatrix} - \lambda_{Z2} \begin{pmatrix} -Y_2 \\ l_2 \\ 0 \end{pmatrix} = 0, \end{aligned} \tag{5.130}$$

where $\bar{H} = RHR^{\mathrm{T}}$. Similary to Equation (5.34), taking inner products between $\dot{q}_i$ and Equation (5.128), $\dot{\boldsymbol{x}}$ and Equation (5.129), and $\boldsymbol{\omega}$ and Equation (5.130), we obtain the relation

$$\sum_{i=1,2} \dot{q}_i^{\mathrm{T}} u_i = \frac{\mathrm{d}}{\mathrm{d}t}(K+P) + c_\varphi \dot{\varphi}^2. \tag{5.131}$$

Finally, it is intersting to note that six vectors associated with f_i, f_2, λ_{Y1}, λ_{Y2}, λ_{Z1} and λ_{Z2} from the fourth term to the ninth term on the right-hand side of Equation (5.130) constitute, together with corresponding six vectors of Equation (5.129), a set of wrench vectors exerted on the three-dimensional rigid object. The last term on the left-hand side of Equation (5.129) is regarded as an external force vector caused by gravity.

First consider the stability of control of blind grasping under weightless circumstances when $g_i(q_i)$ in Equation (5.128) and $Mg(0,1,0)^{\mathrm{T}}$ in Equation (5.129) are missing. The control signal is the same as that in Equation (5.41). Substituting this control signal into Equation (5.128) yields

$$\begin{aligned} H_i(q_i)\ddot{q}_i &+ \left\{\frac{1}{2}\dot{H}_i(q_i) + S_i(q_i,\dot{q}_i) + C_i\right\}\dot{q}_i - (-1)^i J_{0i}^{\mathrm{T}}(q_i)\Delta f_i \boldsymbol{r}_X \\ &-(-1)^i J_{0i}^{\mathrm{T}}(q_i)\frac{f_d}{r_1+r_2}\left\{(Y_1-Y_2)\boldsymbol{r}_Y + (Z_1-Z_2)\boldsymbol{r}_Z\right\} \\ &-\lambda_{Yi}\left\{J_{0i}^{\mathrm{T}}(q_i)\boldsymbol{r}_Y - r_i((-1)^i r_Z(q_{i0})\boldsymbol{e}_i + r_{Zx}\boldsymbol{e}_{0i})\right\} \\ &-\lambda_{Zi}\left\{J_{0i}^{\mathrm{T}}(q_i)\boldsymbol{r}_Z + r_i((-1)^i r_Y(q_{i0})\boldsymbol{e}_i + r_{Yx}\boldsymbol{e}_{0i})\right\} = 0, \ i=1,2, \end{aligned} \tag{5.132}$$

where $\Delta f_i = f_i - f_0$ and f_0 is defined in Equation (5.46). These formulae are the same as those in Equation (5.49). To accompany the closed-loop expression for the finger dynamics, we rewrite Equations (5.129) and (5.130) as the following:

$$M\ddot{\boldsymbol{x}} - (\Delta f_1 - \Delta f_2)\boldsymbol{r}_X + (\lambda_{Y1}+\lambda_{Y2})\boldsymbol{r}_Y + (\lambda_{Z1}+\lambda_{Z2})\boldsymbol{r}_Z = 0, \tag{5.133}$$

$$\begin{aligned} \bar{H}\dot{\boldsymbol{\omega}} &+ \left(\frac{1}{2}\dot{\bar{H}} + S\right)\boldsymbol{\omega} + c_\varphi\begin{pmatrix}\omega_x\\0\\0\end{pmatrix} - \Delta f_1\begin{pmatrix}0\\Z_1\\-Y_1\end{pmatrix} - \Delta f_2\begin{pmatrix}0\\-Z_2\\Y_2\end{pmatrix} \\ -\lambda_{Y1}\begin{pmatrix}Z_1\\0\\l_1\end{pmatrix} &- \lambda_{Y2}\begin{pmatrix}Z_2\\0\\-l_2\end{pmatrix} - \lambda_{Z1}\begin{pmatrix}-Y_1\\-l_1\\0\end{pmatrix} - \lambda_{Z2}\begin{pmatrix}-Y_2\\l_2\\0\end{pmatrix} - \begin{pmatrix}S_X\\S_Y\\S_Z\end{pmatrix} = 0, \end{aligned} \tag{5.134}$$

where

$$S_X = 0, \quad S_Y = f_0(Z_1 - Z_2), \quad S_Z = -f_0(Y_1 - Y_2). \tag{5.135}$$

Then, similarly to the derivation of Equation (5.47), it follows that

$$\frac{\mathrm{d}}{\mathrm{d}t}E = -c_\varphi\dot{\varphi}^2 - \sum_{i=1,2}\dot{q}_i^{\mathrm{T}} C_i \dot{q}_i, \tag{5.136}$$

where

$$E = K + \frac{f_d}{2(r_1 + r_2)} \left\{ (Y_1 - Y_2)^2 + (Z_1 - Z_2)^2 \right\} \tag{5.137}$$

and K is defined as in Equation (5.126).

The overall fingers–object system depicted in Figure 5.2 superficially has 13 DOFs since the pair of fingers has seven joints and the object has three independent translational variables (x, y, z) and three independent angular velocity variables $(\omega_x, \omega_y, \omega_z)$. On the other hand, it has two holonomic constraints $Q_1 = 0$ and $Q_2 = 0$. Further, it is subject to four rolling contact constraints as shown in Equations (5.16) and (5.18). These four constraints are non-holonomic, but they are Pfaffian, that is, they are linear and homogeneous in the velocity variables [components of $\dot{\boldsymbol{X}} = (\dot{\boldsymbol{q}}_1^{\mathrm{T}}, \dot{\boldsymbol{q}}_2^{\mathrm{T}}, \dot{\boldsymbol{x}}^{\mathrm{T}}, \omega_x, \omega_y, \omega_z)^{\mathrm{T}}$]. Hence, in the sense of infinitesimal displacements $\delta\boldsymbol{X}$, these Pfaffian constraints can be written as

$$\boldsymbol{Y}_i^{\mathrm{T}} \delta\boldsymbol{X} = 0, \quad \boldsymbol{Z}_i^{\mathrm{T}} \delta\boldsymbol{X} = 0, \qquad i = 1, 2, \tag{5.138}$$

where $\boldsymbol{Y}_1 = (\boldsymbol{Y}_{q1}^{\mathrm{T}}, 0_4, \boldsymbol{Y}_{\boldsymbol{x}1}^{\mathrm{T}}, Y_{\varphi 1}, Y_{\psi 1}, Y_{\theta 1})^{\mathrm{T}}$, $\boldsymbol{Y}_2 = (0_3, \boldsymbol{Y}_{q2}^{\mathrm{T}}, \boldsymbol{Y}_{\boldsymbol{x}2}^{\mathrm{T}}, Y_{\varphi 2}, Y_{\psi 2}, Y_{\theta 2})^{\mathrm{T}}$ and $\boldsymbol{Z}_1$ and $\boldsymbol{Z}_2$ have similar meanings as treated in the derivation of the variational form described by Equation (5.127). Hence, the total number of DOFs of the fingers–object system is seven. Apparently, under weightlessness the number of finger joints is redundant; in particular some joints of the fingers in the z-axis are redundant. As an extreme case under the circumstances of weightlessness, it may be possible to consider a pair of robot fingers, one of which has a single joint in the z-axis and the other of which has two joints in the x- and z-axes. However, even in this setup with the minimum number of finger joints, the total number of degrees of freedom becomes three. In other words, the scalar function defined by Equation (5.137) cannot be regarded as a Lyapunov function, though it satisfies the Lyapunov-like relation of Equation (5.136).

Notwithstanding the redundancy in the system DOF, it is possible to prove that the closed-loop dynamics of Equations (5.135–5.137) converges asymptotically to the equilibrium manifold satisfying

$$\begin{cases} \dot{\boldsymbol{X}}(t) \to 0, \ Y_1(t) - Y_2(t) \to 0, \ Z_1(t) - Z_2(t) \to 0 \\ f_i(t) \to f_0, \ \lambda_{Yi}(t) \to 0, \ \lambda_{Zi}(t) \to 0 \quad (i = 1, 2) \end{cases} \tag{5.139}$$

as $t \to \infty$. Uniform boundedness of $\dot{\boldsymbol{X}}$, $Y_1 - Y_2$ and $Z_1 - Z_2$ follows immediately from the basic relation of Equation (5.136). Then, uniform boundedness of all multipliers Δf_i, λ_{Yi} and λ_{Zi} $(i = 1, 2)$ can be deduced by applying a similar argument to that presented in Section 5.8. Then, it is possible to see that $\ddot{\boldsymbol{X}}$ becomes bounded uniformly in t. This implies that $\dot{\boldsymbol{X}}(t)$ is uniformly continuous in t. Then, it is important to note that $\dot{q}_i(t)$ $(i = 1, 2)$ and $\dot{\varphi}(t)$ $(= \omega_x(t))$ are square-integrable over $t \in (0, \infty)$. Thus, according to Lemma 2 in Appendix A, it can be concluded that $\dot{q}_i(t) \to 0$ $(i = 1, 2)$ and $\dot{\varphi}(t) \to 0$ as $t \to \infty$. Convergences of $\dot{\psi}(t)$ and $\dot{\theta}(t)$ to zero as $t \to \infty$ can be verified

Table 5.11. Physical parameters of the fingers and object

$l_{11} = l_{21}$	length	0.040 [m]
$l_{12} = l_{22}$	length	0.040 [m]
$l_{13} = l_{23}$	length	0.030 [m]
m_{11}	weight	0.043 [kg]
m_{12}	weight	0.031 [kg]
m_{13}	weight	0.020 [kg]
l_{20}	length	0.000 [m]
m_{20}	weight	0.000 [kg]
m_{21}	weight	0.060 [kg]
m_{22}	weight	0.031 [kg]
m_{23}	weight	0.020 [kg]
I_{XX11}	inertia moment	5.375×10^{-7}[kgm^2]
$I_{YY11} = I_{ZZ11}$	inertia moment	6.002×10^{-6}[kgm^2]
I_{XX12}	inertia moment	3.875×10^{-7}[kgm^2]
$I_{YY12} = I_{ZZ12}$	inertia moment	4.327×10^{-6}[kgm^2]
I_{XX13}	inertia moment	2.500×10^{-7}[kgm^2]
$I_{YY13} = I_{ZZ13}$	inertia moment	1.625×10^{-6}[kgm^2]
I_{XX21}	inertia moment	7.500×10^{-7}[kgm^2]
$I_{YY21} = I_{ZZ21}$	inertia moment	8.375×10^{-6}[kgm^2]
I_{XX22}	inertia moment	3.875×10^{-7}[kgm^2]
$I_{YY22} = I_{ZZ22}$	inertia moment	4.327×10^{-6}[kgm^2]
I_{XX23}	inertia moment	2.500×10^{-7}[kgm^2]
$I_{YY23} = I_{ZZ23}$	inertia moment	1.625×10^{-6}[kgm^2]
$I_{XX} = I_{ZZ}$	inertia moment (object)	1.133×10^{-5}[kgm^2]
I_{YY}	inertia moment (object)	6.000×10^{-6}[kgm^2]
r_0	link radius	0.005 [m]
$r_i (i = 1, 2)$	radius	0.010 [m]
L	base length	0.063 [m]
M	object weight	0.040 [kg]
$l_i (i = 1, 2)$	object width	0.015 [m]
h	object height	0.050 [m]

in a similar argument to that developed through Equations (5.55–5.61) to reach Equation (5.61). Finally, it is possible to confirm by differentiation of Equation (5.20) with respect to t that $\dot{\boldsymbol{x}}(t) \to 0$ as $t \to \infty$. In order to show convergence of $Y_1 - Y_2$, $Z_1 - Z_2$, Δf_i, λ_{Yi} and λ_{Zi} $(i = 1, 2)$ to zero as $t \to \infty$, we can apply a similar argument to that presented in Equations (5.62–5.65) to conclude Equation (5.66).

Finally, it should be remarked that exponential convergence of Equation (5.138) in t can be assured by using a similar approach to that discussed in some cases of planar pinching or by devising a similar proof to that given in Section 5.9.

Table 5.12. Parameters of the control signals

f_d	internal force	1.000 [N]
$c_1 = c_2$	damping coefficient	0.001 [Nms]
c_{20}	damping coefficient	0.006 [Nms]
c_φ	damping coefficient	0.001 [Nms]

To confirm theoretical predictions for convergence of the physical variables of the fingers–object system depicted in Figure 5.2, we show computer simulation results based on the physical parameters of the system model given in Table 5.11. Control gains are given in Table 5.12, where $C_1 = \text{diag}(c_1, c_1, c_1)$ and $C_2 = \text{diag}(c_{20}, c_2, c_2, c_2)$. Note that c_{20} for damping of rotational movement of the first joint of the right-hand finger (corresponding to the thumb) is set considerably larger than c_1 and c_2 for the other joints, whose rotational axes are in the z-direction. The transient behaviours of all the key variables are presented in Figure 5.13. All six velocity variables of the object converge to zero within almost one second. Spinning motion around the x-axis arises at an early stage of the transient process but is very small. Note that in this simulation all the velocity variables are set to zero at the initial time $t = 0$. Initial discrepancies between Y_1 and Y_2 and between Z_1 and Z_2 are given as shown in Figure 5.13, that is, $Y_1(0) - Y_2(0) = 2.0$ [mm] and $Z_1(0) - Z_2(0) = 2.0$ [mm]. Nevertheless, the transient behaviours of $Y_1 - Y_2$, λ_{Yi}, and ω_z are more oscillatory than those of $Z_1 - Z_2$, λ_{Zi} and ω_y.

Next, we discuss the ordinary case of pinching under the effect of gravity by considering the same control signal for blind grasping defined in Equation (5.67). By substituting this control signal into Equation (5.128), we obtain the closed-loop dynamics of the fingers–object system as follows:

$$\begin{aligned} H_i(q_i)\ddot{q}_i &+ \left\{\frac{1}{2}\dot{H}_i(\boldsymbol{q}_i) + S_i(q_i, \dot{q}_i) + C_i\right\}\dot{q}_i - (-1)^i J_{0i}^{\mathrm{T}}(q_i)\Delta f_i \boldsymbol{r}_X \\ &- \Delta\lambda_{Yi}\left\{J_{0i}^{\mathrm{T}}(q_i)\boldsymbol{r}_Y - r_i(-1)^i r_Z(q_{i0})\boldsymbol{e}_i + r_{Zx}\boldsymbol{e}_{0i}\right\} \\ &- \Delta\lambda_{Zi}\left\{J_{0i}^{\mathrm{T}}(q_i)\boldsymbol{r}_Z + r_i(-1)^i r_Y(q_{i0})\boldsymbol{e}_u + r_{Yx}\boldsymbol{e}_{0i}\right\} \\ &- \frac{\Delta Mg}{2} \cdot \frac{\partial y_{0i}}{\partial q_i} - r_i\Delta N_i - r_i\Delta N_{0i} = 0, \qquad i = 1, 2, \end{aligned} \tag{5.140}$$

$$\begin{aligned} M\ddot{\boldsymbol{x}} - (\Delta f_1 - \Delta f_2)\boldsymbol{r}_X &+ (\Delta\lambda_{Y1} + \Delta\lambda_{Y2})\boldsymbol{r}_Y \\ &+ (\Delta\lambda_{Z1} + \Delta\lambda_{Z2})\boldsymbol{r}_Z = 0, \end{aligned} \tag{5.141}$$

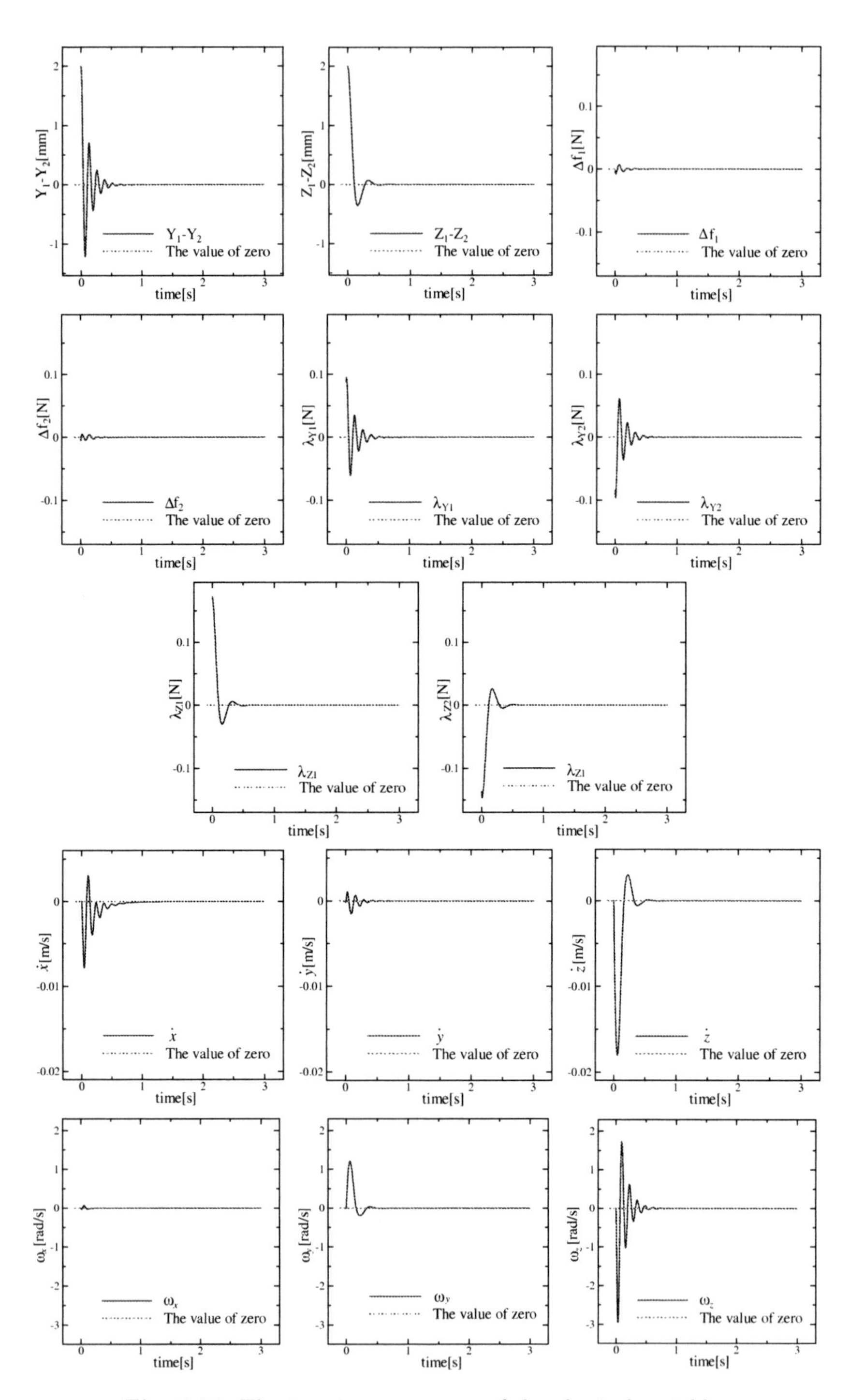

Fig. 5.13. The transient responses of the physical variables

$$\bar{H}\dot{\omega} + \left(\frac{1}{2}\dot{\bar{H}} + S\right)\omega + c_\varphi \begin{pmatrix} \omega_x \\ 0 \\ 0 \end{pmatrix} - \Delta f_1 \begin{pmatrix} 0 \\ Z_1 \\ -Y_1 \end{pmatrix}$$
$$-\Delta f_2 \begin{pmatrix} 0 \\ -Z_2 \\ Y_2 \end{pmatrix} - \Delta\lambda_{Y1} \begin{pmatrix} Z_1 \\ 0 \\ l_1 \end{pmatrix} - \Delta\lambda_{Y2} \begin{pmatrix} Z_2 \\ 0 \\ -l_2 \end{pmatrix}$$
$$-\Delta\lambda_{Z1} \begin{pmatrix} -Y_1 \\ -l_1 \\ 0 \end{pmatrix} - \Delta\lambda_{Z2} \begin{pmatrix} -Y_2 \\ l_2 \\ 0 \end{pmatrix} - \begin{pmatrix} S_X \\ S_Y \\ S_Z \end{pmatrix} = 0, \quad (5.142)$$

where Δf_i, $\Delta\lambda_{Yi}$, $\Delta\lambda_{Zi}$, ΔN_i for $i = 1, 2$, ΔM and ΔN_{02} are defined in Equation (5.77) and N_i ($i = 1, 2$) and N_{02} are also defined in Equation (5.77), $\Delta N_{01} = 0$, and S_X, S_Y and S_Z are defined as follows:

$$\begin{cases} S_X = \dfrac{Mg}{2}\left\{(Z_1 + Z_2)r_{Yy} - (Y_1 + Y_2)r_{Zy}\right\}, \\ S_Y = f_d(Z_1 - Z_2) - \dfrac{Mg}{2}\left\{r_{Xy}(Z_1 + Z_2) + r_{Zy}(l_1 - l_2)\right\}, \\ S_Z = -f_d(Y_1 - Y_2) + \dfrac{Mg}{2}\left\{r_{Xy}(Y_1 + Y_2) + r_{Yy}(l_1 - l_2)\right\}. \end{cases} \quad (5.143)$$

Along a solution trajectory to the closed-loop dynamics of Equations (5.140) to (5.142), the following equality relation follows:

$$\frac{\mathrm{d}}{\mathrm{d}t}(K + W) = -c_\varphi \omega_x^2 - \sum_{i=1,2} \dot{q}_i^{\mathrm{T}} C_i \dot{q}_i, \quad (5.144)$$

where W is defined as Equation (5.88) and K signifies the total kinetic energy described by Equation (5.126). It is quite interesting to compare Equation (5.144) with Equation (5.87), by bearing in mind the difference in modelling physical behaviours of rolling contact constraints about spinning motion of the object around the opposing axis between the contact points. In the former case, it is assumed that rotational movement around the opposing axis is stacked by static friction under the assumption that the object centre of mass remains nearly beneath the opposing axis. This assumption deprives the angular velocity variable ω_x of independence. In the latter case, rotational movements of the object around the x-axis are always accompanied by viscous friction. This does not deprive the velocity ω_x of independence.

We shall show numerical simulation results in Figures 5.14–5.16 by using the same physical setup of the fingers-object system depicted in Figure 5.2, whose parameters are given in Table 5.11. Note that each of the numerical values of Table 5.11 is identical with the corresponding parameter in Table 5.4. However, the simulation results using the control gains given in Table 5.5 lead to considerably slow convergence of most of the physical variables.

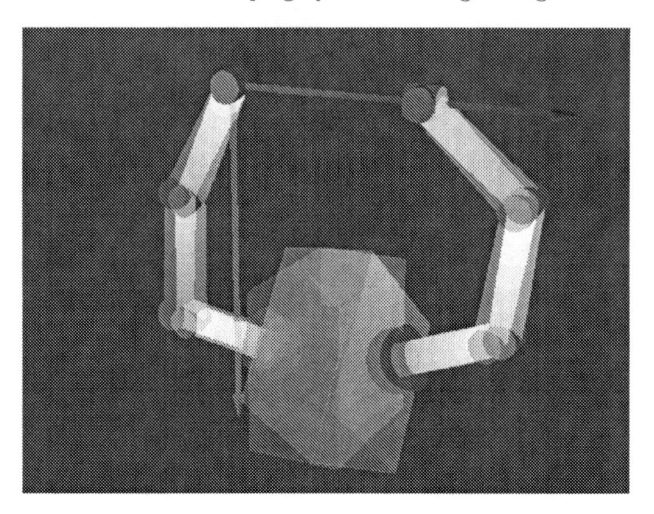

Fig. 5.14. Motions of pinching an object with parallel flat surfaces in 3-D space

Table 5.13. Parameters of the control signals

f_d	internal force	0.300 [N]
$c_1 = c_2$	damping coefficient	0.001 [Nms]
c_{20}	damping coefficient	0.006 [Nms]
c_φ	damping coefficient	4.0×10^{-4} [Nms]

Instead of the control signals in Table 5.5, we show numerical simulation results in Figure 5.15 using the control signal gains given in Table 5.13 and set the gains γ_M, γ_i $(i = 1, 2)$ and γ_0 as in Table 5.5. In particular, the target pushing force parameter f_d had to be decreased to $f_d = 0.3$ [N], which is almost comparable with the numerical value of $Mg/2$. According to Figure 5.16, the speed of convergence of S_X is slow in comparison with that of S_Y or S_Z. The reason is that the early mismatch of the object mass estimator $\hat{M}$ with the true value M induces a large amount of spinning motion as shown in the graph of ω_x in Figure 5.16, which is significant compared with the responses of ω_y and ω_z. Figure 5.14 shows a superimposition of two poses of the overall fingers–object system, the initial pose at $t = 0$ and the final pose satisfying force/torque balance when t tends to infinity. From this figure, we see that the object rotated considerably around the x-axis until establishing force/torque balance. This phenomenon might happen due to the selection of a relatively small damping gain c_φ in Table 5.13. Based on this simulation result, there still remains a lot of interesting problems in the tuning of control gains in relation to the geometry and physical scales of the objects to be grasped.

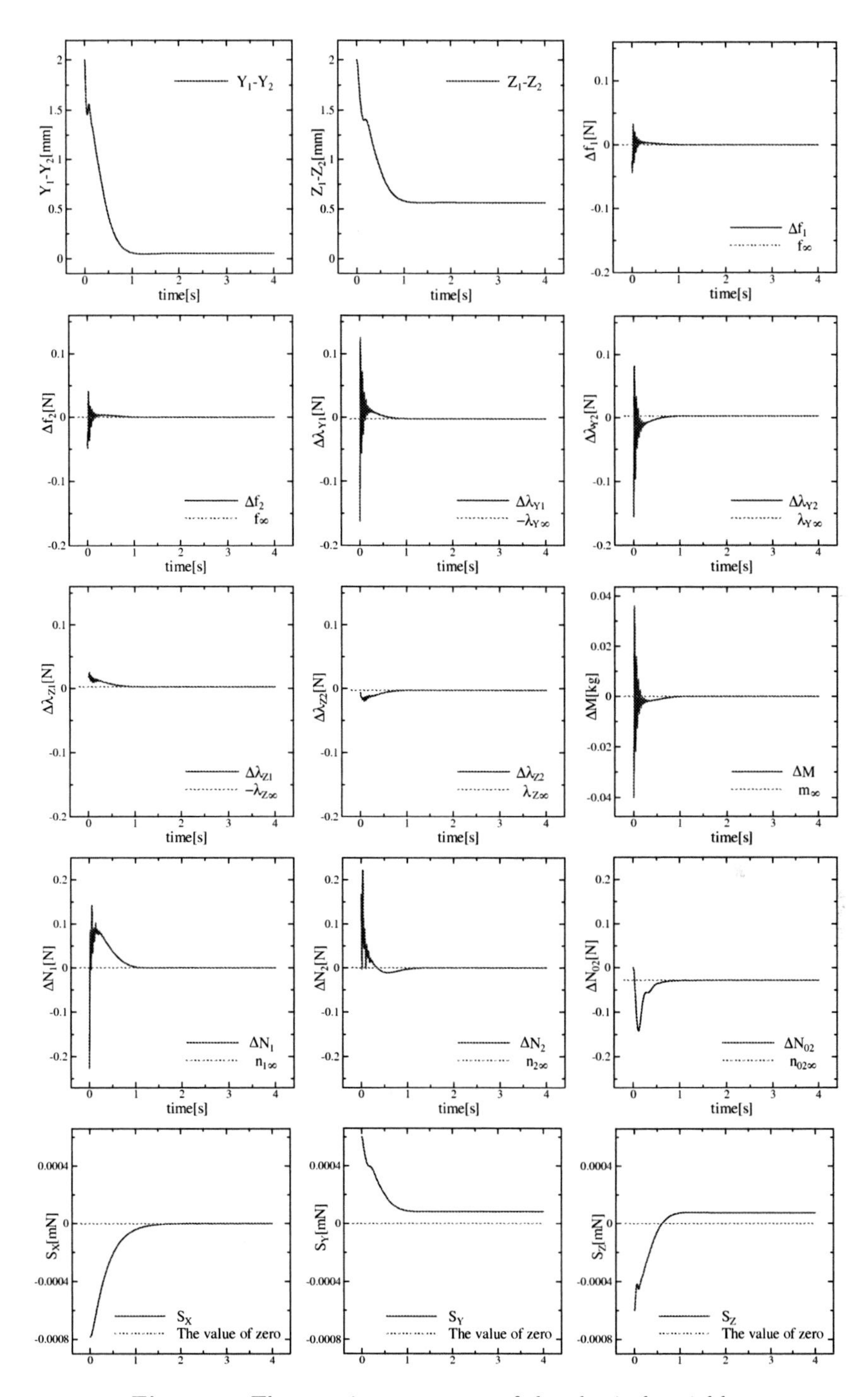

Fig. 5.15. The transient responses of the physical variables

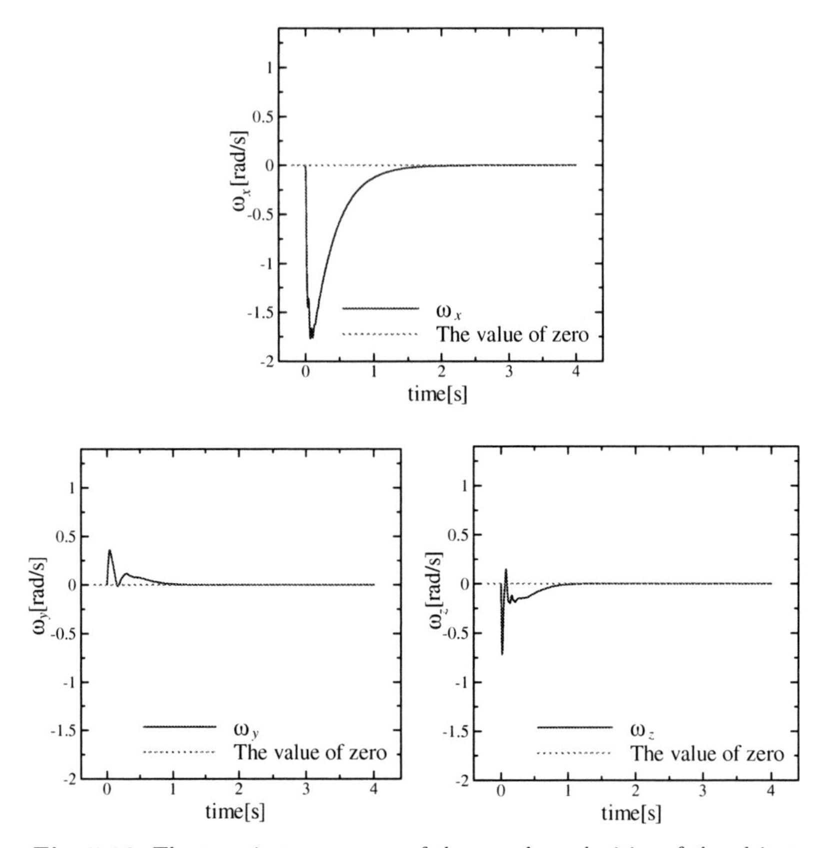

Fig. 5.16. The transient responses of the angular velocities of the object

Theoretical discussions of proofs of convergence of the solution trajectories to the closed-loop dynamics of Equations (5.132–5.134) can be developed in parallel with the mathematical arguments presented in the previous section. The most essential part of the discussions is the derivation of the linear gradient equation for the artificial potential function W in Equation (5.144) defined as Equation (5.88) with respect to q_1, q_2, $\boldsymbol{x}$ and $(\varphi, \psi, \theta)^{\mathrm{T}}$. This gradient equation is coincident with the set of Equations (5.140), (5.141) and (5.142) in which $\dot{q}_i$, $\ddot{q}_i$ $(i = 1, 2)$, $\ddot{\boldsymbol{x}}$, $\boldsymbol{\omega}$ and $\dot{\boldsymbol{\omega}}$ vanish, that is, all the acceleration and velocity terms vanish. Then, what values does the vector $\Delta\boldsymbol{\lambda} = (\Delta f_1, \Delta f_2, \Delta\lambda_{Y1}, \Delta\lambda_{Y2}, \Delta\lambda_{Z1}, \Delta\lambda_{Z2}, \Delta M, \Delta N_1, \Delta N_2, \Delta N_{02})^{\mathrm{T}}$ converge to as $t \to \infty$? According to Equation (5.141), it is possible to notice that Δf_1 and Δf_2 converge to the same certain value f_∞, similarly λ_{Y2} converges to $-\lambda_{Y\infty}$ if λ_{Y1} converges to $\lambda_{Y\infty}$ as $t \to \infty$ and λ_{Z2} converges to $-\lambda_{Z\infty}$ if λ_{Z1} does to $\lambda_{Z\infty}$. Hence, the limiting values of the vector $\Delta\boldsymbol{\lambda}' = (\Delta f_1, \Delta\lambda_{Y1}, \Delta\lambda_{Z1}, \Delta M, \Delta N_1, \Delta N_2, \Delta N_{02})^{\mathrm{T}}$ should satisfy the following linear equation:

$$\Delta \boldsymbol{\lambda}'_\infty = \lim_{t\to\infty} \left((A')^{\mathrm{T}} A'\right)^{-1} (A')^{\mathrm{T}} \boldsymbol{d}', \tag{5.145}$$

where $\Delta\boldsymbol{\lambda}'_\infty = (\Delta f_\infty, \Delta\lambda_{Y\infty}, \Delta\lambda_{Z\infty}, \Delta M_\infty, \Delta N_{1\infty}, \Delta N_{2\infty}, \Delta N_{02\infty})^{\mathrm{T}}$, and

$$A' = \begin{pmatrix} \dfrac{\partial Q_1}{\partial q_1} & \boldsymbol{Y}_{q1} & \boldsymbol{Z}_{q1} & \dfrac{g}{2}\cdot\dfrac{\partial y_{01}}{\partial q_1} & r_1\boldsymbol{e}_1 & 0_3 & 0_3 \\ \dfrac{\partial Q_2}{\partial q_2} & -\boldsymbol{Y}_{q2} & -\boldsymbol{Z}_{q2} & \dfrac{g}{2}\cdot\dfrac{\partial y_{02}}{\partial q_2} & 0_4 & r_2\boldsymbol{e}_2 & r_2\boldsymbol{e}_{02} \\ 0 & Z_1 - Z_2 & -Y_1 + Y_2 & 0 & 0 & 0 & 0 \\ Z_1 + Z_2 & 0 & -l_1 - l_2 & 0 & 0 & 0 & 0 \\ -(Y_1 + Y_2) & l_1 + l_2 & 0 & 0 & 0 & 0 & 0 \end{pmatrix}, \tag{5.146}$$

$$\boldsymbol{d}' = -(0_7, S_X, S_Y, S_Z)^{\mathrm{T}}. \tag{5.147}$$

The Hessian matrix of $\bar{W}$ $(= W - (\gamma_M/2)\Delta M^2)$ with respect to p_1, p_2, q_{20} and $(\varphi, \psi, \theta)^{\mathrm{T}}$ can be calculated in a manner similar to the case when the Hessian in the five-variables problem is obtained in Tables 5.7.

5.12 Supplementary Results

Under the constraints of Equation (5.1) or (5.2) and Equation (5.5), it is possible to derive partial derivatives of $\boldsymbol{r}_X$, $\boldsymbol{r}_Y$ and $\boldsymbol{r}_Z$ with respect to θ and ψ, where

$$\theta(t) = \int_0^t \omega_z(\tau)\,\mathrm{d}\tau, \quad \psi(t) = \int_0^t \omega_y(\tau)\,\mathrm{d}\tau. \tag{5.148}$$

First note that

$$\begin{cases} \dot{\boldsymbol{r}}_Y = \left(\dfrac{\partial \boldsymbol{r}_Y}{\partial\theta}\right)\dot\theta + \left(\dfrac{\partial \boldsymbol{r}_Y}{\partial\psi}\right)\dot\psi = \left(\dfrac{\partial \boldsymbol{r}_Y}{\partial\theta}\right)\omega_z + \left(\dfrac{\partial \boldsymbol{r}_Y}{\partial\psi}\right)\omega_y, \\ \dot{\boldsymbol{r}}_Z = \left(\dfrac{\partial \boldsymbol{r}_Z}{\partial\theta}\right)\omega_z + \left(\dfrac{\partial \boldsymbol{r}_Z}{\partial\psi}\right)\omega_y. \end{cases} \tag{5.149}$$

On the other hand, it follows from Equations (5.5) and (5.2) that

$$\begin{cases} \dot{\boldsymbol{r}}_Y = -\omega_z\boldsymbol{r}_X + \omega_x\boldsymbol{r}_Z = -\omega_z(\boldsymbol{r}_X + \xi_z\boldsymbol{r}_Z) - \omega_y(\xi_y\boldsymbol{r}_Z), \\ \dot{\boldsymbol{r}}_Z = \omega_y\boldsymbol{r}_X - \omega_x\boldsymbol{r}_Y = \omega_y(\boldsymbol{r}_X + \xi_y\boldsymbol{r}_Y) + \omega_z(\xi_z\boldsymbol{r}_Y). \end{cases} \tag{5.150}$$

Comparison of Equation (5.148) with Equation (5.149) leads to

$$\frac{\partial}{\partial\theta}\boldsymbol{r}_Y = -\boldsymbol{r}_X - \xi_z\boldsymbol{r}_Z, \quad \frac{\partial}{\partial\theta}\boldsymbol{r}_Z = \xi_z\boldsymbol{r}_Y, \tag{5.151}$$

$$\frac{\partial}{\partial\psi}\boldsymbol{r}_Y = -\xi_y\boldsymbol{r}_Z, \quad \frac{\partial}{\partial\psi}\boldsymbol{r}_Z = \boldsymbol{r}_X + \xi_y\boldsymbol{r}_Y. \tag{5.152}$$

As to the partial derivatives of $\boldsymbol{r}_X$ in θ and ψ, we see that

$$\begin{cases} \dot{\boldsymbol{r}}_X = \left(\frac{\partial \boldsymbol{r}_X}{\partial \theta}\right)\omega_z + \left(\frac{\partial \boldsymbol{r}_X}{\partial \psi}\right)\omega_y, \\ \dot{\boldsymbol{r}}_X = \omega_z \boldsymbol{r}_Y - \omega_y \boldsymbol{r}_Z, \end{cases} \tag{5.153}$$

which leads to

$$\frac{\partial}{\partial \theta}\boldsymbol{r}_X = \boldsymbol{r}_Y, \quad \frac{\partial}{\partial \psi}\boldsymbol{r}_X = -\boldsymbol{r}_Z. \tag{5.154}$$

Next, we will discuss how to derive the partial derivatives of the inertia matrix $\bar{H}$ of Equation (5.28) with respect to θ or ψ. On account of Equations (5.151) and (5.154), we see that

$$\begin{aligned} \frac{\partial}{\partial \theta}R(t) &= \left(\frac{\partial \boldsymbol{r}_X}{\partial \theta}, \frac{\partial \boldsymbol{r}_Y}{\partial \theta}, \frac{\partial \boldsymbol{r}_Z}{\partial \theta}\right) = (\boldsymbol{r}_Y, -(\boldsymbol{r}_X + \xi_z \boldsymbol{r}_Z), \xi_z \boldsymbol{r}_Y) \\ &= R(t)\Omega_\theta, \end{aligned} \tag{5.155}$$

where Ω_θ denotes the following skew-symmetric matrix:

$$\Omega_\theta = \begin{pmatrix} 0 & -1 & 0 \\ 1 & 0 & \xi_z \\ 0 & -\xi_z & 0 \end{pmatrix}. \tag{5.156}$$

Hence,

$$\begin{aligned} \frac{\partial}{\partial \theta}\bar{H} &= \left(\frac{\partial}{\partial \theta}R\right)HR^{\mathrm{T}} + RH\left(\frac{\partial R}{\partial \theta}\right)^{\mathrm{T}} \\ &= R\left(\Omega_\theta H - H\Omega_\theta\right)R^{\mathrm{T}}. \end{aligned} \tag{5.157}$$

Analogously, we obtain

$$\frac{\partial}{\partial \psi}\bar{H} = R\left(\Omega_\psi H - H\Omega_\psi\right)R^{\mathrm{T}} \tag{5.158}$$

where

$$\Omega_\psi = \begin{pmatrix} 0 & 0 & 1 \\ 0 & 0 & \xi_y \\ -1 & -\xi_y & 0 \end{pmatrix}. \tag{5.159}$$

Thus, the partial derivatives of each entry of $\bar{H}$ with respect to θ and ψ can be found from Equations (5.157) and (5.158). On the other hand, the partial derivatives of ξ_y and ξ_z with respect to θ or ψ can be calculated from the relations

$$\begin{cases} \dfrac{\partial}{\partial \theta}(\boldsymbol{x}_1 - \boldsymbol{x}_2) = (r_1 + r_2)\dfrac{\partial}{\partial \theta}\boldsymbol{r}_X = (r_1 + r_2)\boldsymbol{r}_Y, \\ \dfrac{\partial}{\partial \psi}(\boldsymbol{x}_1 - \boldsymbol{x}_2) = (r_1 + r_2)\dfrac{\partial}{\partial \psi}\boldsymbol{r}_X = -(r_1 + r_2)\boldsymbol{r}_Z. \end{cases} \tag{5.160}$$

Thus, the partial derivatives of each entry h_{ij} $(i, j = 1, 2)$ of H_0 with respect to θ or ψ appearing in Equation (5.33) can be systematically calculated using Equations (5.157), (5.158) and (5.160).

Before closing this chapter, we show how to evaluate the partial derivatives of Y_i and Z_i for $i = 1, 2$ with respect to θ or ψ under the rolling constraints of Equations (5.16) and (5.19). Since those rolling constraints are of a Pfaffian form, Equation (5.16) can be re-interpreted by the analysis of infinitesimally small variation as

$$\begin{cases} \delta Y_1 = \dfrac{\partial Y_1}{\partial \theta}\delta\theta + \left(\dfrac{\partial Y_1}{\partial q_1}\right)^{\mathrm{T}} \delta q_1, \\ \delta Z_1 = \dfrac{\partial Z_1}{\partial \theta}\delta\psi + \left(\dfrac{\partial Z_1}{\partial q_1}\right)^{\mathrm{T}} \delta q_1. \end{cases} \tag{5.161}$$

Comparing these two equations with Equation (5.16) yields

$$\begin{cases} \dfrac{\partial Y_1}{\partial \theta} = r_1, \quad \dfrac{\partial Y_1}{\partial \psi} = 0, \quad \dfrac{\partial Y_1}{\partial q_1} = -r_1 r_{Zz}\boldsymbol{e}_1, \\ \dfrac{\partial Z_1}{\partial \theta} = 0, \quad \dfrac{\partial Z_1}{\partial \psi} = -r_1, \quad \dfrac{\partial Z_1}{\partial q_1} = r_1 r_{Yz}\boldsymbol{e}_1. \end{cases} \tag{5.162}$$

Similarly, we see that

$$\begin{cases} \dfrac{\partial Y_2}{\partial \theta} = -\theta, \quad \dfrac{\partial Y_2}{\partial \psi} = 0 \\ \dfrac{\partial Z_2}{\partial \theta} = 0, \quad \dfrac{\partial Z_2}{\partial \psi} = r_2 \end{cases} \tag{5.163}$$

and the partial derivatives $\partial Y_2/\partial q_2$ and $\partial Z_2/\partial q_2$ are evaluated as in Equation (5.113).

6

Dexterity and Control for Stable Grasping by Soft Fingers

In the previous chapters, stable grasping by a pair of robot fingers interacting rigidly with a rigid object was analysed on the basis of indirect control of rolling contact constraint forces to establish force/torque balance. A class of coordinated control signals based on fingers–thumb opposability was shown to be effective in realising stable grasping in a blind manner.

This chapter extends these results obtained in the case of rigid contacts to the case of robot fingers equipped with soft and deformable finger-ends. The most crucial difference of prehensility between the rigid contact and the soft area contact is that in the former case the stability region of grasping of a thin light object becomes narrow but in the latter case a thinner object with flat surfaces can be grasped securely with a larger stability margin. Dexterity can be enhanced by expansion of the stability margin owing to coordinated regulation of reproducing and damping forces of finger-tip deformations that increase the net DOFs of the system. Another noteworthy difference arising in the latter case is that rolling contact constraints should be treated as non-holonomic. Nevertheless, those constraints can be incorporated into Lagrange's equation of motion of the system accompanied by Lagrange's multipliers. Again, stability analysis becomes applicable with the aid of differetial-geometric concepts such as Riemannian metrics and Morse functions.

6.1 Lumped-Parameterisation of the Behaviours of Soft and Visco-elastic Fingertips

As shown in Figure 6.1 a narrow strip with width $r\mathrm{d}\theta$ and radius $r\sin\theta$ in the contact area produces a reproducing force in the direction to the centre O_{0i} of curvature of the hemisphere with magnitude

$$k(2\pi r\sin\theta)\mathrm{d}\theta \times r(\cos\theta - \cos\theta_0)\cos\theta, \tag{6.1}$$

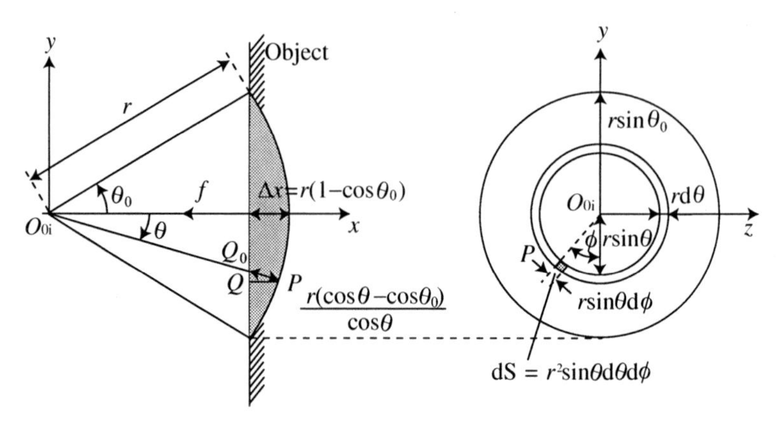

Fig. 6.1. Geometric relations related to lumped parameterisation of fingertip stress distribution into a single reproducing force focussed on the centre of curvature O_{0i}

where k denotes the stiffness parameter of the soft and deformable material per unit area, $(2\pi r^2 \sin\theta)\mathrm{d}\theta$ is the area of the narrow circular strip as shown in Figure 6.1, and

$$\frac{r(\cos\theta - \cos\theta)}{\cos\theta} = \left(r - \frac{r - \Delta x}{\cos\theta}\right)$$
$$= r - \frac{r\cos\theta_0}{\cos\theta} \tag{6.2}$$

denotes the length of deformation at angle θ. Since the total reproducing force with magnitude (6.1) generated from the narrow circular strip contributes to the direction Δx (the arrow denoted by f in Figure 6.1) by $\cos\theta$, the total reproducing force can be expressed as the integral

$$\bar{f} = \int_0^{\theta_0} 2\pi k r^2 \sin\theta(\cos\theta - \cos\theta_0)\mathrm{d}\theta$$
$$= \pi k r^2(1 - \cos\theta_0)^2 = \pi k \Delta x^2. \tag{6.3}$$

This means that the reproducing force produced by the deformed area can be approximately expressed by an increasing function of Δx (the maximum length of displacement). It should be noted that the moment $M = (M_x, M_y, M_z)^{\mathrm{T}}$ around O_{0i} becomes

$$\overrightarrow{O_{0i}P} \times k\overrightarrow{PQ}\mathrm{d}S, \tag{6.4}$$

where $\mathrm{d}S = r^2 \sin\theta\mathrm{d}\theta\mathrm{d}\phi$ (which denotes an infinitesimally small area as shown in Figure 6.1). It is evident that the area integral of Equation (6.4) with respect to $\mathrm{d}S$ over $\phi \in [0, 2\pi]$ and $\theta \in [0, \theta_0]$ vanishes due to the symmetry of the sinusoidal functions $\sin\phi$ and $\cos\phi$ appearing in Equation (6.4). Thus, the

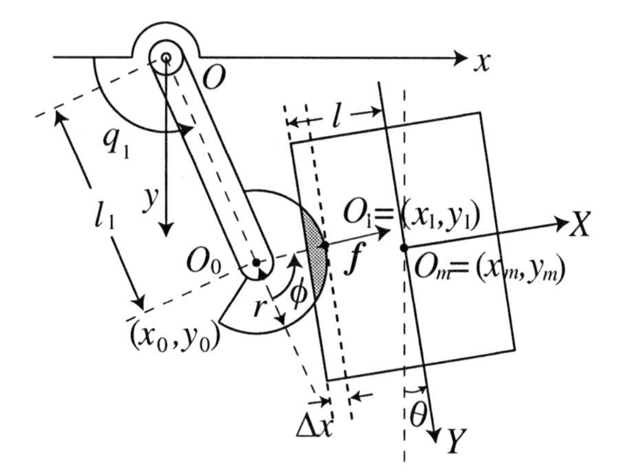

Fig. 6.2. Control for immobilisation of rotational motion of a 2-D object pivoted about the z-axis through O_m by a single-DOF soft finger

moment acting around the point O_{0i} caused by overall deformations of the soft material becomes zero. As to the lumped parametrisation of the distributed viscous forces, a similar argument can be applied. It can be concluded that the viscous force acting on the point O_{0i} can be expressed in the form $\xi_i(\Delta x_i)\Delta\dot{x}_i$, where $\xi_i(\Delta x_i)$ is an increasing function of Δx_i.

6.2 Stabilisation of a 2-D Object by a Single DOF Soft Finger

It is small wonder that we could find the name of Isaac Newton in John Napier's book "Hands". On page 51 of the book, Napier wrote:

> The thumb, the "lesser hand" as Albinus called it, is the most specialized of the digits. Isaac Newton once remarked that, in the absence of any other proof, the thumb alone would convince him of God's existence.

No solid reason for quoting the name of Newton was presented in Napier's book. Hence we attempt to put forward a fact supporting the statement by illustrating the role played by Newtonian mechanics in the simplest model of a robotic thumb in the stabilisation of the physical interaction between the robot thumb and an object.

Consider the simplest testbed problem of control for immobilising a 2-D rigid object rotating around a fixed pivotal axis by a single-DOF finger with a spherical tip made of visco-elastic material. The overall finger–object system is depicted in Figure 6.2, where motion of the finger and object is confined to a horizontal plane. As discussed in Section 3.6 the total DOFs of the system is zero when the finger-end is rigid and contacts rigidly and pointwise with the rigid object (Figure 3.9). In that case, rotational motion of the object cannot be stabilized in a dynamic sense. Differently from that case, such a soft finger with a single joint can stabilise the motion of a 2-D object in a dynamic sense because the net number of DOFs of the system becomes one. However, the rolling constraint must not be treated as a holonomic constraint.

For the sake of the derivation of Lagrange's equation of motion for the finger–object system depicted in Figure 6.2, we derive the kinetic energy K of the overall system and the potential energy P of the fingertip deformation. The former becomes as follows:

$$K = \frac{1}{2} I_1 \dot{q}_1^2 + \frac{1}{2} I \dot{\theta}^2, \tag{6.5}$$

where I_1 denotes the moment of inertia of the finger around the axis through the joint O and I that of the object around O_m. Referring directly to Equation (1.29) and bearing in mind that the reproducing force $f(\Delta x)$ due to finger-tip deformation can be regarded as $-\bar{f}(\Delta x)$ given in Equation (6.3), we can see that the potential energy P of fingertip deformation is described as

$$P(\Delta x) = \int_0^{\Delta x} \bar{f}(\eta) \, \mathrm{d}\eta, \tag{6.6}$$

where $\bar{f}(\eta) = \pi k \eta^2$ as shown in Equation (6.3). In the following, however, we regard the function $\bar{f}(\Delta x)$ as a rapidly increasing function with increasing Δx that is proportional to the stiffness parameter k [N/m^2]. Next, note that the condition of contact between the fingertip and the object should be expressed in the direction normal to the object surface as follows:

$$(r - \Delta x) + l = (x_m - x_0) \cos\theta - (y_m - y_0) \sin\theta \tag{6.7}$$

[see Equation (3.2)]. However, this equality cannot be treated as a holonomic constraint, because Δx is not a fixed constant but a changeable variable, though it is not a component of the generalized coordinates that are taken as $\boldsymbol{X} = (q_1, \theta)^{\mathrm{T}}$. On the other hand, the rolling contact constraint in the direction tangential to the object surface should be expressed as

$$(r - \Delta x) \frac{\mathrm{d}}{\mathrm{d}t} \phi = -\frac{\mathrm{d}}{\mathrm{d}t} Y, \tag{6.8}$$

where Y and ϕ are given by

$$Y = (x_0 - x_m) \sin\theta + (y_0 - y_m) \cos\theta, \tag{6.9}$$

$$\phi = \pi + \theta - q_1 \tag{6.10}$$

[see Equations (3.4) and (3.5)]. It should be remarked that Equation (6.8) cannot be integrated in time t and therefore it must be treated as a non-holonomic constraint. Notwithstanding this fact, Equation (6.8) can be regarded in a form written in terms of infinitesimally small variation $\delta \boldsymbol{X}$ as follows:

$$\left\{(r-\Delta x)\frac{\partial \phi}{\partial \boldsymbol{X}^{\mathrm{T}}}+\frac{\partial Y}{\partial \boldsymbol{X}^{\mathrm{T}}}\right\}\delta \boldsymbol{X}=0. \tag{6.11}$$

Thus, the variational form for the Lagrangian $L = K - P$ with external forces of control input u, damping $\xi(\Delta x)\Delta\dot{x}$, and the non-holonomic constraint of Equation (6.11) can be expressed as follows:

$$\int_{t_1}^{t_2}\left[\delta L-\xi(\Delta x)\Delta\dot{x}\frac{\partial \Delta x}{\partial \boldsymbol{X}^{\mathrm{T}}}\delta\boldsymbol{X}+u^{\mathrm{T}}\delta q_1 \right.$$
$$\left. +\lambda\left\{(r-\Delta x)\frac{\partial \phi}{\partial \boldsymbol{X}^{\mathrm{T}}}+\frac{\partial Y}{\partial \boldsymbol{X}^{\mathrm{T}}}\right\}\delta\boldsymbol{X}\right]\mathrm{d}t=0 \tag{6.12}$$

where λ expresses a Lagrange multiplier introduced correspondingly to the non-holonomic constraint of Equation (6.8). Applying the variational principle to the above form it follows that

$$I_1\ddot{q}_1+\left\{\bar{f}(\Delta x)+\xi(\Delta x)\Delta\dot{x}\right\}\frac{\partial \Delta x}{\partial q_1}-\lambda\left\{(r-\Delta x)\frac{\partial\phi}{\partial q_1}+\frac{\partial Y}{\partial q_1}\right\}=u, \tag{6.13}$$

$$I\ddot{\theta}-\left\{\bar{f}(\Delta x)+\xi(\Delta x)\Delta\dot{x}\right\}\frac{\partial \Delta x}{\partial \theta}-\lambda\left\{(r-\Delta x)\frac{\partial\phi}{\partial \theta}+\frac{\partial Y}{\partial \theta}\right\}=0. \tag{6.14}$$

Since it follows that

$$\begin{cases}\dfrac{\partial Y}{\partial\theta}=-(r-\Delta x)-l, \quad \dfrac{\partial\phi}{\partial\theta}=1, \quad \dfrac{\partial\Delta x}{\partial\theta}=Y,\\[2mm] \dfrac{\partial Y}{\partial q_1}=J_0^{\mathrm{T}}(q_1)\boldsymbol{r}_Y, \quad \dfrac{\partial\phi}{\partial q_1}=-1, \quad \dfrac{\partial\Delta x}{\partial q_1}=J_0^{\mathrm{T}}(q_1)\boldsymbol{r}_X-Y,\\[2mm] J_0^{\mathrm{T}}(q_1)=\dfrac{\partial(x_0,y_0)}{\partial q_1}, \quad \boldsymbol{r}_X=(\cos\theta,-\sin\theta)^{\mathrm{T}}, \quad \boldsymbol{r}_Y=(\sin\theta,\cos\theta).\end{cases} \tag{6.15}$$

Equation (6.13) and (6.14) are reduced to

$$I_1\ddot{q}_1+\left\{\bar{f}(\Delta x)+\xi(\Delta x)\Delta\dot{x}\right\}J_0^{\mathrm{T}}(q_1)\boldsymbol{r}_X$$
$$-\lambda\left\{(r-\Delta x)-J_0^{\mathrm{T}}(q_1)\boldsymbol{r}_Y\right\}=u, \tag{6.16}$$
$$I\ddot{\theta}-\left\{\bar{f}(\Delta x)+\xi(\Delta x)\Delta\dot{x}\right\}Y+l\lambda=0, \tag{6.17}$$

respectively. From this equation we see that the tangential constraint force with the magnitude λ emerges at the centre of the contact area in the direction tangential to the object surface.

Now, applying a similar argument given in Chapter 3 for the dynamics of the overall finger-object system of Figure 6.2, we show stability of the closed-loop dynamics when the following control signal is used:

$$u=-c\dot{q}_1-(f_d/r)J_0^{\mathrm{T}}(q_1)\begin{pmatrix}x_0-x_m\\ y_0-y_m\end{pmatrix}. \tag{6.18}$$

This control signal is the same as that of Equation (3.30), though in this case the finger has a single joint whereas in the case of the system shown in Figure 3.1 the finger has three joints. In a similar manner to the derivation of Equation (3.42), we see that

$$\begin{pmatrix} x_0 - x_m \\ y_0 - y_m \end{pmatrix} = -(r - \Delta x + l)\boldsymbol{r}_X + Y\boldsymbol{r}_Y. \tag{6.19}$$

Hence, substituting Equation (6.18) into Equation (6.16) yields

$$I_1\ddot{q}_1 + c_1\dot{q}_1 + \left\{\bar{f}(\Delta x) + \xi(\Delta x)\Delta\dot{x} - \frac{r - \Delta x + l}{r} f_d\right\} J_0^{\mathrm{T}}(q_1)\boldsymbol{r}_X$$
$$+\lambda(r - \Delta x) - \left(\lambda - \frac{f_d}{r}Y\right) J_0^{\mathrm{T}}(q_1)\boldsymbol{r}_Y = 0. \tag{6.20}$$

Multiplying Equation (6.20) by $\dot{q}_1$ and Equation (6.17) by $\dot{\theta}$ yields

$$\frac{\mathrm{d}}{\mathrm{d}t}\left\{\frac{1}{2}(I_1\dot{q}_1^2 + I\dot{\theta}^2) + P(\Delta x) + \frac{f_d}{2r}\left(Y^2 + (l + r - \Delta x)^2\right)\right\}$$
$$= -c\dot{q}_1^2 - \xi(\Delta x)\Delta\dot{x}^2. \tag{6.21}$$

If we define

$$E = \frac{1}{2}(I_1\dot{q}_1^2 + I\dot{\theta}^2) + P(\Delta x) + \frac{f_d}{2r}\left\{Y^2 + (l + r - \Delta x)^2\right\} \tag{6.22}$$

then Equation (6.21) is rewritten in the form

$$\frac{\mathrm{d}}{\mathrm{d}t}E = -c\dot{q}_1^2 - \xi(\Delta x)\Delta\dot{x}^2. \tag{6.23}$$

Evidently, E attains its minimum value E_m when $\dot{q}_1 = 0$, $\dot{\theta} = 0$, $Y = 0$ and $\Delta x = \Delta x_d$, where Δx_d denotes the value of Δx satisfying

$$\bar{f}(\Delta x_d) = \left(1 + \frac{l}{r}\right) f_d - \frac{f_d}{r}\Delta x_d \tag{6.24}$$

because, at $\Delta x = \Delta x_d$, $\partial E/\partial \Delta x = 0$ (see Figure 6.3). Then, it is possible to rewrite Equation (6.23) as

$$\frac{\mathrm{d}}{\mathrm{d}t}\bar{E} = -c\dot{q}_1^2 - \xi(\Delta x)\Delta\dot{x}^2. \tag{6.25}$$

where

$$\bar{E} = E - E_m \geq 0. \tag{6.26}$$

Note that $\bar{E}$ is positive definite with respect to the state vector $\boldsymbol{X} = (q_1, \theta)^{\mathrm{T}}$ and $\dot{\boldsymbol{X}} = (\dot{q}_1, \dot{\theta})^{\mathrm{T}}$. Thus, $\bar{E}$ in Equation (6.25) plays a role of a Lyapunov function. By applying Lemma 2 of Appendix A, we conclude that, as $t \to \infty$,

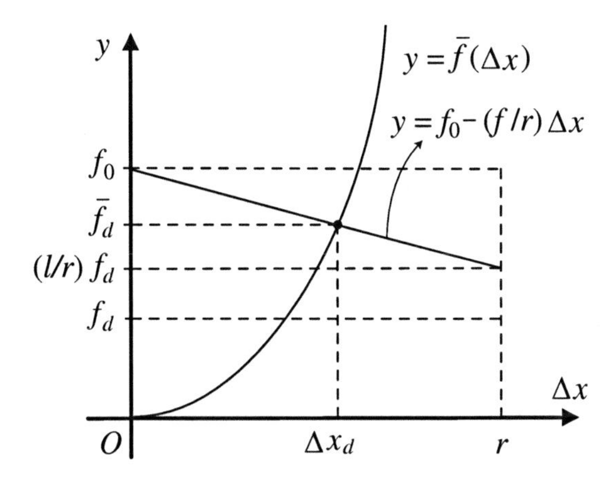

Fig. 6.3. The pressing force $\bar{f}(\Delta x_d)$ realising the force/torque balance is determined by the crossing point of the curve $y = \bar{f}(\Delta x)$ and the straight line $y = f_0 - (f_d/r)\Delta x$, where $f_0 = (1 + l/r) f_d$

$$\dot{q}_1 \to 0 \quad \text{and} \quad \Delta\dot{x} \to 0. \tag{6.27}$$

On the other hand, Equation (6.8) implies that

$$(r - \Delta x)(\dot{\theta} - \dot{q}_1) = \dot{\theta}\{(r - \Delta x) + l\} - \dot{x}_0 \sin\theta - \dot{y}_0 \cos\theta \tag{6.28}$$

from which it follows that

$$\dot{\theta} = \frac{1}{l}\{\dot{x}_0 \sin\theta + \dot{y}_0 \cos\theta - (r - \Delta x)\dot{q}_1\}. \tag{6.29}$$

This shows that $\dot{\theta} \to 0$ as $t \to \infty$ owing to Equation (6.27). Thus, applying a similar argument to that developed in Section 3.1, we can conclude that as $t \to \infty$

$$Y \to 0, \quad \Delta x \to \Delta x_d, \quad \lambda \to 0. \tag{6.30}$$

In addition to the control signal introduced in Equation (6.18), it is possible to append another term $-r\hat{N}_0$ with a similar meaning to the term $-r\hat{N}_0 e$ in Equation (3.103), which was discussed fully in Section 3.5. Hence, let

$$\begin{aligned} \hat{N}_0(t) &= \hat{N}_0(0) + \gamma_0^{-1} \int_0^t r\dot{q}_1(\tau)\,\mathrm{d}\tau \\ &= \hat{N}_0(0) + (r/\gamma_0)(q_1(t) - q_1(0)) \end{aligned} \tag{6.31}$$

and

$$u = -c\dot{q}_1 - (f_d/r)J_0^{\mathrm{T}}(q_1)\begin{pmatrix} x_0 - x_m \\ y_0 - y_m \end{pmatrix} - r\hat{N}_0. \tag{6.32}$$

Table 6.1. Physical parameters for the case of one DOF

m_1	link mass	0.025[kg]
I_1	inertia moment	3.333×10^{-6}[kg · m^2]
l_1	link length	0.040[m]
r	radius	0.010[m]
M	object mass	0.009[kg]
h	object length	0.050[m]
w	object width	0.030[m]
I	object inertia moment	3.000×10^{-6}[kg · m^2]
l	object length	0.020[m]
k	stiffness	3.000×10^5[N/m^2]
c_Δ	viscosity	1000.0[Ns/m^2]

Table 6.2. Parameters of the control signals

f_d	internal force	0.250[N]
c	damping coefficient	0.001[msN]
γ_0	regressor gain	0.001
$\hat{N}_0(0)$	initial estimated value	0.0
γ_λ	CSM gain	3000.0

Convergence of the closed-loop dynamics when the control signal of Equation (6.32) is substituted into Equation (6.16) to the equilibrium state can be proved in a similar way to that given in Section 6.5.

To confirm this theoretical prediction, we shall show a numerical simulation result obtained by a model of the finger–object system whose physical parameters are given in Table 6.1. In the table, the viscosity c_Δ stands for a constant governing the relation

$$\xi(\Delta x)\Delta\dot{x} = c_\Delta(\Delta x)^2\Delta\dot{x}, \tag{6.33}$$

that is, the viscosity affecting the finger-end O_{01} in the direction $-\boldsymbol{f}$ is proportional to the square of Δx similarly to the stiffness. The parameters of the control signal of Equation (6.32) and the initial value for $\hat{N}_0$ are shown in Table 6.2. We show the transient responses of all the physical variables in Figure 6.4. In this case, Y and λ do not converge to zero as $t \to \infty$ but converge to some constant values very quickly within 0.15 [s]. The result for $\alpha + \theta$ shown in Figure 6.3 implies that the sum of two forces originating at the centre of area contact with magnitude $\bar{f}(\Delta x)$ normal to the object surface and the magnitude λ tangent to the object surface should be directed toward the pivot point O_m. The details of this observation about the condition

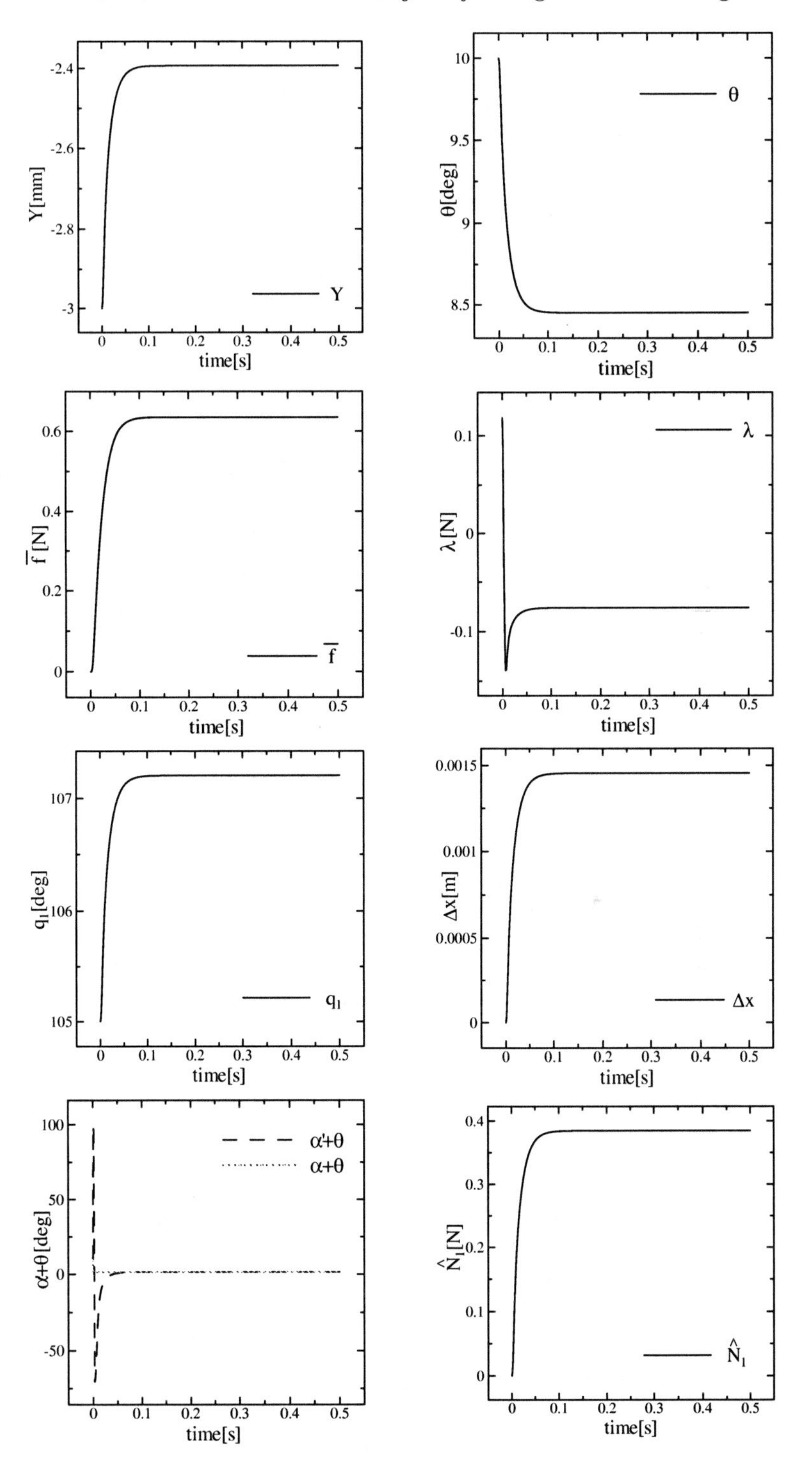

Fig. 6.4. The transient responses of the physical variables when the control input of Equation (6.32) is used

Table 6.3. Physical parameters for dual single-DOF fingers

$m_{11} = m_{21}$	link mass	0.025[kg]
$I_{11} = I_{21}$	inertia moment	3.333×10^{-6}[kg · m^2]
$l_{11} = l_{21}$	link length	0.040[m]
$r_1 = r_2$	radius	0.010[m]
M	object mass	0.009[kg]
h	object length	0.050[m]
$w = (l_1 + l_2)$	object width	0.030[m]
d	object height	0.010[m]
I	object inertia moment	2.550×10^{-6}[kg · m^2]
$l_1 = l_2$	object length	0.015[m]
k	stiffness	3.000×10^5[N/m^2]
c_Δ	viscosity	1000.0[Ns/m^2]

of force/torque balance is the same as in the rigid contact case discussed in Section 3.6 (in particular, refer to Equation (3.132).

6.3 Stable Grasping of a 2-D Object by Dual Single-DOF Soft Fingers

The thumb itself is rather big and fat in comparison with the other digits such as the index and middle fingers. It is strong and therefore can generate a large pressing force on an object. It plays a vital role in finger-pressure therapy. Nevertheless, the thumb can be dexterous through deliberate exercises as seen in our everyday life. In fact, nowadays, even elderly people use it to push buttons on a cell phone while grasping it by other four digits.

The thumb and index (or middle) finger can be used for prehensing a small, light, and thin object easily and quickly based upon the fingers–thumb opposability. The dexterity of precision prehension of a thin object on the basis of such opposability is deeply indebted to the softness of the thumb and finger pads. In this section, we shall analyse the stability of the precision prehension of a small, light, thin object by a pair of dual single-DOF fingers with spherical fingertips that are soft and visco-elastic. Before this, we show some computer simulation results of a 2-D object with ordinary width $l = l_1 + l_2$, $l_1 = l_2 = 0.015$ [m] being grasped by a pair of dual single-DOF fingers whose physical parameters are shown in Table 6.3. A schematic of the fingers–object system is quite similar to Figure 2.10, where in this case the fingertip spheres are soft and deformable. Firstly we show the transient responses of the physical variables of the system in Figure 6.5, where the following control

Table 6.4. Parameters of the control signals

f_d	internal force	0.250[N]
$c_1 = c_2$	damping coefficient	0.001[msN]
γ_f	CSM gain	1500.0
γ_λ	CSM gain	3000.0

signal is employed:

$$u_i = -c_i\dot{q}_i + (-1)^i \frac{f_d}{r_1 + r_2} J_{0i}^{\mathrm{T}}(q_i) \begin{pmatrix} x_{01} - x_{02} \\ y_{01} - y_{02} \end{pmatrix}, \quad i = 1, 2. \tag{6.34}$$

Control gains are given in Table 6.4. The second term on the right-hand side of Equation (6.34) stands for exertion of the pressing force $\boldsymbol{F}_i$ $(i = 1, 2)$ on the object in the direction shown in Figure 6.6. In this case, Lagrange's equation of motion for the overall finger–object system can easily be derived in a similar manner to in the previous section for the system of Figure 6.6 [also refer to the derivation of Equations (2.52) and (2.53)]. This results in the equations:

$$\begin{aligned} &I_i\ddot{q}_i - (-1)^i f_i J_{0i}^{\mathrm{T}}(q_i)\boldsymbol{r}_X \\ &\quad +\lambda_i \left\{(r_i - \Delta x_i) - J_{0i}^{\mathrm{T}}(q_i)\boldsymbol{r}_Y\right\} = u_i, \quad i = 1, 2, \end{aligned} \tag{6.35}$$

$$M\ddot{\boldsymbol{x}} - R_\theta(f_1 - f_2, -(\lambda_1 + \lambda_2))^{\mathrm{T}} = 0, \tag{6.36}$$

$$I\ddot{\theta} - f_1 Y_1 + f_2 Y_2 + l_1\lambda_1 - l_2\lambda_2 = 0, \tag{6.37}$$

where $\boldsymbol{x} = (x, y)^{\mathrm{T}}$, $\boldsymbol{x}_{0i} = (x_{0i}, y_{0i})^{\mathrm{T}}$, $J_{0i}(q_i) = \partial(x_{0i}, y_{0i})/\partial q_i$ and

$$f_i(\Delta x_i, \Delta\dot{x}_i) = \bar{f}_i(\Delta x_i) + \xi_i(\Delta x_i)\Delta\dot{x}_i \tag{6.38}$$

and $\boldsymbol{r}_X = (\cos\theta, -\sin\theta)^{\mathrm{T}}$, $\boldsymbol{r}_Y = (\sin\theta, \cos\theta)^{\mathrm{T}}$ and $R_\theta = (\boldsymbol{r}_X, \boldsymbol{r}_Y)$. It should also be remarked that the maximum displacements Δx_i $(i = 1, 2)$ of deformation arising at the centres of the contact areas are dependent on the position state variables according to

$$\Delta x_i = r_i + l_i + (-1)^i(\boldsymbol{x} - \boldsymbol{x}_{0i})^{\mathrm{T}}\boldsymbol{r}_X, \quad i = 1, 2 \tag{6.39}$$

and similarly it follows that

$$\begin{cases} \boldsymbol{x}_i = \boldsymbol{x}_{0i} - (-1)^i(r_i - \Delta x_i)\boldsymbol{r}_X, \quad i = 1, 2 \\ \boldsymbol{x} = \boldsymbol{x}_1 + l_1\boldsymbol{r}_X - Y_1\boldsymbol{r}_Y = \boldsymbol{x}_2 - l_2\boldsymbol{r}_X - Y_2\boldsymbol{r}_Y \\ Y_i = (\boldsymbol{x}_{0i} - \boldsymbol{x})^{\mathrm{T}}\boldsymbol{r}_Y, \quad i = 1, 2 \end{cases} \tag{6.40}$$

and

$$-(r_i - \Delta x_i)\frac{\mathrm{d}}{\mathrm{d}t}\left(\frac{3\pi}{2} - (-1)^i\theta - q_i\right) = \frac{\mathrm{d}}{\mathrm{d}t}Y_i, \quad i = 1, 2. \tag{6.41}$$

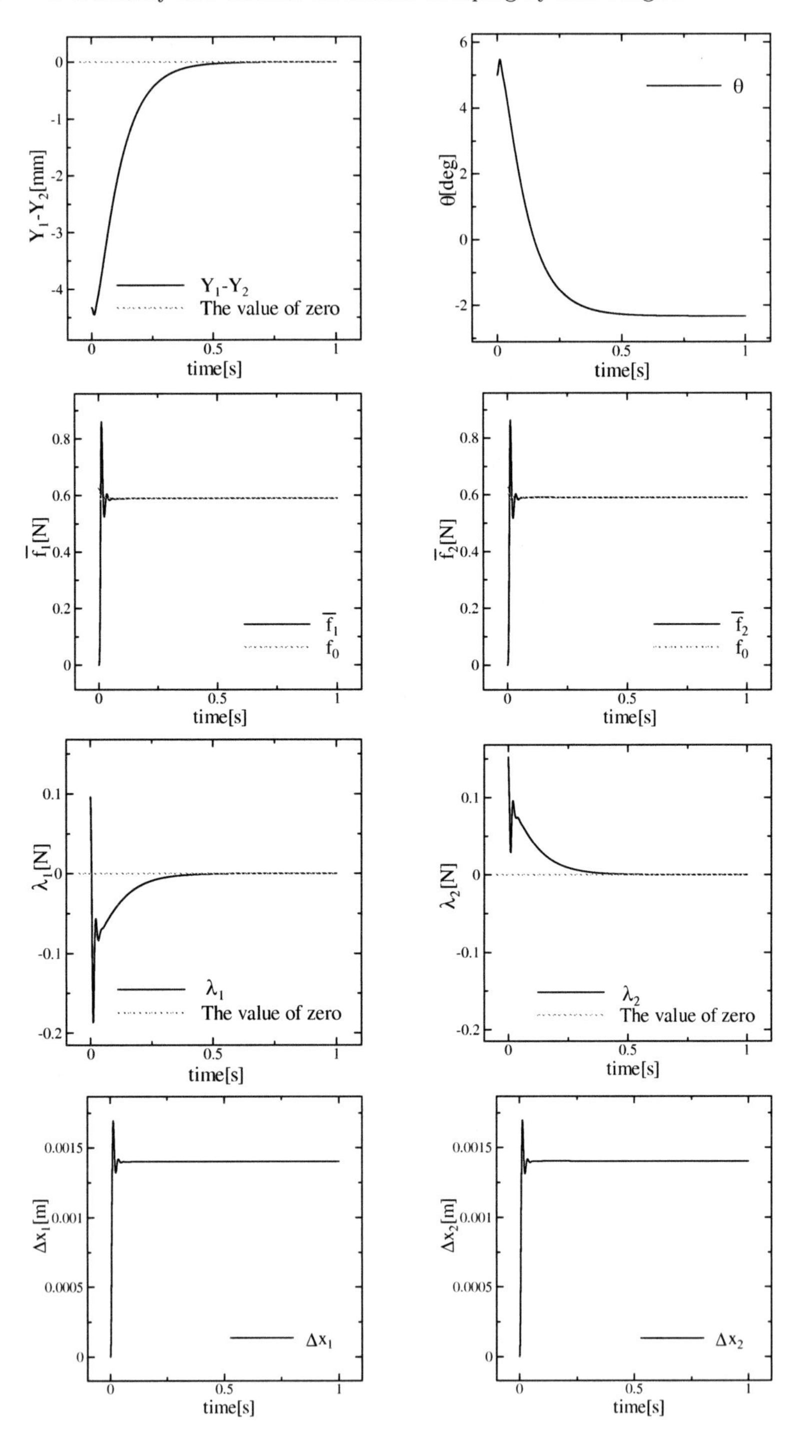

Fig. 6.5. The transient responses of the physical parameters when the control signals of Equation (6.34) are employed

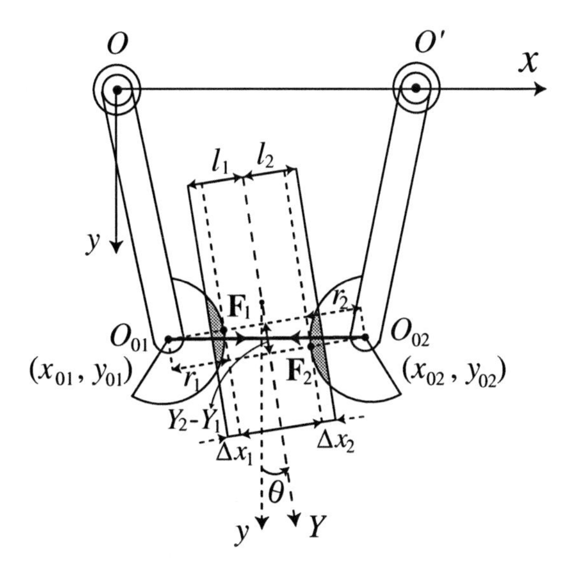

Fig. 6.6. Precision prehension of a 2-D rigid object by means of a pair of single-DOF robot fingers with soft fingertips

The last equations for $i = 1, 2$ express non-holonomic constraints due to rolling of the fingertips and object surfaces.

According to Figure 6.5, $Y_1 - Y_2$ converges asymptotically to zero as $t \to \infty$ and, coincidently, the constraint forces λ_1 and λ_2 converge to zero as $t \to \infty$ with the same speed of convergence as that of $Y_1 - Y_2$. In contrast, Δx_1 and Δx_2 together with $\bar{f}_1(\Delta x_1)$ and $\bar{f}_2(\Delta x_2)$ quickly converge to their corresponding constants within 0.2 [s]. In this simulation, we use the same characteristics of fingertip visco-elasticity, that is,

$$\begin{cases} \bar{f}_1(\Delta x) = \bar{f}_2(\Delta x) = k\Delta x^2 \\ \xi_1(\Delta x)\Delta\dot{x} = \xi_2(\Delta x)\Delta\dot{x} = c_\Delta(\Delta x)^2\Delta\dot{x} \end{cases} \tag{6.42}$$

and the constants k and c_Δ are specified in Table 6.3. As seen from Figure 6.5, both Δx_1 and Δx_2 converge to the same value Δx_d and $\bar{f}_1(\Delta x)$ and $\bar{f}_2(\Delta x)$ converge to the same value $\bar{f}(\Delta x_d)$. In what follows, however, we will show that, even if the parameters of fingertip stiffness and viscosity are different for the left and right fingers, $\bar{f}_1(\Delta x_1)$ and $\bar{f}_2(\Delta x_2)$ must converge asymptotically to the same common value $\bar{f}_d$ as $t \to \infty$. To do this, let us show the closed-loop dynamics of robot fingers when Equation (6.34) is substituted into Equation (6.35), which results in the form

$$\begin{aligned} I_i\ddot{q}_i + c_i\dot{q}_i - (-1)^i\Delta f_i J_{0i}^{\mathrm{T}}(q_i)\boldsymbol{r}_X \\ +\Delta\lambda_i\left\{(r_i - \Delta x_i) - J_{0i}^{\mathrm{T}}(q_i)\boldsymbol{r}_Y\right\} - N_i = 0, \quad i = 1, 2, \end{aligned} \tag{6.43}$$

where $N_i = (-1)^i \frac{(r_i - \Delta x_i)}{r_1 + r_2} f_d(Y_1 - Y_2)$ and

$$\begin{cases} \Delta f_i = f_i + \dfrac{f_d}{r_1 + r_2}(\boldsymbol{x}_{01} - \boldsymbol{x}_{02})^{\mathrm{T}} \boldsymbol{r}_X, \\ \Delta \lambda_i = \lambda_i + (-1)^i \dfrac{f_d}{r_1 + r_2}(\boldsymbol{x}_{01} - \boldsymbol{x}_{02})^{\mathrm{T}} \boldsymbol{r}_Y. \end{cases} \tag{6.44}$$

In relation to this, it is convenient to rewrite the dynamics of the object expressed by Equations (6.36) and (6.37) equivalently in the following forms:

$$M\ddot{\boldsymbol{x}} - R_\theta(\Delta f_1 - \Delta f_2,\ -\Delta\lambda_1 - \Delta\lambda_2)^{\mathrm{T}} = 0, \tag{6.45}$$

$$I\ddot{\theta} - \Delta f_1 Y_1 + \Delta f_2 Y_2 + l_1 \Delta\lambda_1 - l_2 \Delta\lambda_2 + S = 0, \tag{6.46}$$

where

$$S = -f_d\left(1 - \frac{\Delta x_1 + \Delta x_2}{r_1 + r_2}\right)(Y_1 - Y_2). \tag{6.47}$$

It is then easy to derive Lyapunov's relation by calculating

$$\sum_{i=1,2} \dot{q}_i \times (6.43) + \dot{\boldsymbol{x}}^{\mathrm{T}} \times (6.45) + \dot{\theta} \times (6.46),$$

which results in

$$\frac{\mathrm{d}}{\mathrm{d}t}E = \sum_{i=1,2} -\left\{c_i \dot{q}_i^2 + \xi(\Delta x_i)\Delta \dot{x}_i^2\right\}, \tag{6.48}$$

where

$$\begin{cases} K = \dfrac{1}{2}\left\{I_1 \dot{q}_1^2 + I_2 \dot{q}_2^2 + M\dot{x}^2 + M\dot{y}^2 + I\dot{\theta}^2\right\}, \\ P = \displaystyle\sum_{i=1,2} \int_0^{\Delta x_i} \bar{f}_i(\eta)\,\mathrm{d}\eta + \frac{f_d}{2(r_1 + r_2)}(Y_1 - Y_2)^2, \\ E = K + P. \end{cases} \tag{6.49}$$

In the derivation of S in Equation (6.46), we use the relations

$$\begin{cases} (\boldsymbol{x}_{01} - \boldsymbol{x}_{02})^{\mathrm{T}} \boldsymbol{r}_Y = Y_1 - Y_2 \\ -(\boldsymbol{x}_{01} - \boldsymbol{x}_{02})^{\mathrm{T}} \boldsymbol{r}_X = l_1 + l_2 + r_1 + r_2 - (\Delta x_1 + \Delta x_2) \end{cases} \tag{6.50}$$

and

$$\frac{f_d}{r_1 + r_2}(\boldsymbol{x}_{01} - \boldsymbol{x}_{02}) = -\frac{l(\Delta x_1 + \Delta x_2)}{r_1 + r_2} f_d \boldsymbol{r}_X + \frac{Y_1 - Y_2}{r_1 + r_2} f_d \boldsymbol{r}_Y \tag{6.51}$$

that can easily be obtained from Equation (6.40). Since E is positive definite with respect to $\boldsymbol{X} = (q_1, q_2, x, y, \theta)^{\mathrm{T}}$ and $\dot{\boldsymbol{X}}$ under the non-holonomic constraints of Equation (6.41) and the relation of Equation (6.39), E in Equation (6.48) plays the role of a Lyapunov function. Hence, Lemma 2 stated in Appendix A concludes that, as $t \to \infty$,

Table 6.5. Parameters of the control signals

f_d	internal force	0.25[N]
$c_1 = c_2$	damping coefficient	0.001[msN]
$\gamma_i(i = 1, 2)$	regressor gain	0.001
$\hat{N}_i(0)(i = 1, 2)$	initial estimate value	0.0
γ_f	CSM gain	1500.0
γ_λ	CSM gain	3000.0

$$\dot{q}_i(t) \to 0, \quad \Delta\dot{x}_i(t) \to 0, \quad i = 1, 2. \tag{6.52}$$

Then, applying a similar argument to that developed in the previous section, we conclude

$$\Delta f_i \to 0 \quad \text{and} \quad \Delta\lambda_i \to 0, \quad i = 1, 2 \tag{6.53}$$

as $t \to \infty$. In particular, $\Delta f_i \to 0$ and $\Delta\dot{x}_i \to 0$ as $t \to \infty$ imply that

$$\bar{f}_i(\Delta x_i) - \frac{f_d}{r_1 + r_2}\left\{(l_1 + l_2 + r_1 + r_2) - (\Delta x_1 + \Delta x_2)\right\} \to 0, \quad i = 1, 2. \tag{6.54}$$

This shows that the constants Δx_{di} $(i = 1, 2)$ should be determined so that they satisfy the following two equations simultaneously:

$$\begin{cases} \bar{f}_1(\Delta x_{d1}) = \left(1 + \dfrac{l_1 + l_2 - \Delta x_{d1} - \Delta x_{d2}}{r_1 + r_2}\right) f_d, \\ \bar{f}_2(\Delta x_{d2}) = \left(1 + \dfrac{l_1 + l_2 - \Delta x_{d1} - \Delta x_{d2}}{r_1 + r_2}\right) f_d, \end{cases} \tag{6.55}$$

from which it can be concluded that both reproducing forces $\bar{f}_1(\Delta x_1)$ and $\bar{f}_2(\Delta x_2)$ must converge to the same value of $\bar{f}_d$ in such a way that

$$\bar{f}_1(\Delta x_1) \text{ and } \bar{f}_2(\Delta x_2) \to \bar{f}_d = \left(1 + \frac{l_1 + l_2 - \Delta x_{d1} - \Delta x_{d2}}{r_1 + r_2}\right) f_d \tag{6.56}$$

as $t \to \infty$.

We conducted another computer simulation for the setup of Figure 6.6 by using the control signal

$$u_i = -c_i\dot{q}_i + (-1)^i \frac{f_d}{r_1 + r_2} J_{0i}^{\mathrm{T}}(q_i)(\boldsymbol{x}_{01} - \boldsymbol{x}_{02}) - r_i\hat{N}_i, \quad i = 1, 2, \tag{6.57}$$

where

$$\hat{N}_i(t) = \hat{N}_i(0) + \gamma_i^{-1} r_i \left\{q_i(t) - q_i(0)\right\}, \quad i = 1, 2. \tag{6.58}$$

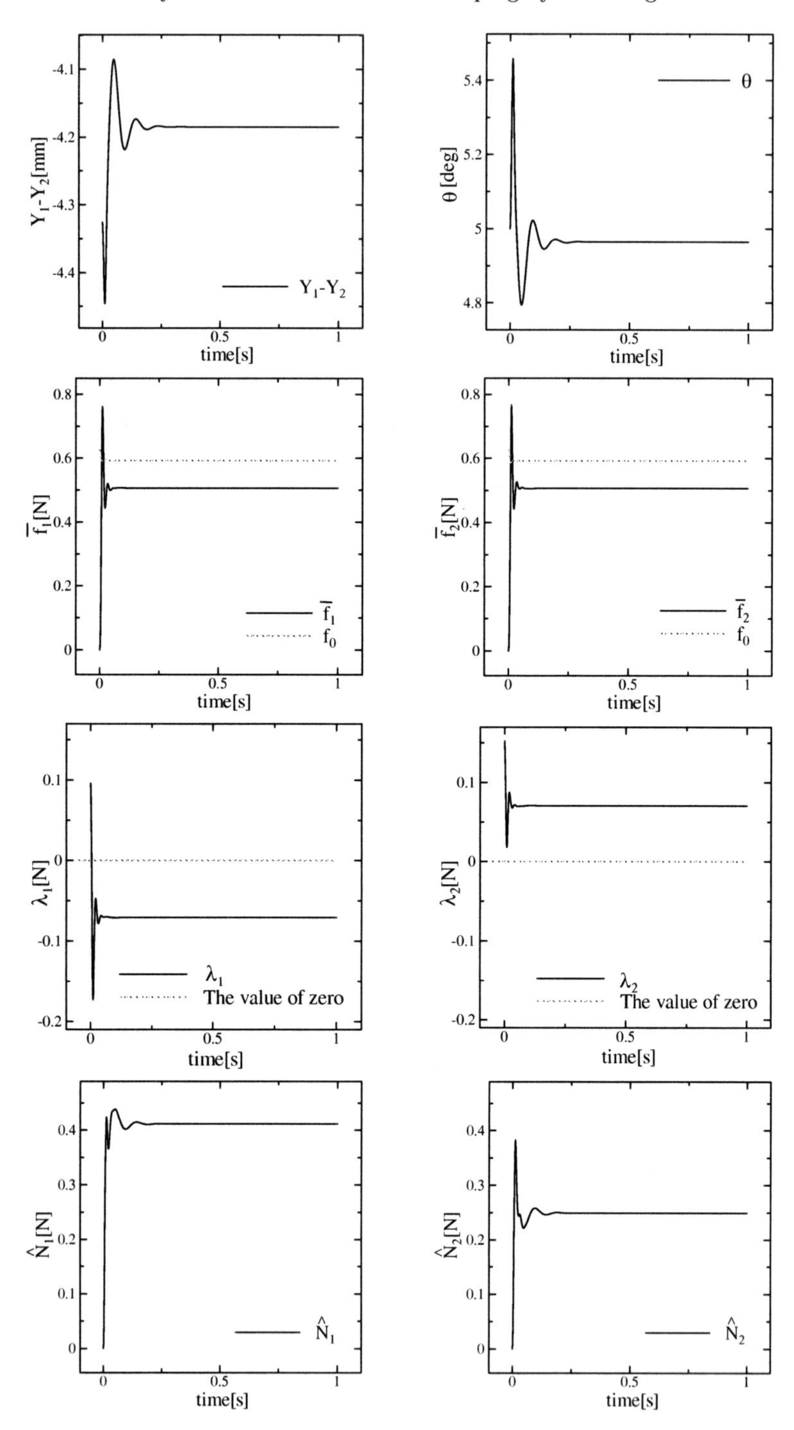

Fig. 6.7. The transient responses of the physical parameters when the control signals of Equation (6.57) are employed

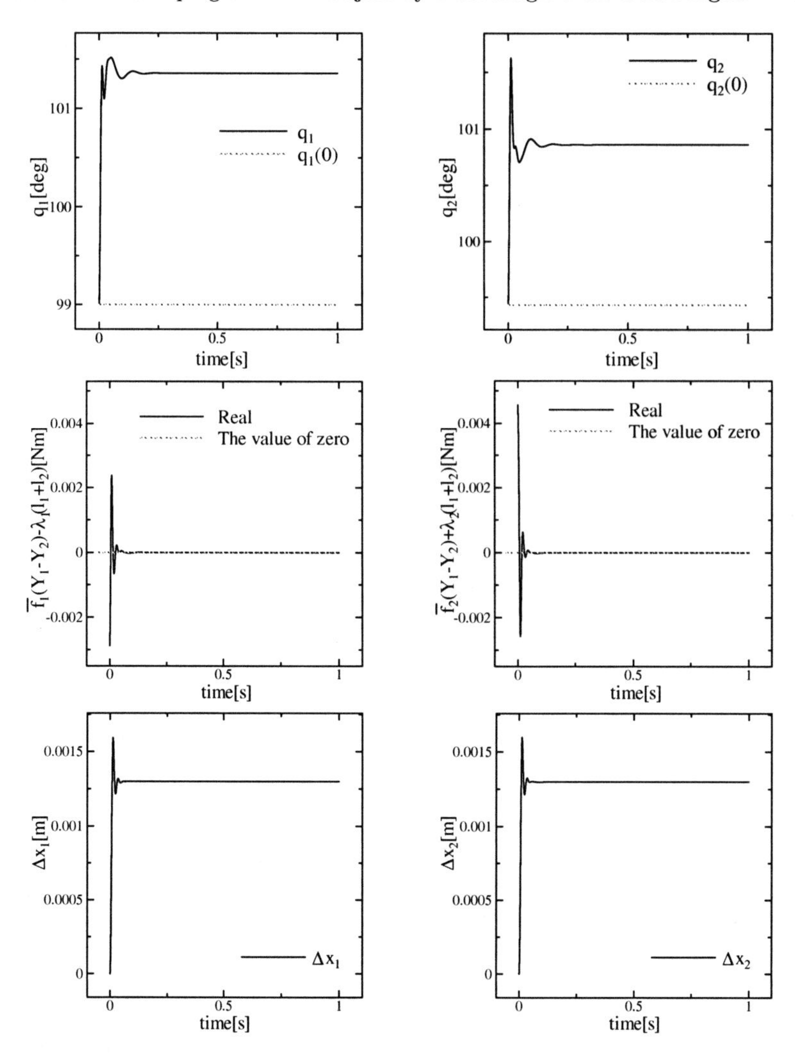

Fig. 6.8. The transient responses of the physical variables when the same control signals as those in Figure 6.7 are employed

The control gains used in this simulation are given in Table 6.5. We show the transient responses of the physical variables in Figures 6.7 and 6.8. Comparing Figure 6.7 with Figure 6.5, the convergence speeds of λ_i $(i = 1, 2)$ to their corresponding constants have been drastically improved. Accompanying this improvement in convergences of λ_i $(i = 1, 2)$, the convergence performance of the variables $Y_1 - Y_2$ and θ is also improved in this case, though $Y_1 - Y_2$ does not converge to zero. Nevertheless, force/torque balance is established in the sequel in such a way that

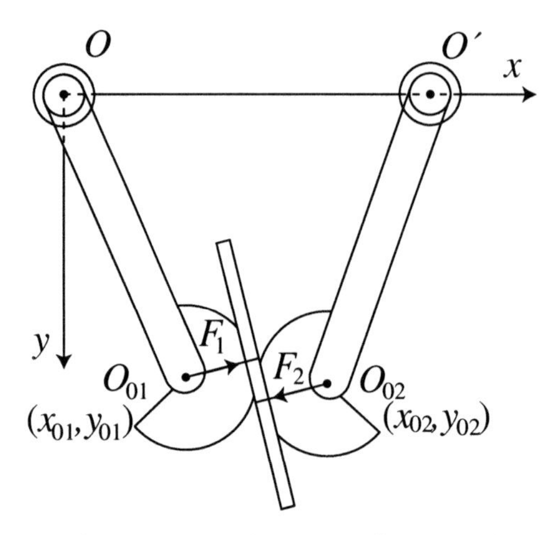

Fig. 6.9. Prehension of a thin object by a pair of single-DOF fingers with soft tips

$$\bar{f}_i(Y_1 - Y_2) - \lambda_i(l_1 + l_2) \to 0, \quad i = 1, 2 \tag{6.59}$$

as $t \to \infty$, as seen in Figure 6.8. The maximum displacements Δx_1 and Δx_2 of the soft fingertips also converge quickly to their corresponding constant values Δx_{d1} and Δx_{d2}, respectively, which must be determined so as to satisfy Equations (6.55) and (6.56).

Let us now return to the problem posed in the second paragraph of this section, namely the problem of the stability of grasping a thin light object with a pair of single-DOF dual fingers with soft fingertips as sketched in Figure 6.9. When the thickness $l_1 + l_2$ of the object becomes small, the attractor region of the equilibrium state corresponding to the state of force/torque balance may shrink. In fact, in the case of the testbed problem discussed in Section 6.2, the smaller width l of the object may result in a larger angular velocity $\dot{\theta}$ of rotational motion of the object owing to Equation (6.29). In the case of grasping a thin object as in Figure 6.9, the constraint Equations (6.39) and (6.41) lead to the relation

$$\begin{aligned}\dot{\theta} = \frac{1}{l_1 + l_2} \{ &-(r_1 - \Delta x_1)\dot{q}_1 + (r_2 - \Delta x_2)\dot{q}_2 \} \\ &+ (\dot{x}_{01} + \dot{x}_{02}) \sin\theta + (\dot{y}_{01} + \dot{y}_{02}) \cos\theta \} ,\end{aligned} \tag{6.60}$$

which can be derived in a similar manner to Equation (2.76). Hence, if $l_1 + l_2$ becomes small, large rotational movements of the object may result. Notwithstanding this theoretical observation, the control signals of Equation (6.57) can still be applied for stable precision prehension of a thin light object as in Figure 6.9. We show a computer simulation result in Figure 6.10 when the object is a paper card with thickness $l_1 + l_2 = 0.001$ [m] and mass $M = 0.0003$ [kg]. The parameters of the fingers and the card are given in Table 6.6 and the

Table 6.6. Physical parameters for dual single-DOF fingers

$m_{11} = m_{21}$	link mass	0.025[kg]
$I_{11} = I_{21}$	inertia moment	3.333×10^{-6}[kg · m^2]
$l_{11} = l_{21}$	link length	0.040[m]
$r_1 = r_2$	radius	0.010[m]
M	object mass	0.0003[kg]
h	object length	0.050[m]
$w = (l_1 + l_2)$	object width	0.001[m]
d	object height	0.010[m]
I	object inertia moment	7.003×10^{-8}[kg · m^2]
$l_1 = l_2$	object length	0.0005[m]
k	stiffness	2300.0[N/m^2]
c_Δ	viscosity	30.0[Ns/m^2]

Table 6.7. Parameters of the control signals

f_d	internal force	0.1[N]
$c_1 = c_2$	damping coefficient	0.001[msN]
$\gamma_i(i = 1, 2)$	regressor gain	0.001
$\hat{N}_i(0)(i = 1, 2)$	initial estimate value	0.000[N]

control gains are given in Table 6.7. Comparing Figure 6.10 with Figures 6.7 and 6.8, we see that the convergences of Δx_1 and Δx_2 to their corresponding constants become slow due to the choice of a smaller $f_d = 0.1$ [N] in the case of Figure 6.10. As predicted theoretically, transient movements of $Y_1 - Y_2$ and θ in Figure 6.10 are inflational at an early stage after the instant $t = 0$ of grasping compared with in Figures 6.7 and 6.8. Nevertheless, the transient motions of $Y_1 - Y_2$ and θ quickly cease after around 0.25 [s] and converge to their corresponding constants that attain the state of force/torque balance, though the magnitude $|Y_1 - Y_2|$ remains large relative to the object thickness. This convergence might become possible because it is assumed in the simulation that no slipping arises between the fingertips and object surfaces. We should bear in mind that, in the case of human pinching of a thin and light object, control of slip between the finger-tips and the object becomes more dominant than that of rolling if the initial value of $|Y_1 - Y_2|$ is relatively large. A theoretical proof that validates the convergence of the overall closed-loop dynamics to the state of force/torque balance seems difficult without assuming the additional presence of viscous-like forces induced by fingertip deformations at both contact points in the directions tangent to the object

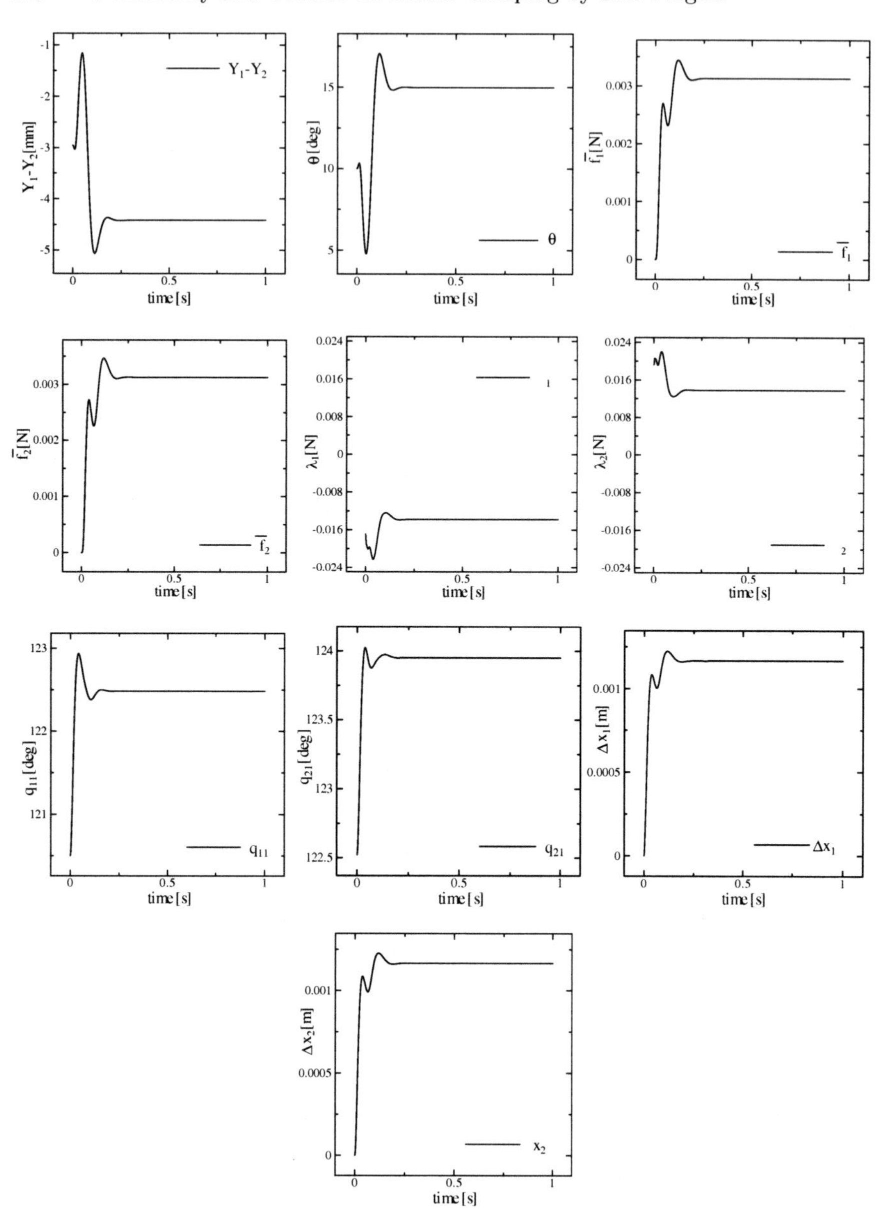

Fig. 6.10. The transient responses of the physical variables in the case of manipulation of a thin object by a pair of robot fingers with soft ends

surfaces. This point will be discussed again in the last paragraph of the next section.

6.4 Convergence to Force/Torque Balance by a Pair of Soft Fingers

This section concentrates on the mathematical proof of asymptotic convergence of closed-loop dynamics of the overall fingers–object system to the equilibrium state of force/torque balance when the control signals of Equation (6.34) are employed and the object thickness $l_1 + l_2$ is of order of the fingertip radius r_i $(i = 1, 2)$.

First, we spell out in the following the closed-loop dynamics of the overall fingers–object system when the control signals of Equation (6.34) are substituted into Equation (6.35):

$$I_i\ddot{q}_i + c_i\dot{q}_i - (-1)^i J_{0i}^{\mathrm{T}}(q_i)\left\{ f_i \boldsymbol{r}_X - \frac{l(\Delta x_1 + \Delta x_2)}{r_1 + r_2} f_d \boldsymbol{r}_X + \frac{Y_1 - Y_2}{r_1 + r_2} f_d \boldsymbol{r}_Y \right\}$$
$$+ \lambda_i \left\{ (r_i - \Delta x_i) - J_{0i}^{\mathrm{T}}(q_i)\boldsymbol{r}_Y \right\} = 0, \quad i = 1, 2, \tag{6.61}$$

$$M\ddot{\boldsymbol{x}} - R_\theta \begin{pmatrix} f_1 - f_2 \\ -(\lambda_1 + \lambda_2) \end{pmatrix} = 0, \tag{6.62}$$

$$I\ddot{\theta} - f_1 Y_1 + f_2 Y_2 + l_1\lambda_1 - l_2\lambda_2 = 0. \tag{6.63}$$

At the same time, we note that from Equation (6.51) it follows that

$$\|\boldsymbol{x}_{01} - \boldsymbol{x}_{02}\|^2 = l^2(\Delta x_1 + \Delta x_2) + (Y_1 - Y_2)^2, \tag{6.64}$$

where we define the length l as a function of the magnitude of $\Delta x_1 + \Delta x_2$ as follows:

$$\begin{aligned} l(\Delta x_1 + \Delta x_2) &= l_1 + l_2 + (r_1 - \Delta x_1) + (r_2 - \Delta x_2) \\ &= l_1 + l_2 + r_1 + r_2 - (\Delta x_1 + \Delta x_2). \end{aligned} \tag{6.65}$$

On the other hand, it is easy to check and confirm that the sum of $\dot{q}_i$ times Equation (6.61) for $i = 1, 2$, $\dot{\boldsymbol{x}}$ times Equation (6.62), and $\dot{\theta}$ times Equation (6.63) yields

$$\frac{\mathrm{d}}{\mathrm{d}t}\left\{ K + P + \frac{f_d}{2(r_1 + r_2)}\|\boldsymbol{x}_{01} - \boldsymbol{x}_{02}\|^2 + \sum_{i=1,2} \frac{\gamma_i}{2}\hat{N}_i^2 \right\}$$
$$= - \sum_{i=1,2} \left\{ c_i\dot{q}_i^2 + \xi_i(\Delta x_i)\Delta\dot{x}_i^2 \right\}, \tag{6.66}$$

where

$$P = \sum_{i=1,2} \int_0^{\Delta x_i} \bar{f}_i(\eta)\,\mathrm{d}\eta. \tag{6.67}$$

The symbol K signifies the total kinetic energy of the overall fingers–object system defined in Equation (6.49).

Next, by using $\boldsymbol{X} = (q_1, q_2, r^{-1}x, r^{-1}y, \theta)^{\mathrm{T}}$, Δf_i and $\Delta\lambda_i$ $(i = 1, 2)$ defined by Equation (6.44), where r stands for a scale factor, we rewrite Equations (6.61–6.63) in a vector–matrix form:

$$\begin{aligned} H\ddot{\boldsymbol{X}} + C\dot{\boldsymbol{X}} + \sum_{i=1,2} \xi_i(\Delta x_i)\Delta\dot{x}_i \frac{\partial \Delta x_i}{\partial \boldsymbol{X}} \\ -A(\boldsymbol{X})\Delta\bar{\boldsymbol{f}} + B(\boldsymbol{X})\Delta\boldsymbol{\lambda} - (Y_1 - Y_2) f_d \boldsymbol{e} = 0, \end{aligned} \tag{6.68}$$

where $H = \mathrm{diag}(I_1, I_2, r^2 M, r^2 M, I)$, $C = \mathrm{diag}(c_1, c_2, 0, 0, 0)$ and

$$A(\boldsymbol{X}) = \left(\frac{\partial \Delta x_1}{\partial \boldsymbol{X}}, \frac{\partial \Delta x_2}{\partial \boldsymbol{X}}\right) = \begin{pmatrix} -J_{01}^{\mathrm{T}}(q_1)\boldsymbol{r}_X & 0 \\ 0 & J_{02}^{\mathrm{T}}(q_2)\boldsymbol{r}_X \\ r\cos\theta & -r\cos\theta \\ -r\sin\theta & r\sin\theta \\ Y_1 & -Y_2 \end{pmatrix}, \tag{6.69}$$

$$\begin{aligned} B(\boldsymbol{X}) &= \left(-(r_1 - \Delta x_1)\frac{\partial \phi}{\partial \boldsymbol{X}} - \frac{\partial Y_1}{\partial \boldsymbol{X}}, -(r_2 - \Delta x_2)\frac{\partial \phi_2}{\partial \boldsymbol{X}} - \frac{\partial Y_2}{\partial \boldsymbol{X}}\right) \\ &= \begin{pmatrix} (r_1 - \Delta x_1) - J_{01}^{\mathrm{T}}(q_1)\boldsymbol{r}_Y & 0 \\ 0 & (r_2 - \Delta x_2) - J_{02}^{\mathrm{T}}(q_2)\boldsymbol{r}_Y \\ r\sin\theta & r\sin\theta \\ r\cos\theta & r\cos\theta \\ l_1 & -l_2 \end{pmatrix}, \end{aligned} \tag{6.70}$$

$$\boldsymbol{e} = \left(-\frac{r_1 - \Delta x_1}{r_1 + r_2}, \frac{r_2 - \Delta x_2}{r_1 + r_2}, 0, 0, \frac{r_1 + r_2 - \Delta x_1 - \Delta x_2}{r_1 + r_2}\right)^{\mathrm{T}}, \tag{6.71}$$

$$\Delta\bar{\boldsymbol{f}} = (\Delta\bar{f}_1, \Delta\bar{f}_2)^{\mathrm{T}}, \quad \Delta\boldsymbol{\lambda} = (\Delta\lambda_1, \Delta\lambda_2)^{\mathrm{T}}. \tag{6.72}$$

We remark that, by referring to Equation (6.51), four contact forces as components of $\Delta\bar{\boldsymbol{f}}$ and $\Delta\boldsymbol{\lambda}$ can be written explicitly in the form:

$$\begin{aligned} \Delta\bar{f}_i &= \bar{f}_i + \frac{f_d}{r_1 + r_2}(\boldsymbol{x}_{01} - \boldsymbol{x}_{02})^{\mathrm{T}}\boldsymbol{r}_X \\ &= \bar{f}_i(\Delta x_i) - \frac{l(\Delta x_1 + \Delta x_2)}{r_1 + r_2} f_d, \quad i = 1, 2, \end{aligned} \tag{6.73}$$

$$\begin{aligned} \Delta\lambda_i &= \lambda_i + (-1)^i \frac{f_d}{r_1 + r_2}(\boldsymbol{x}_{01} - \boldsymbol{x}_{02})^{\mathrm{T}}\boldsymbol{r}_Y \\ &= \lambda_i + (-1)^i \frac{Y_1 - Y_2}{r_1 + r_2} f_d, \quad i = 1, 2. \end{aligned} \tag{6.74}$$

It is also important to remark that, according to Equation (6.55), we can conveniently define the potential functions of fingertip deformation in the following way:

$$\Delta P(\delta x_i) = \int_0^{\delta x_i} \left\{\bar{f}_i(\Delta x_{id} + \eta) - \bar{f}_d\right\} \mathrm{d}\eta, \quad i = 1, 2, \tag{6.75}$$

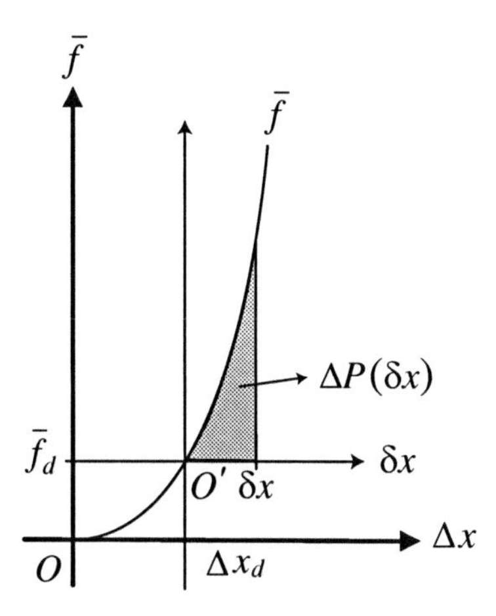

Fig. 6.11. A potential function $\Delta P(\delta x)$ induced from the reproducing force $\bar{f}(\Delta x)$ relative to a fixed value $\Delta x = \Delta x_d$, which is defined as the area by Equation (6.75)

where $\delta x_i = \Delta x_i - \Delta x_{id}$ and $\bar{f}_d = \bar{f}_1(\Delta x_{d1}) = \bar{f}_2(\Delta x_{d2}) = \{1 + l(\Delta x_{d1} + \Delta x_{d2})/(r_1 + r_2)\} f_d$ [see Equations (6.55) and (6.56)]. As shown in Figure 6.11, the function $\Delta P_i(\delta x_i)$ is positive definite in δx_i as far as $0 \le \Delta x_{di} + \delta x_i < r_i$ $(i = 1, 2)$.

$$E = K + \Delta P_1(\delta x_1) + \Delta P_2(\delta x_2) + \frac{f_d(Y_1 - Y_2)^2}{2(r_1 + r_2)} \tag{6.76}$$

then it is possible to derive the relation along a solution trajectory to the closed-loop dynamics of Equation (6.68)

$$\frac{\mathrm{d}}{\mathrm{d}t} E = -\sum_{i=1,2} \left\{ c_i \dot{q}_i^2 + \xi_i(\Delta x_i) \delta \dot{x}_i^2 \right\}. \tag{6.77}$$

As discussed earlier, the closed-loop system of Equation (6.68) or equivalently Equations (6.61–6.63) has three DOFs because it has five independent position variables q_1, q_2, x, y and θ and two constraints

$$-(r_i - \Delta x_i)\delta\phi_i = \delta Y_i, \quad i = 1, 2, \tag{6.78}$$

which corresponds to Equation (6.41), where $\phi_i = 3\pi/2 - (-1)^i\theta - q_i$, $i = 1, 2$. On the other hand, the energy function E defined by Equation (6.76) is positive definite in $\dot{\boldsymbol{X}}$ and $\boldsymbol{X}$ under the constraints of Equation (6.78), which has a unique minimum $E = 0$ at $Y_1 - Y_2 = 0$, $\delta x_1 = 0$ and $\delta x_2 = 0$.

Now, we prove the exponential convergence of the closed-loop dynamics of Equation (6.68) to the state of force/torque balance, *i.e.*,

$$Y_1 - Y_2 \to 0, \quad \Delta f_i \to 0, \quad \lambda_i \to 0 \qquad (i = 1, 2) \tag{6.79}$$

as $t \to \infty$. Note that $\Delta f_i \to 0$ as $t \to \infty$ implies that $\delta x_i (= \Delta x_i - \Delta x_{di}) \to 0$ as $t \to \infty$ for $i = 1, 2$, where the Δx_{di} $(i = 1, 2)$ can be determined uniquely so that they satisfy Equation (6.55). Similarly to the proof of exponential stability given in Section 2.7, we consider the pseudo-inverse of 4×5 matrix

$$U_Y = [A(\boldsymbol{X}), -B(\boldsymbol{X})]^{\mathrm{T}} \tag{6.80}$$

that is defined as

$$U_Y^+ = U_Y^{\mathrm{T}}(U_Y U_Y^{\mathrm{T}})^{-1}. \tag{6.81}$$

Then, if we define the 5×5 projection matrix

$$P_Y = I_5 - U_Y^+ U_Y \tag{6.82}$$

then it follows that

$$P_Y U_Y^{\mathrm{T}} = P_Y [A(\boldsymbol{X}), -B(\boldsymbol{X})] = 0. \tag{6.83}$$

Similarly, if we introduce

$$\begin{cases} U_{f1} = \left[\dfrac{\partial \Delta x_2}{\partial \boldsymbol{X}}, -B(\boldsymbol{X}), \boldsymbol{e} \right]^{\mathrm{T}}, \\ U_{f2} = \left[\dfrac{\partial \Delta x_1}{\partial \boldsymbol{X}}, -B(\boldsymbol{X}), \boldsymbol{e} \right]^{\mathrm{T}}, \\ P_{f1} = I_5 - U_{f1}^+ U_{f1}, \quad P_{f2} = I_5 - U_{f2}^+ U_{f2}, \end{cases} \tag{6.84}$$

where U_{fi} denotes the pseudo-inverse of U_{fi} for $i = 1, 2$, then it follows that

$$\begin{cases} P_{f1} \dfrac{\partial \Delta x_2}{\partial \boldsymbol{X}} = 0, \quad P_{f1} B(\boldsymbol{X}) = 0, \quad P_{f1} \boldsymbol{e} = 0, \\ P_{f2} \dfrac{\partial \Delta x_1}{\partial \boldsymbol{X}} = 0, \quad P_{f2} B(\boldsymbol{X}) = 0, \quad P_{f2} \boldsymbol{e} = 0. \end{cases} \tag{6.85}$$

We are now in a position to prove the exponential convergence of the trajectory $(\boldsymbol{X}(t), \dot{\boldsymbol{X}}(t))$ of the closed-loop dynamics of Equation (6.68) to the equilibrium state $(\boldsymbol{X}_\infty, \dot{\boldsymbol{X}} = 0)$ as $t \to \infty$, where at $\boldsymbol{X} = \boldsymbol{X}_\infty$ the scalar function E of Equation (6.76) is minimised so that $E(\boldsymbol{X}_\infty) = 0$, that is, $\delta x_1 = 0$, $\delta x_2 = 0$ and $Y_1 - Y_2 = 0$, and at the same time $\Delta \bar{\boldsymbol{f}} = 0$ and $\Delta \boldsymbol{\lambda} = 0$. To do this, it is convenient to modify E in such a way that

$$V_\alpha = E - \alpha \dot{\boldsymbol{X}}^{\mathrm{T}} H P_{Y\infty} \boldsymbol{e}_\infty \frac{Y_1 - Y_2}{\gamma (r_1 + r_2)} - \alpha \sum_{i=1,2} \dot{\boldsymbol{X}}^{\mathrm{T}} H P_{fi\infty} \boldsymbol{e}_{i\infty} \frac{\delta x_i}{\sqrt{\gamma_i}}, \tag{6.86}$$

where $\alpha > 0$ is a parameter and we define

$$\gamma = \boldsymbol{e}_\infty^{\mathrm{T}} P_{Y\infty} \boldsymbol{e}_\infty, \quad \gamma_i = \boldsymbol{e}_{i\infty}^{\mathrm{T}} P_{fi\infty} \boldsymbol{e}_{i\infty}, \quad i = 1, 2 \tag{6.87}$$

and

$$\boldsymbol{e}_i = \partial \delta x_i / \partial \boldsymbol{X}, \quad i = 1, 2 \tag{6.88}$$

and use the symbols $P_{Y\infty}$, $P_{fi\infty}$, $\boldsymbol{e}_\infty$ and $\boldsymbol{e}_{i\infty}$ to denote the convergent values of P_Y, P_{fi}, $\boldsymbol{e}$ and $\boldsymbol{e}_i$ when $t \to \infty$ and $\boldsymbol{X}$ converges to $\boldsymbol{X}_\infty$. We also rewrite Equation (6.68) conveniently in the following form:

$$H\ddot{\boldsymbol{X}} + C\dot{\boldsymbol{X}} + \sum_{i=1,2} \{\xi_i(\Delta x_{id})\boldsymbol{e}_{i\infty}\} \Delta \dot{x}_i - A(\boldsymbol{X}_\infty)\Delta \bar{\boldsymbol{f}}$$
$$+ B(\boldsymbol{X}_\infty)\Delta \boldsymbol{\lambda} - (Y_1 - Y_2) f_d \boldsymbol{e}_\infty + \boldsymbol{h}(\boldsymbol{X} - \boldsymbol{X}_\infty) = 0, \tag{6.89}$$

where

$$\boldsymbol{h}(\boldsymbol{X} - \boldsymbol{X}_\infty) = \sum_{i=1,2} \{\xi_i(\Delta x_i)\boldsymbol{e}_i - \xi_i(\Delta x_{id})\boldsymbol{e}_{i\infty}\} \Delta \dot{x}_i - \{A(\boldsymbol{X}) - A(\boldsymbol{X}_\infty)\} \Delta \bar{\boldsymbol{f}}$$
$$+ \{B(\boldsymbol{X}) - B(\boldsymbol{X}_\infty)\} \Delta \boldsymbol{\lambda} - (Y_1 - Y_2) f_d (\boldsymbol{e} - \boldsymbol{e}_\infty) = 0. \tag{6.90}$$

Similarly to the derivation of Equation (2.130) in Section 2.7, it is possible to see that

$$\dot{\boldsymbol{X}}^{\mathrm{T}} H P_{Y\infty} \boldsymbol{e}_\infty \frac{Y_1 - Y_2}{(r_1 + r_2)\gamma} \leq \frac{2\lambda_M(H)\dot{\boldsymbol{X}}^{\mathrm{T}} H \dot{\boldsymbol{X}}}{f_d (r_1 + r_2)\gamma} + \frac{f_d (Y_1 - Y_2)^2}{8(r_1 + r_2)} \tag{6.91}$$

$$\dot{\boldsymbol{X}}^{\mathrm{T}} H P_{fi\infty} \boldsymbol{e}_{i\infty} \frac{\delta x_i}{\sqrt{\gamma_i}} \leq \frac{\lambda_M(H)\dot{\boldsymbol{X}}^{\mathrm{T}} H \dot{\boldsymbol{X}}}{2k_0} + \frac{k_0}{2}(\delta x_i)^2, \quad i = 1, 2, \tag{6.92}$$

where $\lambda_M(H)$ denotes the maximum eigenvalue of the matrix H. Since at this stage it is difficult to continue the argument in a generic way, we specialise the proof on the basis of configuration-dependent values of the fingers–object system whose kinematic parameters and control gains are provided in Tables 6.3 and 6.4, respectively. In this specialised case, it is easy to check that

$$\frac{2\lambda_M(H)}{f_d(r_1 + r_2)} = O(10^{-3}). \tag{6.93}$$

Further, we assume that $\bar{f}_i(\Delta x)$ for each i (= 1 or 2) is expressed by Equation (6.3) with $k = 3.0 \times 10^5$ [N/m^2]. Then, it is possible to check that

$$\frac{\lambda_M(H)}{2k_0} \leq O(10^{-7}), \quad \frac{k_0}{2}(\delta x_i)^2 \leq \frac{1}{8} P_i(\delta x_i) \tag{6.94}$$

provided that k_0 is set as $k_0 = 15\pi$ and the maximum length of deformation Δx_i eventually lies in an interval 0.5×10^{-3} [m] $\leq \Delta x_{di} \leq 5.0 \times 10^{-3}$ [m]. Then, if we assume that

$$\boldsymbol{e}_\infty^{\mathrm{T}} P_{Y\infty} \boldsymbol{e}_\infty \geq 0.2, \quad \sqrt{\boldsymbol{e}_{i\infty} P_{fi\infty} \boldsymbol{e}_{i\infty}} \geq O(10^{-3}), \quad i = 1, 2 \tag{6.95}$$

then it is possible to see that

$$\dot{\boldsymbol{X}}^{\mathrm{T}} H P_{Y\infty} \frac{Y_1 - Y_2}{(r_1 + r_2)\gamma} \leq \frac{K}{16} + \frac{f_d (Y_1 - Y_2)^2}{8(r_1 + r_2)}, \tag{6.96}$$

$$\begin{aligned} \sum_{i=1,2} \dot{\boldsymbol{X}}^{\mathrm{T}} H \left\{ P_{fi\infty} \boldsymbol{e}_{i\infty} \frac{\delta x_i}{\sqrt{\gamma_i}} \right\} &\leq \sum_{i=1,2} \left\{ \frac{K}{32} + \frac{1}{8} P_i(\delta x_i) \right\} \\ &= \frac{K}{16} + \frac{1}{8} \left\{ P_1(\delta x_1) + P_2(\delta x_2) \right\}. \end{aligned} \tag{6.97}$$

Thus, substituting these two equations into Equation (6.68), we obtain

$$\left(1 - \frac{\alpha}{4}\right) E \leq V_\alpha \leq \left(1 + \frac{\alpha}{4}\right) E. \tag{6.98}$$

We next evaluate the derivative of V_α in time t in such a way that

$$\begin{aligned} \dot{V}_\alpha = \dot{E} - \alpha \ddot{\boldsymbol{X}}^{\mathrm{T}} H &\left\{ P_{Y\infty} \boldsymbol{e}_\infty \frac{Y_1 - Y_2}{(r_1 + r_2)\gamma} + \sum_{i=1,2} P_{fi\infty} \boldsymbol{e}_\infty \frac{\delta x_i}{\sqrt{\gamma_i}} \right\} \\ -\alpha \dot{X}^{\mathrm{T}} H &\left\{ P_{Y\infty} \boldsymbol{e}_\infty \frac{\dot{Y}_1 - \dot{Y}_2}{(r_1 + r_2)\gamma} + \sum_{i=1,2} P_{fi\infty} \boldsymbol{e}_{i\infty} \frac{\delta \dot{x}_i}{\sqrt{\gamma_i}} \right\}. \end{aligned} \tag{6.99}$$

Referring to the relations of Equation (6.85) and substituting Equations (6.89) and (6.77) into Equation (6.99), we obtain

$$\begin{aligned} \dot{V}_\alpha = &- \sum_{i=1,2} (c_i \dot{q}_i^2 + \xi_i(\Delta x_i) \Delta \dot{x}_i^2) - \alpha \frac{f_d (Y_1 - Y_2)^2}{r_1 + r_2} - \alpha \sum_{i=1,2} \Delta \bar{f}_i \delta x_i \\ &+ \alpha (C \dot{\boldsymbol{X}})^{\mathrm{T}} \left\{ P_{Y\infty} \boldsymbol{e}_\infty \frac{Y_1 - Y_2}{(r_1 + r_2)\gamma} + \sum_{i=1,2} P_{fi\infty} \boldsymbol{e}_{i\infty} \frac{\delta x_i}{\sqrt{\gamma_i}} \right\} \\ &+ \sum_{i=1,2} \alpha \xi_i (\Delta x_{di}) \Delta \dot{x}_i \delta x_i \\ &- \alpha \dot{\boldsymbol{X}}^{\mathrm{T}} H \left\{ P_{Y\infty} \boldsymbol{e}_\infty \frac{\dot{Y}_1 - \dot{Y}_2}{(r_1 + r_2)\gamma} + \sum_{i=1,2} P_{fi\infty} \boldsymbol{e}_{i\infty} \frac{\Delta \dot{x}_i}{\sqrt{\gamma_i}} \right\} \\ &+ \alpha \boldsymbol{h}^{\mathrm{T}} (\boldsymbol{X} - \boldsymbol{X}_\infty) \left\{ P_{Y\infty} \boldsymbol{e}_\infty \frac{Y_1 - Y_2}{(r_1 + r_2)\gamma} + \sum_{i=1,2} P_{fi\infty} \boldsymbol{e}_{i\infty} \frac{\delta x_i}{\sqrt{\gamma_i}} \right\}. \end{aligned} \tag{6.100}$$

Similarly to the derivation of Equation (2.130) in Section 2.7, it is possible to see that

$$(C \dot{\boldsymbol{X}})^{\mathrm{T}} P_{Y\infty} \boldsymbol{e}_\infty \frac{Y_1 - Y_2}{(r_1 + r_2)\gamma} \leq \frac{f_d (Y_1 - Y_2)^2}{2(r_1 + r_2)} + \frac{c_1 \dot{q}_1^2 + c_2 \dot{q}_2^2}{2 f_d (r_1 + r_2)} \tag{6.101}$$

if c_i $(i = 1, 2)$ are chosen so as to satisfy

$$\frac{c_i}{f_d(r_1 + r_2)} = 0.05 \sim 0.2, \quad i = 1, 2. \tag{6.102}$$

In fact, according to Table 6.4, the choices of $c_i = 0.001$ [msN] and $f_d = 0.25$ [N] together with the assumption $\gamma \geq 0.2$ in Equation (6.95) yield

$$\frac{c_i}{f_d(r_1 + r_2)\gamma} \leq 1.0, \tag{6.103}$$

which makes Equation (6.101) reduce to

$$(C\dot{\boldsymbol{X}})^{\mathrm{T}} P_Y \boldsymbol{e}_\infty \frac{Y_1 - Y_2}{(r_1 + r_2)\gamma} \leq \frac{1}{2}(c_1\dot{q}_1^2 + c_2\dot{q}_2^2) + \frac{f_d(Y_1 - Y_2)^2}{2(r_1 + r_2)}. \tag{6.104}$$

Similary, it is possible to see that

$$(C\dot{\boldsymbol{X}})^{\mathrm{T}} P_{fi\infty} \boldsymbol{e}_{fi\infty} \frac{\delta x_i}{\sqrt{\gamma_i}} \leq \frac{c_1^2\dot{q}_1^2 + c_2\dot{q}_2^2}{2\beta} + \frac{\beta(\delta x_i)^2}{2} \tag{6.105}$$

with an appropriate parameter $\beta > 0$. Then, by choosing $\beta = O(10^{-2})$, it is possible to see that

$$(C\dot{\boldsymbol{X}})^{\mathrm{T}} P_{fi\infty} \boldsymbol{e}_{fi\infty} \frac{\delta x_i}{\sqrt{\gamma_i}} \leq \frac{1}{16}(c_1\dot{q}_1^2 + c_2\dot{q}_2^2) + \frac{1}{8}P_i(\delta x_i), \quad i = 1, 2. \tag{6.106}$$

Next, it is necessary to note that differentiation of both equalities of Equation (6.50) with respect to t leads to

$$\begin{cases} \dot{Y}_1 - \dot{Y}_2 = (\dot{\boldsymbol{x}}_{01} - \dot{\boldsymbol{x}}_{02})^{\mathrm{T}} \boldsymbol{r}_Y + (\boldsymbol{x}_{01} - \boldsymbol{x}_{02})^{\mathrm{T}} \dot{\theta} \boldsymbol{r}_Y \\ \qquad\qquad = (\dot{\boldsymbol{x}}_{01} - \dot{\boldsymbol{x}}_{02})^{\mathrm{T}} \boldsymbol{r}_Y - l(\Delta x_1 + \Delta x_2)\dot{\theta}, \\ \Delta \dot{x}_i = (-1)^i \left\{ (\dot{\boldsymbol{x}} - \dot{\boldsymbol{x}}_{0i})^{\mathrm{T}} \boldsymbol{r}_X + Y_i \dot{\theta} \right\}, \quad i = 1, 2. \end{cases} \tag{6.107}$$

By using these relations, it is possible to confirm that the last term on the right-hand side of Equation (6.99) is bounded from above in such a manner that

$$-\dot{\boldsymbol{X}}^{\mathrm{T}} H \left\{ P_{Y\infty} \boldsymbol{e}_\infty \frac{\dot{Y}_1 - \dot{Y}_2}{(r_1 + r_2)r} + \sum_{i=1,2} P_{fi\infty} \boldsymbol{e}_{i\infty} \frac{\Delta \dot{x}_i}{\sqrt{\gamma_i}} \right\}$$
$$\leq O(K) \leq \frac{1}{8}\left(c_i\dot{q}_1^2 + c_2\dot{q}_2\right), \tag{6.108}$$

where K denotes the total kinetic energy and the last inequality follows from a similar argument to that given in the paragraph including Equations (2.140) and (2.141). Now, note that the non-linear remaining term $\boldsymbol{h}(\boldsymbol{X} - \boldsymbol{X}_\infty)$ defined in Equation (6.90) is already quadratic in components of $\Delta \boldsymbol{X}$ $(= \boldsymbol{X} - \boldsymbol{X}_\infty)$

and $\Delta\dot{x}_i$ $(i = 1, 2)$, $\boldsymbol{X} \to \boldsymbol{X}_\infty$ (and hence $Y_1 - Y_2 \to 0$ and $\delta x_i \to 0$ for $i = 1, 2$) and $\dot{\boldsymbol{X}} \to 0$ as $t \to \infty$, and E is decreasing with increase of t as long as $\dot{\boldsymbol{X}} \neq 0$. This implies that there exists a small positive number δ_0 such that for any $(\boldsymbol{X}(0), \dot{\boldsymbol{X}}(0))$ satisfying $E(\boldsymbol{X}(0), \dot{\boldsymbol{X}}(0)) \leq \delta_0$ it follows that

$$\begin{aligned} &\boldsymbol{h}^{\mathrm{T}}(\boldsymbol{X} - \boldsymbol{X}_\infty)\left\{P_{Y\infty}\boldsymbol{e}_\infty\frac{Y_1 - Y_2}{(r_1 + r_2)\gamma} + \sum_{i=1,2} P_{fi\infty}\boldsymbol{e}_{i\infty}\frac{\delta x_i}{\sqrt{\gamma_i}}\right\} \\ &\leq \sum_{i=1,2}\left\{\xi_i(\Delta x_i)\Delta\dot{x}_i^2 + \frac{1}{8}c_i\dot{q}_i^2\right\} + \frac{f_d(Y_1 - Y_2)^2}{8(r_1 + r_2)} + \sum_{i=1,2}\frac{1}{8}P_i(\delta x_i). \end{aligned} \tag{6.109}$$

Thus, substituting Equations (6.104), (6.106), (6.108) and (6.109) into Equation (6.100), we can show that

$$\begin{aligned} \dot{V}_\alpha \leq &-\left(1 - \frac{7\alpha}{8}\right)(c_1\dot{q}_1^2 + c_2\dot{q}_2^2) \\ &-\frac{3\alpha}{8}\cdot\frac{f_d(Y_1 - Y_2)^2}{(r_1 + r_2)} - \frac{\alpha}{2}\left\{P_1(\delta x_1) + P_2(\delta x_2)\right\}, \end{aligned} \tag{6.110}$$

where we used the inequality $\Delta\bar{f}_i\delta x_i \geq 2P_i(\delta x_i)$ for $i = 1, 2$. Particularly if we choose $\alpha = 1.0$ and remark that $(1/8)(c_1\dot{q}_1^2 + c_2\dot{q}_2^2)$ is larger than or equal to the total kinetic energy K under the equality conditions of Equations (6.39) to (6.41), it is possible to conclude that

$$\dot{V}_1 = -\frac{3}{4}E. \tag{6.111}$$

This inequality, by using Equation (6.98) and setting $\alpha = 1.0$, is reduced to

$$\dot{V}_1 \leq -\frac{3}{5}V_1, \tag{6.112}$$

from which it is concluded that

$$\begin{aligned} E(t) &\leq \frac{4}{3}V_1(t) \leq \frac{4}{3}V_1(0)\mathrm{e}^{-0.6t} \leq \frac{4}{3}\cdot\frac{5}{4}E(0)\mathrm{e}^{-0.6t} \\ &= \frac{5}{3}E(0)\mathrm{e}^{-0.6t} \end{aligned} \tag{6.113}$$

Exponential convergence of the closed-loop dynamics to the equilibrium state of force/torque balance when the control signals of Equation (6.57) are used can also be proved theoretically in a similar manner to the discussion given above.

Before closing this section, we remark that the control signals of Equation (6.34) based on fingers–thumb opposition may work well even in the case of precision prehension of a thin light object if the control gains f_d and c_i $(i = 1, 2)$ are carefully chosen and the fingertip stiffness and damping parameters k and c_0 are smaller. As an illustrative example, we conducted numerical

Table 6.8. Physical parameters for dual single-DOF fingers

$m_{11} = m_{21}$	link mass	0.025[kg]
$I_{11} = I_{21}$	inertia moment	3.333×10^{-6}[kg · m^2]
$l_{11} = l_{21}$	link length	0.040[m]
$r_1 = r_2$	radius	0.010[m]
M	object mass	0.0003[kg]
h	object length	0.050[m]
$w = (l_1 + l_2)$	object width	0.001[m]
d	object height	0.010[m]
I	object inertia moment	7.003×10^{-8}[kg · m^2]
$l_1 = l_2$	object length	0.0005[m]
k	stiffness	1.0×10^4[N/m^2]
c_Δ	viscosity	5.0[Ns/m^2]

Table 6.9. Parameters of the control signals

f_d	internal force	0.1[N]
$c_1 = c_2$	damping coefficient	0.003[msN]
γ_f	CSM gain	1500.0
γ_λ	CSM gain	3000.0

simulation of pinching motion of a pair of single-DOF fingers whose kinematic parameters are given in Table 6.8. In Figure 6.12 we show the transient responses of the physical variables of the fingers–object system under the choice of control parameters given in Table 6.9. In this case, the damping factors c_i for the finger joints are chosen larger than those in Table 6.5. Even though these choices of $c_i = 0.003$ [msN] and $f_d = 0.1$ [N] do not satisfy the conditions of Equation (6.102), all the physical variables converge to their corresponding constant values that reflect the state of force/torque balance. In fact, we see from Figure 6.12 that $Y_1 - Y_2 \to 0$, $\bar{f}_1 - \bar{f}_2 \to 0$, $\lambda_1 \to 0$ and $\lambda_2 \to 0$ as $t \to \infty$. It is quite interesting to notice that the transient responses of the physical variables $Y_1 - Y_2$, θ, λ_1 and λ_2 are oscillatory at initial time $t = 0$, though they eventually converge exponentially to their corresponding constant values. In contrast, the variables Δx_i and $\bar{f}_i$ for $i = 1, 2$ converge quickly and smoothly to their constants as $t \to \infty$, as seen in Figure 6.12. This simulation result suggests that there must arise viscous-like forces induced by rollings of the deformed fingertips against the rigid object and they work as a whole at the contact areas in directions tangential to the object surfaces. If such viscous-like forces in the directions tangential to the object surfaces were introduced in the model of the dynamics of the overall fingers–object system, oscillatory phenonomena of $Y_1 - Y_2$, θ and λ_i $(i = 1, 2)$ might disappear provided that

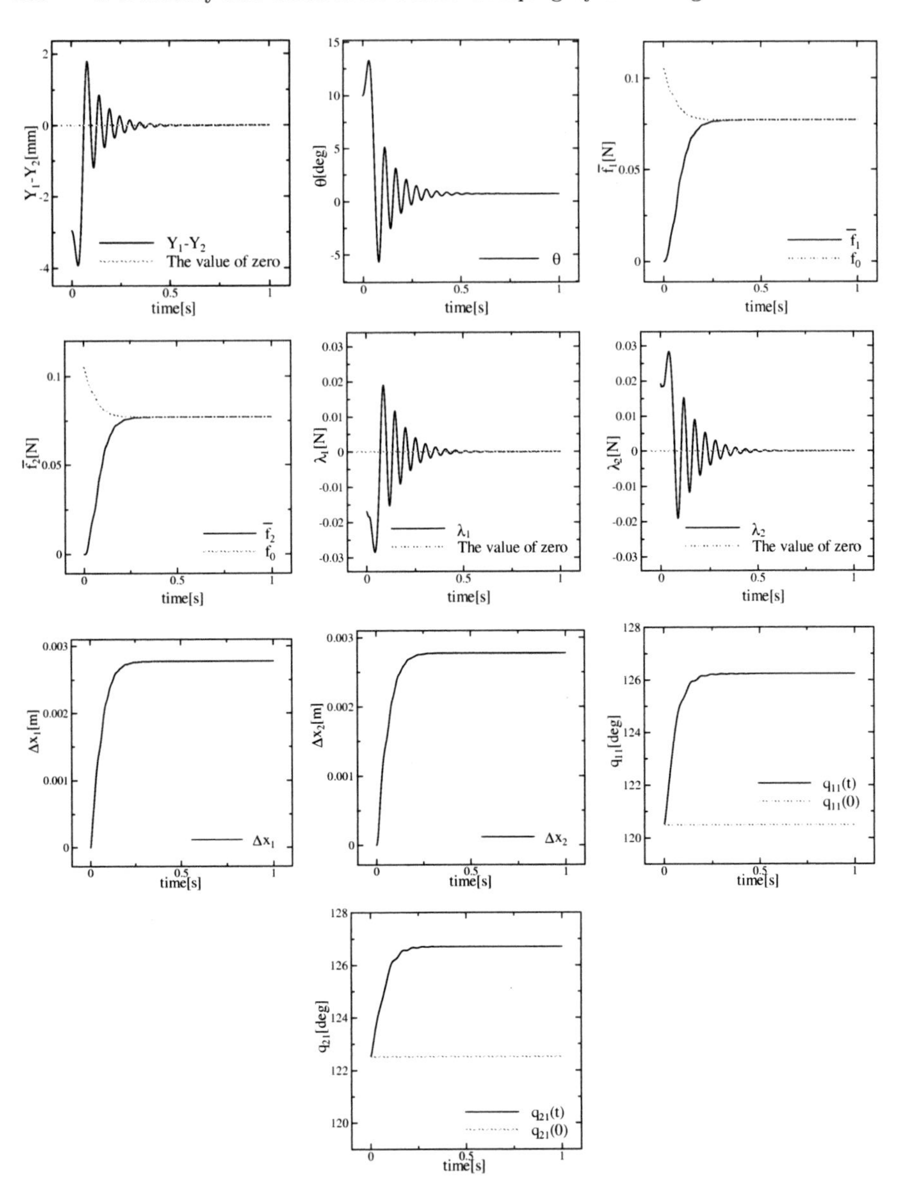

Fig. 6.12. The transient responses of pinching by dual single-DOF soft fingers

the damping parameters c_i $(i = 1, 2)$ for the finger joints could be readjusted. Certainly, a theoretical proof of the exponential convergence of the closed-loop dynamics to the state of force/torque balance would be devised in a similar but extended way to that presented above.

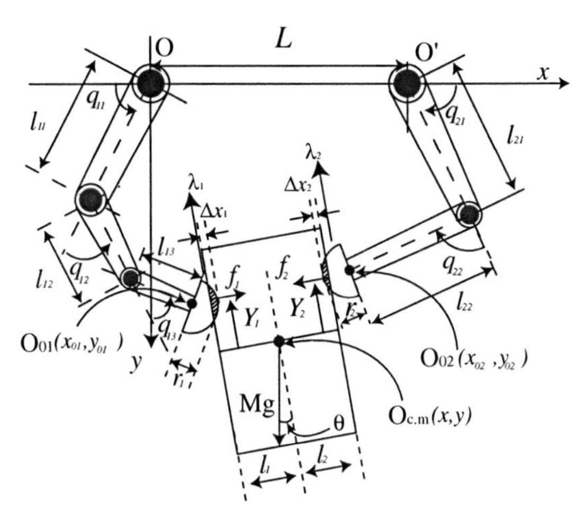

Fig. 6.13. Two robot fingers pinching an object with parallel flat surfaces under gravity.

6.5 Precision Prehension of a 2-D Object under Gravity

Let us consider the stability of grasping a 2-D object by a pair of multi-joint fingers with soft spherical fingertips under gravity. First we derive Lagrange's equation of motion of the overall fingers–object system depicted in Figure 6.13. It is assumed that all rotational motions of finger joints and the object together with translational motion of the object are confined in the vertical plane (xy-plane in Figure 6.13) directly affected by the gravity. We also assume that lumped parameterisation of the contact forces caused by deformation of fingertip material gives rise to the form

$$f_i(\Delta x_i, \Delta \dot{x}_i) = \bar{f}_i(\Delta x_i) + \xi_i(\Delta x_i)\Delta \dot{x}_i, \quad i = 1, 2 \tag{6.114}$$

and $\bar{f}_i(\Delta x_i)$ can be well approximated as in Equation (6.114) and $\xi_i(\Delta x_i)$ are also proportional to Δx_i^2 for $i = 1, 2$, respectively. In this case, the total kinetic energy and potential energy are expressed as follows:

$$K = \frac{1}{2}\left\{ \sum_{i=1,2} \dot{\boldsymbol{q}}_i^{\mathrm{T}} H_i(\boldsymbol{q}_i)\dot{\boldsymbol{q}}_i + M\|\dot{\boldsymbol{x}}\|^2 + I\dot{\theta}^2 \right\}, \tag{6.115}$$

$$P = P_1 + P_2 - Mgy + \sum_{i=1,2} \int_0^{\Delta x_i} \bar{f}_i(\eta)\,\mathrm{d}\eta, \tag{6.116}$$

where $\boldsymbol{q}_1 = (q_{11}, q_{12}, q_{13})^{\mathrm{T}}$, $\boldsymbol{q}_2 = (q_{21}, q_{22})^{\mathrm{T}}$, $\boldsymbol{x} = (x, y)^{\mathrm{T}}$, and H_i and I denote the moments of inertia of the fingers $i = 1, 2$ and the object, M denotes the mass of the object, P_i the potential energy for finger i ($i = 1, 2$) and $-Mgy$

denotes that of the object. Rolling contact constraints can be expressed by focussing on the relative velocities at the centre of each contact area in the direction tangential to the object surface. Therefore, similarly to Equation (6.41), we consider the following non-holonomic constraints:

$$-(r_i - \Delta x_i)\frac{\mathrm{d}}{\mathrm{d}t}\left(\frac{3\pi}{2} - (-1)^i\theta - \boldsymbol{q}_i^{\mathrm{T}}\boldsymbol{e}_i\right) = \frac{\mathrm{d}}{\mathrm{d}t}Y_i, \quad i = 1, 2, \tag{6.117}$$

where $\boldsymbol{e}_1 = (1, 1, 1)^{\mathrm{T}}$ and $\boldsymbol{e}_2 = (1, 1)^{\mathrm{T}}$. Then, in a similar derivation to that of Equations (6.13) and (6.14) and Equations (6.35–6.37), we obtain Lagrange's equations of motion for the system in the following form:

$$\begin{aligned} H_i(\boldsymbol{q}_i)\ddot{\boldsymbol{q}}_i &+ \left(\frac{1}{2}\dot{H}_i(\boldsymbol{q}_i) + S_i(\boldsymbol{q}_i, \dot{\boldsymbol{q}}_i)\right)\dot{\boldsymbol{q}}_i - (-1)^i f_i J_{0i}^{\mathrm{T}}(\boldsymbol{q}_i)\boldsymbol{r}_X \\ &+\lambda_i \left\{(r_i - \Delta x_i)\boldsymbol{e}_i - J_{0i}^{\mathrm{T}}(\boldsymbol{q}_i)\boldsymbol{r}_Y\right\} + \boldsymbol{g}_i(\boldsymbol{q}_i) = \boldsymbol{u}_i, \end{aligned} \tag{6.118}$$

$$M\ddot{\boldsymbol{x}} - R_\theta(f_1 - f_2, -\lambda_1 - \lambda_2)^{\mathrm{T}} - (0, Mg)^{\mathrm{T}} = 0, \tag{6.119}$$

$$I\ddot{\theta} - f_1 Y_1 + f_2 Y_2 + l_1\lambda_1 - l_2\lambda_2 = 0, \tag{6.120}$$

where $J_{0i}^{\mathrm{T}}(\boldsymbol{q}_i) = \partial \boldsymbol{x}_{0i}^{\mathrm{T}}/\partial \boldsymbol{q}_i$ $(i = 1, 2)$. Then, it is obvious that the input–output pair $\boldsymbol{u} = (\boldsymbol{u}_1^{\mathrm{T}}, \boldsymbol{u}_2^{\mathrm{T}})^{\mathrm{T}}$, $\dot{\boldsymbol{q}} = (\dot{\boldsymbol{q}}_1^{\mathrm{T}}, \dot{\boldsymbol{q}}_2^{\mathrm{T}})^{\mathrm{T}}$ concerning the overall dynamics described above satisfies the relation

$$\begin{aligned} &\int_0^t (\dot{\boldsymbol{q}}_1^{\mathrm{T}}\boldsymbol{u}_1 + \dot{\boldsymbol{q}}_2^{\mathrm{T}}\boldsymbol{u}_2)\,\mathrm{d}\tau \\ &\qquad = E(t) - E(0) - \int_0^t \sum_{i=1,2} \xi(\Delta x_i(\tau))\Delta\dot{x}_i^2(\tau)\,\mathrm{d}\tau, \end{aligned} \tag{6.121}$$

where $E(t)$ denotes the value of the total energy $E = K + P$ at time t.

Now we will consider the same control signals as those proposed in Section 4.3 [see Equation (4.49)] for stable grasping of a 2-D object in a blind manner in the case of rigid contact. We repeat it in the following:

$$\begin{aligned} \boldsymbol{u}_i = \boldsymbol{g}_i(\boldsymbol{q}_i) - c_i\dot{\boldsymbol{q}}_i &+ (-1)^i\frac{f_d}{r_1 + r_2}J_{0i}^{\mathrm{T}}(\boldsymbol{q}_i)(\boldsymbol{x}_{01} - \boldsymbol{x}_{02}) \\ &-\frac{\hat{M}g}{2}\left(\frac{\partial y_{0i}}{\partial \boldsymbol{q}_i}\right) - r_i\hat{N}_i\boldsymbol{e}_i, \quad i = 1, 2, \end{aligned} \tag{6.122}$$

where

$$\hat{M}(t) = \hat{M}(0) + \int_0^t \frac{g\gamma_M^{-1}}{2} \sum_{i=1,2} \left(\frac{\partial y_{0i}}{\partial \boldsymbol{q}_i}\right)^{\mathrm{T}} \dot{\boldsymbol{q}}_i(\tau)\,\mathrm{d}\tau$$

$$= \hat{M}(0) + \frac{g\gamma_M^{-1}}{2} \sum_{i=1,2} \{y_{0i}(t) - y_{0i}(0)\}, \tag{6.123}$$

$$\hat{N}_i(t) = \hat{N}_i(0) - \int_0^t \gamma_{Ni}^{-1} r_i \boldsymbol{e}_i^{\mathrm{T}} \dot{\boldsymbol{q}}_i(\tau)\,\mathrm{d}\tau$$

$$= \hat{N}_i(0) - \gamma_{Ni}^{-1} r_i \boldsymbol{e}_i^{\mathrm{T}} \{\boldsymbol{q}_i(t) - \boldsymbol{q}_i(0)\}, \quad i = 1, 2. \tag{6.124}$$

In the design of the control signals, we implicitly assume that all joint angles of fingers can be measured by optical encoders mounted in the joint actuators and thereby the effect of gravity at the finger joints can be compensated. At the same time, all other terms of $\boldsymbol{u}_i$ together with $\hat{M}(t)$ and $\hat{N}_i(t)$ $(i = 1, 2)$ can be easily calculated by using only the finger kinematics and measurements of the finger joint angles. In other words, there is no need to use prior knowledge of the object or sensing data by visual, force or tactile sensing. There is no need to assume that the object side surfaces are parallel. In fact, the same control signals as Equation (6.122) can be applied for other 2-D object whose side surfaces are not parallel, as shown in Figure 6.14. However, the dynamics of Lagrange's equation of motion of the system in the case of Figure 6.14 differ from Equation (6.118–6.120). In fact, similarly to the derivation of Table 5.4, we obtain the following forms of the dynamics for the overall fingers–object system of Figure 6.14:

$$H_i \ddot{\boldsymbol{q}}_i + \left(\frac{1}{2}\dot{H}_i + S_i\right)\dot{\boldsymbol{q}}_i - (-1)^i f_i J_{0i}^{\mathrm{T}} \boldsymbol{r}_{Xi}$$
$$+\lambda_i \left\{(r_i - \Delta x_i)\boldsymbol{e}_i - J_{0i}^{\mathrm{T}} \boldsymbol{r}_{Yi}\right\} + \boldsymbol{g}(\boldsymbol{q}_i) = \boldsymbol{u}_i, \tag{6.125}$$

where

$$\boldsymbol{r}_{Xi} = \begin{pmatrix} \cos(\theta + (-1)^i\theta_0) \\ -\sin(\theta + (-1)^i\theta_0) \end{pmatrix}, \quad \boldsymbol{r}_{Yi} = \begin{pmatrix} \sin(\theta + (-1)^i\theta_0) \\ \cos(\theta + (-1)^i\theta_0) \end{pmatrix} \tag{6.126}$$

and

$$M\ddot{\boldsymbol{x}} + R_{\theta-\theta_0}\begin{pmatrix} f_1 \\ \lambda_1 \end{pmatrix} - R_{\theta+\theta_0}\begin{pmatrix} f_2 \\ -\lambda_2 \end{pmatrix} - Mg\begin{pmatrix} 0 \\ 1 \end{pmatrix} = 0, \tag{6.127}$$

$$I\ddot{\theta} - f_1 Y_1 + f_2 Y_2 + \lambda_1 l_1 - \lambda_2 l_2 = 0. \tag{6.128}$$

Before discussing theoretically the effectiveness of such control signals and the proof of the convergence of the closed-loop dynamics to a state of force/torque balance, we will show a computer simulation result for the overall system depicted in Figure 6.14. The physical paramters of the fingers and the object are given in Table 6.10 and the control gains together with the initial value for $\hat{M}$ are given in Table 6.11. We set $\hat{N}_i(0) = 0$ for $i = 1, 2$.

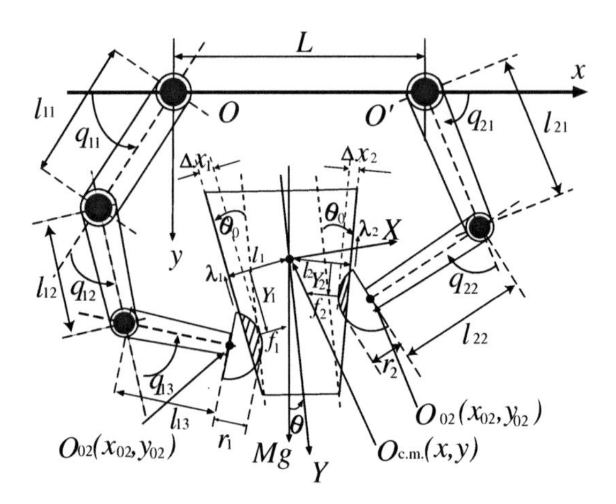

Fig. 6.14. Two robot fingers pinching an object with non-parallel flat surfaces under gravity

We show the transient responses of the key variables in Figure 6.15. Similarly to the case of precision prehension in the case of rigid fingers discussed in Chapter 4, the physical model of the overall fingers–object system in Figure 6.14 is redundant in DOF and therefore all physical variables $\Delta\bar{f}'_i$, $\Delta\lambda'_i$, $\Delta N'_i$ $(i = 1, 2)$, ΔM and S' converge to zero as $t \to \infty$, as seen in Figure 6.15. Here, physical variables mentioned above are defined as follows:

$$\begin{aligned}\Delta\bar{f}'_i &= f_i + (-1)^i \frac{Mg}{2} \sin(\theta + (-1)^i\theta_0) \\ &\quad - \frac{f_d}{r_1 + r_2} \left\{ l'_w \cos\theta_0 + (-1)^i d' \sin\theta_0 \right\}, \quad i = 1, 2, \end{aligned} \tag{6.129}$$

$$\begin{aligned}\Delta\lambda'_i &= \lambda_i - \frac{Mg}{2} \cos(\theta + (-1)^i\theta_0) \\ &\quad - \frac{f_d}{r_1 + r_2} \left\{ l'_w \sin\theta_0 + (-1)^i d' \cos\theta_0 \right\}, \quad i = 1, 2, \end{aligned} \tag{6.130}$$

$$\Delta M = \hat{M} - M, \tag{6.131}$$

$$\Delta N'_i = \hat{N}_i - (1 - \Delta x_i / r_i) N_i, \quad i = 1, 2, \tag{6.132}$$

$$N'_i = -\left\{ \frac{f_d}{r_1 + r_2} (l'_w \sin\theta_0 - (-1)^i d' \cos\theta_0) + \frac{Mg}{2} \cos(\theta + (-1)^i\theta_0 \right\}, \tag{6.133}$$

$$d' = (x_{01} - x_{02}) \sin\theta + (y_{01} - y_{02}) \cos\theta, \tag{6.134}$$

$$l'_w = -(x_{01} - x_{02}) \cos\theta + (y_{01} - y_{02}) \sin\theta, \tag{6.135}$$

$$Y'_1 - Y'_2 = (Y_1 - Y_2) \cos\theta_0 - (l_1 - l_2) \sin\theta_0, \tag{6.136}$$

Table 6.10. Physical parameters for the pair of fingers with three-DOFs and two-DOFs

$l_{11} = l_{21}$	length	0.065 [m]
l_{12}	length	0.039 [m]
l_{13}	length	0.026 [m]
l_{22}	length	0.065 [m]
$m_{11} = m_{21}$	weight	0.045 [kg]
m_{12}	weight	0.025 [kg]
m_{13}	weight	0.015 [kg]
m_{22}	weight	0.040 [kg]
$I_{11} = I_{21}$	inertia moment	1.584×10^{-5}[kgm^2]
I_{12}	inertia moment	3.169×10^{-6}[kgm^2]
I_{13}	inertia moment	8.450×10^{-7}[kgm^2]
I_{22}	inertia moment	1.408×10^{-5}[kgm^2]
r_1	radius	0.010 [m]
r_2	radius	0.020 [m]
L base	length	0.063 [m]
M object	weight	0.040 [kg]
l_1	object width	0.013 [m]
l_2	object width	0.023 [m]
h	object height	0.050 [m]
I	inertia moment	1.248×10^{-5}[kgm^2]
θ_0	object inclination angle	-15.00[deg]
k_i(i=1,2)	stiffness	3.000×10^5[N/m^2]
$c_{\Delta 1}$	viscosity	1000[Ns/m^2]
$c_{\Delta 2}$	viscosity	500.0[Ns/m^2]

Table 6.11. Parameters of the control signals & initial value of the estimator

f_d	internal force	1.0 [N]
c_i i=1,2	damping coefficient	0.006 [Nms/rad]
γ_M	regressor gain	0.01
γ_{Ni} i=1,2	regressor gain	0.001
$M(0)$	initial value	0.010[kg]

$$S' = -f_d \left\{ l'_w \left(\frac{r_1 - r_2 - \Delta x_1 + \Delta x_2}{r_1 + r_2} \right) \sin\theta_0 + d' \left(1 - \frac{\Delta x_1 + \Delta x_2}{r_1 + r_2} \right) \cos\theta_0 \right\} - \frac{Mg}{2} N', \tag{6.137}$$

$$N' = \sum_{i=1,2} \left\{ Y_1 \sin(\theta + (-1)^i \theta_0 + (-1)^i l_i \cos(\theta + (-1)^i \theta_0) \right\}. \tag{6.138}$$

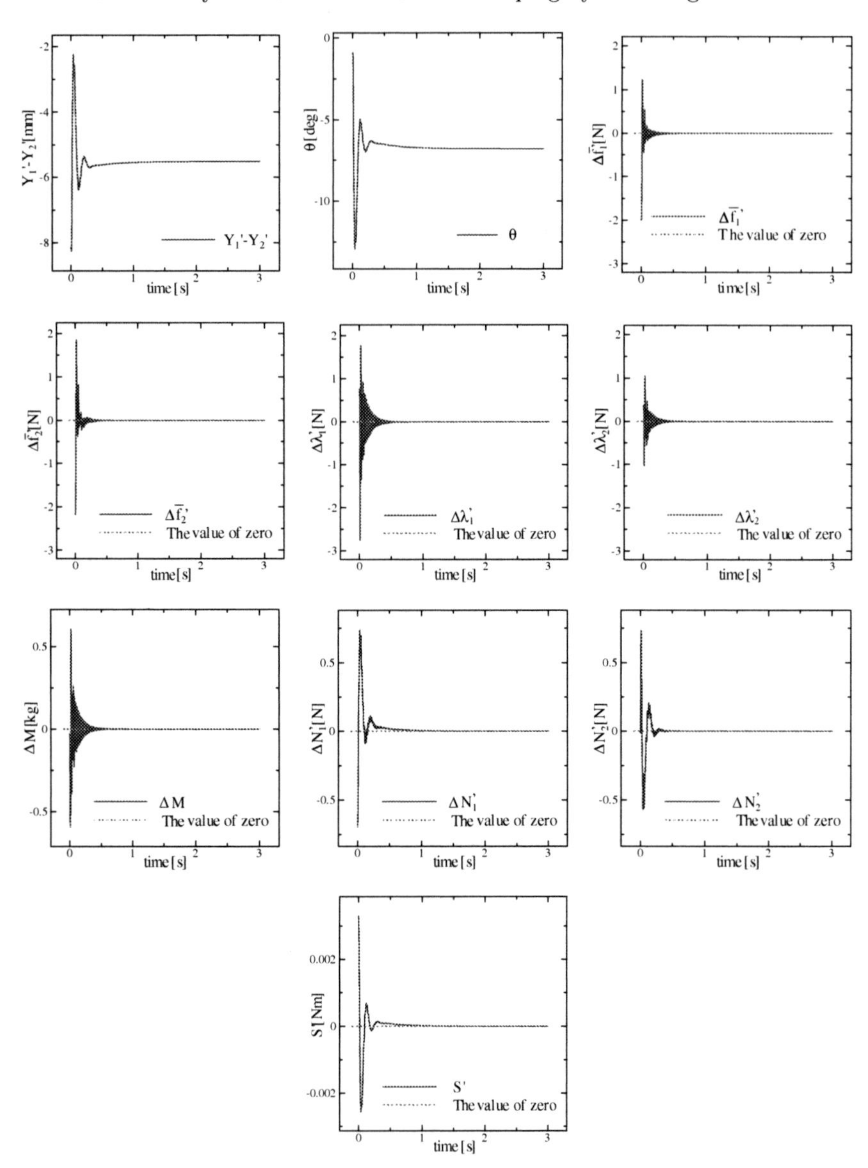

Fig. 6.15. The transient responses of the physical variables

The physical meanings of all these variables have been provided in Section 4.3 and compare well with those given in Table 4.3. It is also interesting to notice from Figure 6.15 that the physical variables with physical units of force and torque, that is, $\Delta f_i'$ $(i = 1, 2)$, $\Delta\lambda_i'$ $(i = 1, 2)$ and ΔM, are quite oscillatory just after manoeuvring the system, though such oscillations decay quickly. The true cause of such oscillatory phenomena of the force and torque variables is the large mismatch between the initial guess $\hat{M}(0)$ of the object mass and its true value M and inadequate gain tuning for γ_M in the object mass estimator. It should be remarked that this simulation starts at $t = 0$ under the condition that $\Delta x_1 = \Delta x_2 = 0$ and hence $f_1(\Delta x_1) = f_2(\Delta x_2) = 0$ and $\dot{\boldsymbol{q}}_i(0) = 0$ $(i = 1, 2)$, $\dot{\boldsymbol{x}}(0) = 0$, and $\dot{\theta}(0) = 0$. Therefore, at the instant just after $t = 0$, Δf_i $(i = 1, 2)$ jumps to a negative value around -2.0 [N] but $\bar{f}_i(t)$ remains positive after this instance as seen in (c) and (d) of Figure 6.15.

Theoretical analysis of the stability of the closed-loop dynamics of the overall fingers–object system depicted in Figure 6.15 can be carried out in parallel with the arguments developed in Sections 4.5 and 4.6 by devising adequate modifications for the treatment of the visco-elastic characteristics of the soft fingertips, as already treated in the simplest case given in the previous section. One of the most important differences of grasping by means of a pair of rigid fingers and with soft fingers is that the attractor region of a certain state of force/torque balance is inclined to shrink when the grasped object becomes thinner and lighter, as remarked on in the previous section. It can be remarked that, in the design of coordinated control signals for stable grasping of such a thin and light object, the term for estimation of the object mass can be omitted and therefore the analysis of the behaviour of the object mass estimator is unnecessary. In the next section, some results of numerical simulation of grasping of thin and light objects by means of a pair of 2-D and 3-D robot fingers with soft fingertips will be exhibited, where the oscillatory phenomena in the force and torque variables will disappear.

6.6 Prehension of a 3-D Object by a Pair of Soft Fingers

As discussed in Chapter 5, even in the case of stable grasping of a 3-D object with parallel flat surfaces by means of a pair of soft fingers, the same structure of coordinated control signals as shown in Equation (6.122) can be used. We consider a setup of two robot fingers with soft hemispherical fingertips as shown in Figure 6.16, which is a soft-fingertip version of the setup shown in Figure 5.2. Then, the coordinated control signals based on fingers–thumb opposition are defined as follows:

$$\begin{aligned} \boldsymbol{u}_i = \boldsymbol{g}_i(q_i) - c_i\dot{q}_i + (-1)^i \frac{f_d}{r_1 + r_2} J_i^{\mathrm{T}}(q_i)(\boldsymbol{x}_{01} - \boldsymbol{x}_{02}) \\ - \frac{\hat{M}g}{2}\frac{\partial y_{0i}}{\partial q_i} - r_i\hat{N}_i\boldsymbol{e}_i - r_i\hat{N}_{0i}\boldsymbol{e}_{0i}, \end{aligned} \tag{6.139}$$

where $\boldsymbol{e}_1 = (1, 1)^{\mathrm{T}}$, $\boldsymbol{e}_2 = (0, 1, 1)^{\mathrm{T}}$, $\boldsymbol{e}_{01} = 0$ and $\boldsymbol{e}_{02} = (1, 0, 0)^{\mathrm{T}}$, and $\hat{M}(t)$, $\hat{N}(t)$ $(i = 1, 2)$, and $\hat{N}_{0i}$ (for $i = 2$ only) are defined in the same forms as in Equations (5.68), (5.69) and (5.70) respectively. Although the theoretical analysis of the closed-loop dynamics has not yet been tackled, computer

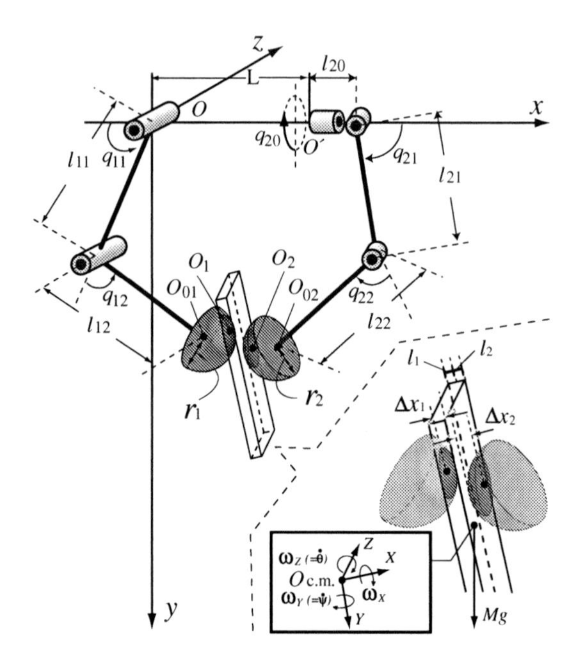

Fig. 6.16. Two robot fingers pinching a 3-D thin object with parallel flat surfaces under gravity

Table 6.12. Physical paramaters (in the case of 3-D grasp)

$l_{11}=l_{21}$	length	0.040 [m]
$l_{12}=l_{22}$	length	0.040 [m]
l_{20}	length	0.000 [m]
$m_{11}=m_{21}$	weight	0.045 [kg]
$m_{12}=m_{22}$	weight	0.035 [kg]
m_{13}	weight	0.020 [kg]
$I_{xx11}=I_{xx21}$	inertia moment	5.625×10^{-7}[kgm^2]
$I_{yy11}=I_{yy21}$	inertia moment	1.613×10^{-5}[kgm^2]
$I_{zz11}=I_{zz21}$	inertia moment	1.613×10^{-5}[kgm^2]
$I_{xx12}=I_{xx21}$	inertia moment	4.375×10^{-7}[kgm^2]
$I_{yy12}=I_{yy21}$	inertia moment	1.254×10^{-5}[kgm^2]
$I_{yy22}=I_{yy22}$	inertia moment	1.254×10^{-5}[kgm^2]
r_0	link radius	0.005 [m]
r_i(i=1,2)	radius	0.01 [m]
L	base length	0.063 [m]
M	object weight	6.667×10^{-3} [kg]
l_i(i=1,2)	object width	2.500×10^{-3} [m]
h	object height	0.050 [m]
k_i (i=1,2)	stiffness	3.000×10^{5}[N/m^2]
$c_{\Delta i}$(i=1,2)	viscosity	1000[Ns/m^2]

Table 6.13. Parameters of the control signals and initial values

f_d	0.100 [N]	c_i (i=1,2)	0.006
c_{q20}	0.006	γ_M	0.001
γ_{Ni} (i=1,2)	0.001	γ_{N0}	0.001
$M(0)$	0.000 [kg]	$\hat{N}_1(0)$	0.000 [N]
$\hat{N}_2(0)$	0.000 [N]	$\hat{N}_0(0)$	0.000 [N]

Table 6.14. Physical paramaters

M	object weight	1.000×10^{-3} [kg]
l_i (i=1,2)	object width	5.000×10^{-4} [m]
h	object height	0.050 [m]
k_i	stiffness parameter	2300.0[N/m^2]
$c_{\Delta i}$ (i=1,2)	viscosity parameter	50.00[Ns/m^2]

Table 6.15. Parameters of the control signals and initial values

f_d	0.050 [N]	c_i (i=1,2)	0.004
c_{q20}	0.004	γ_M	0.01
γ_{Ni} (i=1,2)	0.0004	γ_{N0}	0.0004
$M(0)$	0.000 [kg]	$\hat{N}_1(0)$	0.000 [N]
$\hat{N}_2(0)$	0.000 [N]	$\hat{N}_0(0)$	0.000 [N]

simulations can be carried out in line with treatments of reproducing forces of fingertip deformations instead of contact constraint forces normal to the object surfaces as discussed in the previous section. We show two numerical simulation results based on the physical parameters of the pair of robot fingers and the object given in Table 6.12. We show firstly a numerical simulation result for pinching motion of the system of a pair of fingers and an object whose physical parameters are given in Table 6.12. In this first simulation, the object thickness and weight were set as follows:

$$l_1 + l_2 = 5.0 \text{ [mm]}, \quad M = 6.667 \text{ [g]}. \tag{6.140}$$

The parameters of the control signals of Equation (6.139) and the initial data for the estimators $\hat{M}$, $\hat{N}_i$ $(i = 1, 2)$ and $\hat{N}_{02}$ $(= \hat{N}_0)$ are given in Table 6.13. We show the transient responses of the concerned physical variables in Figure 6.17. In comparison with the responses in the case of a pair of robot fingers with rigid fingertips shown in Figure 5.8, the transient behaviour of the object mass estimator $\hat{M}(t)$ is rather smooth, but that of $Y_1 - Y_2$ together with the constraint forces λ_{Yi} and λ_{Zi} $(i = 1, 2)$ is oscillatory before converging to a constant value.

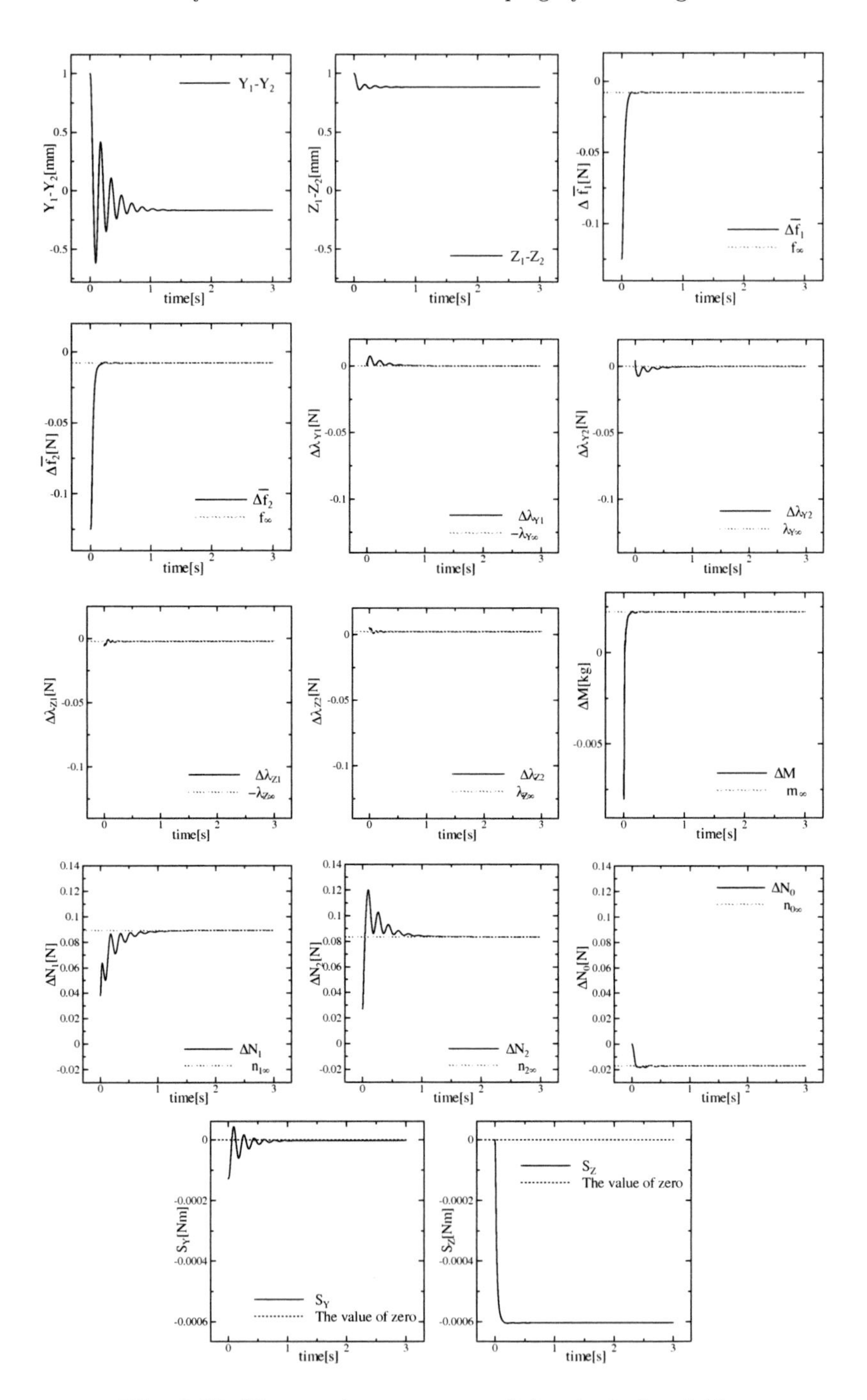

Fig. 6.17. The transient responses of the physical variables

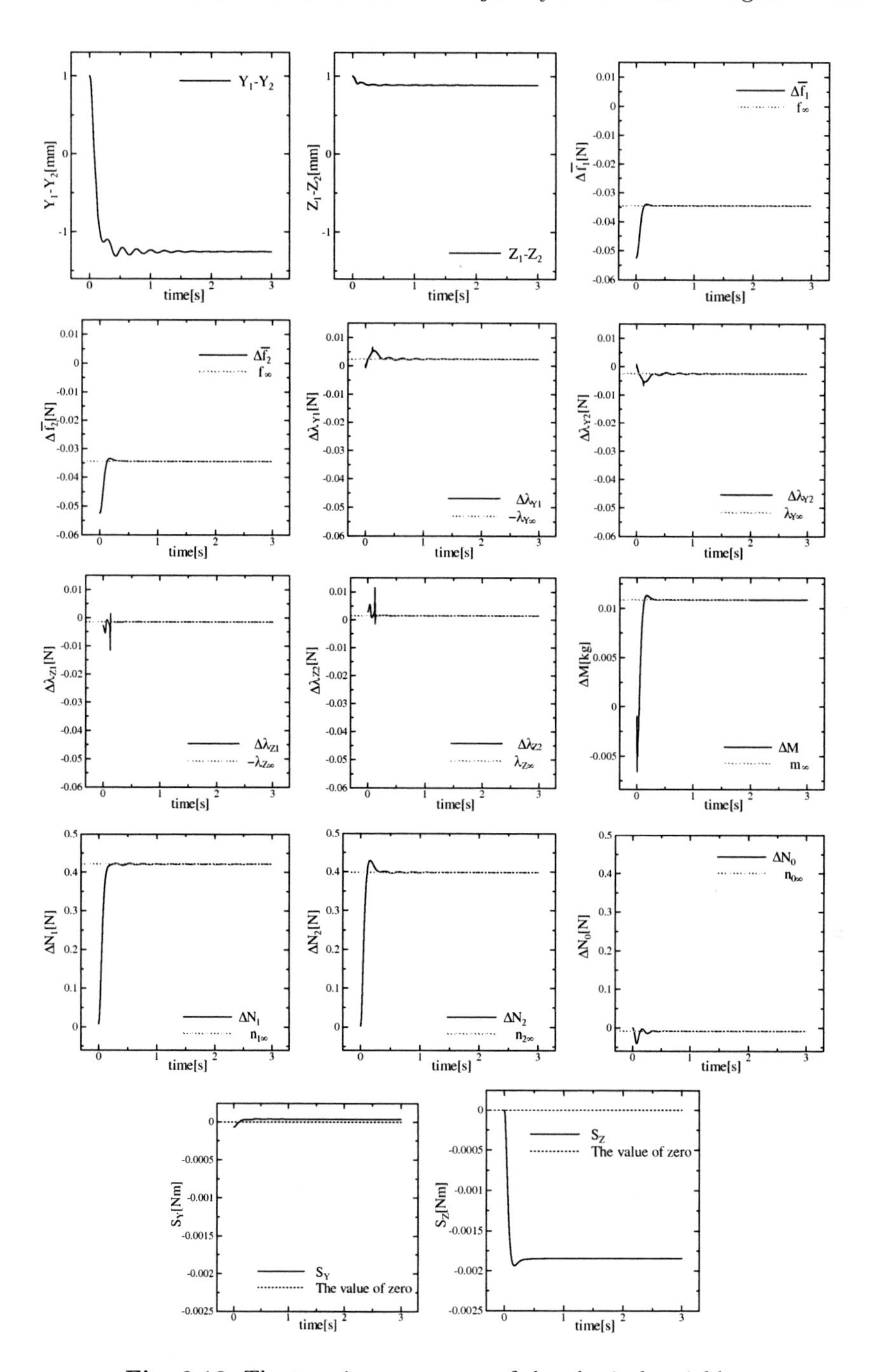

Fig. 6.18. The transient responses of the physical variables

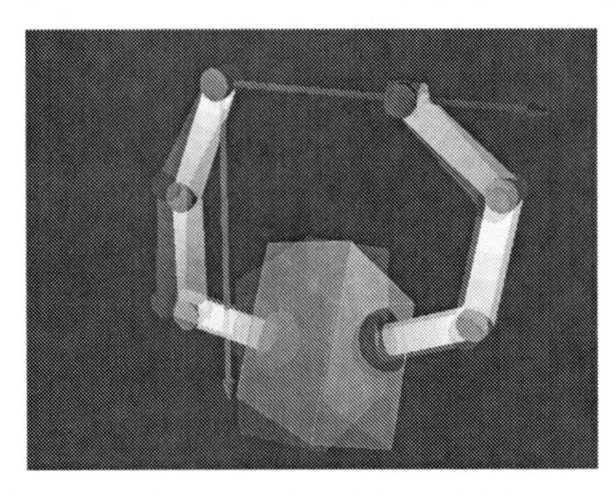

Fig. 6.19. Motions of pinching an object with parallel flat surfaces in 3-D space

We show another simulation result on pinching motion in Figure 6.18 when the object is thinner and lighter as given in Table 6.14, *i.e.*,

$$l_1 + l_2 = 1.0 \text{ [mm]}, \quad M = 1.0 \text{ [g]}. \tag{6.141}$$

In this case, the control parameters and initial conditions for the estimators $\hat{M}$, $\hat{N}_i$ $(i = 1, 2)$ and $\hat{N}_{02}$ are set as in Table 6.15. The transient behaviours of the physical variables in this case are shown in Figure 6.18, in which each transient response behaves like the corresponding result in Figure 6.17. It should be remarked or rather emphasised that in the latter simulation the stiffness parameter k_i $(i = 1, 2)$ for the elasticity of the fingertip material had to be chosen sufficiently small in order to obtain smooth convergence of physical variables as shown in Figure 6.18. In fact, note that $k_i = 3.0 \times 10^5$ [N/m^2] and $c_{\Delta \mathrm{i}} = 1000$ [Ns/m^2] in the former case but $k_i = 2.3 \times 10^3$ [N/m^2] and $c_{\Delta \mathrm{i}} = 50$ [Ns/m^2] in the latter case. In the latter case, the fourth term of the estimator $\hat{M}$ in the control signals of Equation (6.139) might be discarded for practical situations of controlling the motion of the pinching task.

6.7 Dynamics of a Full–Variables Model for 3-D Grasping by Soft Fingers

In this final section, we will show the full dynamics of the physical interaction between a pair of soft fingers and a rigid object with parallel surfaces as shown in Figure 6.19 without assuming non-occurence of spinning motion of the object around the opposing axis. In this case, the finger mechanisms are the same as Figure 5.2, but the fingertips are soft and visco-elastic as in Section 6.1. Then, the Lagrangian of the system can be described as

$$L = K - P, \tag{6.142}$$

where

$$K = \frac{1}{2}\sum_{i=1,2}\dot{q}_i^{\mathrm{T}} H_i(q_i)\dot{q}_i + \frac{1}{2}M\|\dot{\boldsymbol{x}}\|^2 + \frac{1}{2}\boldsymbol{\omega} RHR^{\mathrm{T}}\boldsymbol{\omega}, \tag{6.143}$$

$$P = P_1(q_1) + P_2(q_2) - Mgy + \sum_{i=1,2}\int_0^{\Delta x_i} \bar{f}_i(\eta)\,\mathrm{d}\eta. \tag{6.144}$$

Note that the kinetic energy of Equation (6.143) is the same as that of Equation (5.126) and the potential energy of Equation (6.144) includes the sum of potential energies of fingertip deformation for both fingers. In this case, the geometric relations among the positions of the centres of fingertip spheres are denoted by $\boldsymbol{x}_{0i}$ or O_{0i} $(i = 1, 2)$, the positions of the centres of the area contacts denoted by $\boldsymbol{x}_i$ or O_i $(i = 1, 2)$, the position $\boldsymbol{x}$ of the object mass centre $O_{\mathrm{c.m.}}$ is given as follows instead of by Equations (5.7), (5.8) and (5.9):

$$\begin{cases} \boldsymbol{x}_i = \boldsymbol{x}_{0i} - (-1)^i(r_i - \Delta x_i)\boldsymbol{r}_X \\ \boldsymbol{x} = \boldsymbol{x}_{0i} - (-1)^i(r_i - \Delta x_i + l_i)\boldsymbol{r}_X - Y_i\boldsymbol{r}_Y - Z_i\boldsymbol{r}_Z \end{cases} \quad i = 1, 2, \tag{6.145}$$

At the same time, rolling contact constraints should be expressed, instead of by Equations (5.16) and (5.19), as follows:

$$\begin{cases} (r_1 - \Delta x_1)\{\omega_z - r_{Zz}\dot{p}_1\} = \dot{Y}_1, \\ (r_1 - \Delta x_1)\{-\omega_y + r_{Yz}\dot{p}_1\} = \dot{Z}_1, \end{cases} \tag{6.146}$$

$$\begin{cases} (r_2 - \Delta x_2)\{-\omega_z + (r_{Zz}\cos q_{20} - r_{Zy}\sin q_{20})\dot{p}_2 + r_{Zx}\dot{q}_{20}\} = \dot{Y}_2, \\ (r_2 - \Delta x_2)\{\omega_y - (r_{Yz}\cos q_{20} - r_{Yy}\sin q_{20})\dot{p}_2 - r_{Yx}\dot{q}_{20}\} = \dot{Z}_2. \end{cases} \tag{6.147}$$

Then, the variational principle of the form

$$\begin{aligned} \int_{t_0}^{t_1} \delta L\,\mathrm{d}t = \int_{t_0}^{t_1} &- \sum_{i=1,2}\left\{u_i^{\mathrm{T}}\delta q_i + (\lambda_{Yi}\boldsymbol{Y}_i^{\mathrm{T}} + \lambda_{Zi}\boldsymbol{Z}_i^{\mathrm{T}})\delta\boldsymbol{X}\right\}\mathrm{d}t \\ &+ \int_{t_0}^{t_1}\Bigg\{c_\varphi\omega_x\delta\varphi + c_\psi\omega_y\delta\psi + c_\theta\omega_z\delta\theta \\ &\qquad + \sum_{i=1,2}\xi_i(\Delta x_i)\Delta\dot{x}_i\frac{\partial\Delta x_i}{\partial\boldsymbol{X}^{\mathrm{T}}}\delta\boldsymbol{X}\Bigg\}\,\mathrm{d}t \end{aligned} \tag{6.148}$$

can be applied for the system, where $\boldsymbol{Y}_1 = (\boldsymbol{Y}_{q1}^{\mathrm{T}}, 0_4, \boldsymbol{Y}_{\boldsymbol{x}1}^{\mathrm{T}}, Y_{\varphi 1}, Y_{\psi 1}, Y_{\theta 1})^{\mathrm{T}}$ and $\boldsymbol{Y}_2$, $\boldsymbol{Z}_1$, $\boldsymbol{Z}_2$ express similar meanings, and $\boldsymbol{X} = (q_1^{\mathrm{T}}, q_2^{\mathrm{T}}, \boldsymbol{x}^{\mathrm{T}}, \varphi, \psi, \theta)^{\mathrm{T}}$. This yields the following set of Lagrange equations:

$$\begin{aligned} H_i(q_i)\ddot{q}_i + \left\{\frac{1}{2}\dot{H}_i(q_i) + S_i(q_i, \dot{q}_i)\right\}\dot{q}_i - (-1)^i f_i J_{0i}^{\mathrm{T}}(q_i)\boldsymbol{r}_X \\ -\lambda_{Yi}\boldsymbol{Y}_{qi} - \lambda_{Zi}\boldsymbol{Z}_{qi} + g_i(q_i) = u_i, \quad i = 1, 2, \end{aligned} \tag{6.149}$$

Table 6.16. Parameters of the control signals

f_d	internal force	1.000 [N]
$c_1 = c_2$	damping coefficient	0.001 [Nms]
c_{20}	damping coefficient	0.006 [Nms]
γ_M	regressor gain	0.050 $[\mathrm{m}^2/\mathrm{kgs}^2]$
$\gamma_{Ni}(i = 1, 2)$	regressor gain	5.000×10^{-4} $[\mathrm{s}^2/\mathrm{kg}]$
γ_{N02}	regressor gain	5.000×10^{-4} $[\mathrm{s}^2/\mathrm{kg}]$

$$M\ddot{\boldsymbol{x}} - (f_1 - f_2)\boldsymbol{r}_X + (\lambda_{Y1} + \lambda_{Y2})\boldsymbol{r}_Y + (\lambda_{Z1} + \lambda_{Z2})\boldsymbol{r}_Z - Mg\begin{pmatrix}0\\1\\0\end{pmatrix} = 0, \tag{6.150}$$

$$\bar{H}\dot{\boldsymbol{\omega}} + \left(\frac{1}{2}\dot{\bar{H}} + S\right)\boldsymbol{\omega} + C_\omega\boldsymbol{\omega} - f_1\begin{pmatrix}0\\Z_1\\-Y_1\end{pmatrix} - f_2\begin{pmatrix}0\\-Z_2\\Y_2\end{pmatrix} - \lambda_{Y1}\begin{pmatrix}Z_1\\0\\l_1\end{pmatrix} - \lambda_{Y2}\begin{pmatrix}Z_1\\0\\-l_2\end{pmatrix} - \lambda_{Z1}\begin{pmatrix}-Y_1\\-l_1\\0\end{pmatrix} - \lambda_{Z2}\begin{pmatrix}-Y_2\\l_2\\0\end{pmatrix} = 0, \tag{6.151}$$

where $C_\omega = \mathrm{diag}(c_\varphi, c_\psi, c_\theta)$ and

$$f_i = \bar{f}_i(\Delta x_i) + \xi_i(\Delta x_i)\Delta\dot{x}_i, \quad i = 1, 2. \tag{6.152}$$

Note that Equations (6.149), (6.150) and (6.151) can be respectively compared with Equations (5.128), (5.129) and (5.130) in the case of rigid fingers. It should be remarked that, in Equation (6.149), $\boldsymbol{Y}_{qi}$ and $\boldsymbol{Z}_{qi}$ $(i = 1, 2)$ are defined in Table 5.1 but in this case r_i in Equations (T-5) and (T-6) should be replaced with $r_i - \Delta x_i$ for $i = 1, 2$.

Even though spinning motion around the opposition axis may arise, the same control signal of Equation (6.139) can be applied for stabilisation of the system toward the force/torque balance. We will show first one simulation result based on numerical solutions of the closed-loop dynamics of Equations (6.149–6.151) when the "blind grasp" control signals of Equation (6.139) are substituted into Equation (6.149). The control gains of the signals are given as in Table 6.16 and the physical parameters of the fingers and object are presented in Table 6.17. The viscosities for the rollings are given in the last part of Table 6.17. Figure 6.19 shows the superposition of the initial configuration of the fingers–object system with the final configuration as time tends to infinity. The details of the transient responses of physical variables are given in Figures 6.20 and 6.21. It is interesting to note that the magnitudes of the normal forces $\bar{f}_i(\Delta x_i)$ $(i = 1, 2)$ converge quickly to their corresponding constant values together with maximum deformations Δx_i $(i = 1, 2)$ without

Table 6.17. Physical parameters of the fingers and object

$l_{11} = l_{21}$	length	0.040 [m]
$l_{12} = l_{22}$	length	0.040 [m]
$l_{13} = l_{23}$	length	0.030 [m]
m_{11}	weight	0.043 [kg]
m_{12}	weight	0.031 [kg]
m_{13}	weight	0.020 [kg]
l_{20}	length	0.000 [m]
m_{20}	weight	0.000 [kg]
m_{21}	weight	0.060 [kg]
m_{22}	weight	0.031 [kg]
m_{23}	weight	0.020 [kg]
I_{XX11}	inertia moment	5.375×10^{-7}[kgm^2]
$I_{YY11} = I_{ZZ11}$	inertia moment	6.002×10^{-6}[kgm^2]
I_{XX12}	inertia moment	3.875×10^{-7}[kgm^2]
$I_{YY12} = I_{ZZ12}$	inertia moment	4.327×10^{-6}[kgm^2]
I_{XX13}	inertia moment	2.500×10^{-7}[kgm^2]
$I_{YY13} = I_{ZZ13}$	inertia moment	1.625×10^{-6}[kgm^2]
I_{XX21}	inertia moment	7.500×10^{-7}[kgm^2]
$I_{YY21} = I_{ZZ21}$	inertia moment	8.375×10^{-6}[kgm^2]
I_{XX22}	inertia moment	3.875×10^{-7}[kgm^2]
$I_{YY22} = I_{ZZ22}$	inertia moment	4.327×10^{-6}[kgm^2]
I_{XX23}	inertia moment	2.500×10^{-7}[kgm^2]
$I_{YY23} = I_{ZZ23}$	inertia moment	1.625×10^{-6}[kgm^2]
$I_{XX} = I_{ZZ}$	inertia moment(object)	1.133×10^{-5}[kgm^2]
I_{YY}	inertia moment(object)	6.000×10^{-6}[kgm^2]
r_0	link radius	0.005 [m]
$r_i (i = 1, 2)$	radius	0.010 [m]
L	base length	0.063 [m]
M	object weight	0.001 [kg]
$l_i (i = 1, 2)$	object width	0.015 [m]
h	object height	0.050 [m]
$k_i (i = 1, 2)$	stiffness	3.000×10^5[N/m^2]
$c_{\Delta i} (i = 1, 2)$	viscosity	1000.0[Ns/m^2]
c_φ	viscosity	0.001 [Nms]
c_ψ	viscosity	5.0×10^{-4} [Nms]
c_θ	viscosity	5.0×10^{-4} [Nms]

oscillatory phenomena. It should be remarked that meanings of Δf_i, $\Delta\lambda_{yi}$, $\Delta\lambda_{zi}$, ΔM, ΔN_i, ΔN_{02} can be found in Equations (5.76) and S_X in Equation (5.143) but N_i $(i = 1, 2)$, N_{02}, S_Y and S_Z should become

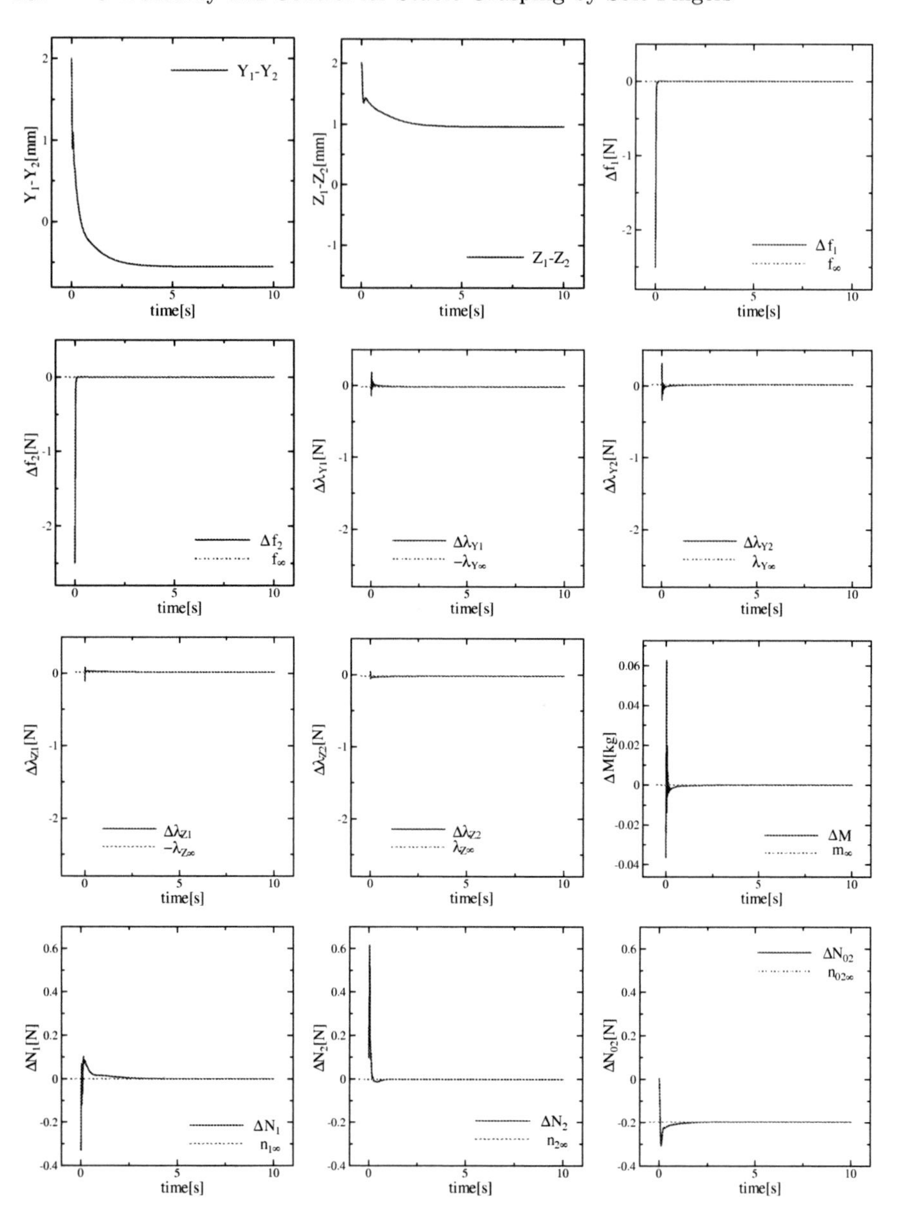

Fig. 6.20. The transient responses of the physical variables

$$
\begin{aligned}
N_i = & \frac{f_d(r_i - \Delta x_i)}{(r_1 + r_2) r_i} \left\{ (Y_1 - Y_2) r_Z(q_{i0}) - (Z_1 - Z_2) r_Y(q_{i0}) \right\} \\
& -(-1)^i \frac{(r_i - \Delta x_i) M g}{2 r_i} \left\{ r_{Yy} r_Z(q_{i0}) - r_{Zy} r_Y(q_{i0}) \right\}, \ i = 1, 2, \quad (6.153)
\end{aligned}
$$

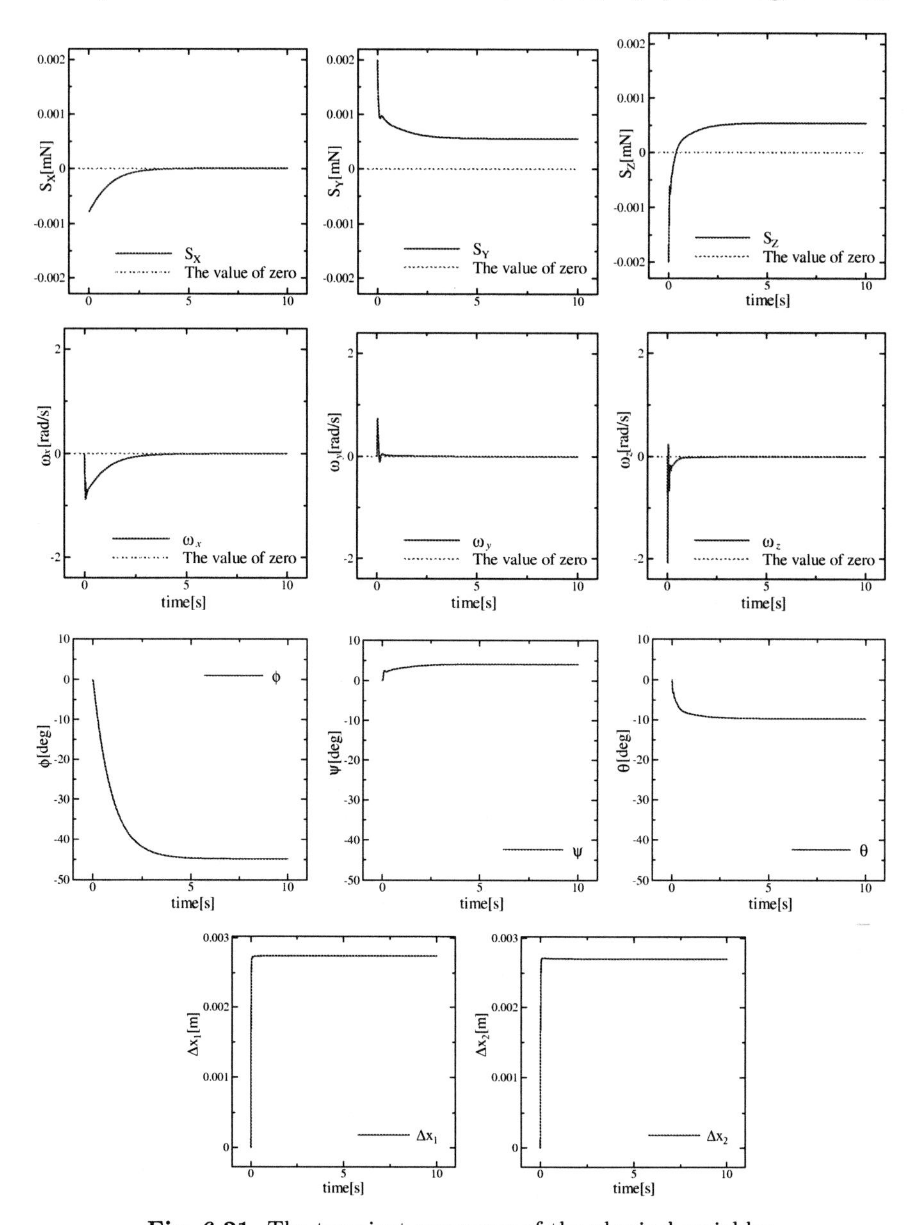

Fig. 6.21. The transient responses of the physical variables

$$
N_{02} = (-1)^i \frac{f_d(r_2 - \Delta x_2)}{(r_1 + r_2)r_2} \{(Y_1 - Y_2)r_{Zx} - (Z_1 - Z_2)r_{Yx}\}
- \frac{(r_2 - \Delta x_2)Mg}{2r_2}(r_{Yy}r_{Zx} - r_{Zy}r_{Yx}), \tag{6.154}
$$

Table 6.18. Parameters of the control signals

f_d	internal force	0.020 [N]
$c_1 = c_2$	damping coefficient	3.000×10^{-4} [Nms]
c_{20}	damping coefficient	0.001 [Nms]
γ_M	regressor gain	0.050 [m^2/kgs^2]
$\gamma_{Ni}(i = 1, 2)$	regressor gain	5.000×10^{-4} [s^2/kg]
γ_{N02}	regressor gain	5.000×10^{-4} [s^2/kg]

$$S_Y = f_d \left(1 - \frac{\Delta x_1 + \Delta x_2}{r_1 + r_2}\right)(Z_1 - Z_2) \\ -\frac{Mg}{2}\left\{r_{Xy}(Z_1 + Z_2) + r_{Zy}(l_1 - l_2)\right\}, \tag{6.155}$$

$$S_Z = -f_d \left(1 - \frac{\Delta x_1 + \Delta x_2}{r_1 + r_2}\right)(Y_1 - Y_2) \\ +\frac{Mg}{2}\left\{r_{Xy}(Y_1 + Y_2) + r_{Yy}(l_1 - l_2)\right\}. \tag{6.156}$$

However, the convergences of the physical variables $Y_1 - Y_2$ and $Z_1 - Z_2$ becomes slow relatively to the rapid growth of the angular velocity ω_x as shown in Figure 6.21 inducing a large spinning motion around the opposition axis as seen in Figure 6.19. We see from Figure 6.21 that the variable $\varphi(t)$ changes from $\varphi(0) = 0$ [deg] to about $\varphi(\infty) = -45$ [deg]. Nevertheless, increasing the viscosity c_φ for rotational motion around the x-axis may degrade the speed of convergence, as predicted through computer simulation trials. In this simulation, the object width l $(= l_1 + l_2)$ is not small in comparison with the fingertip radius r_i $(i = 1, 2)$. Therefore, even if $c_\theta = c_\psi = 0$, it is possible to confirm that solutions to the closed-loop dynamics converge to a state of force/torque balance.

In the case that the object to be grasped is very thin and light, the viscosities c_θ and c_ψ are crucial to some extent for the convergence of the solution trajectories of the closed-loop dynamics. We show another simulation result based on the use of the control gains given in Table 6.18, with an object size $l_1 + l_2 = 1.0 \times 10^{-3}$ [m] and object mass $M = 1.0 \times 10^{-3}$ [kg]. We set in this case $c_\varphi = 1.0 \times 10^{-3}$ [Nms] and $c_\psi = c_\theta = 1.0 \times 10^{-4}$ [Nms]. Figure 6.22 shows a superposition of the initial pose of the fingers–object system with its final pose as t tends to infinity and the force/torque balance is attained. In this simulation we intentionally used the object mass estimator though the true mass is so small that it can be reasonably neglected in a practical situation. Figure 6.23 shows transient behaviours of principal physical variables. As shown in the graph of ΔM, around at the beginning of manoeuvring of the system $\hat{M}(t)$ becomes negative, which may cause oscillatory phenomena about the variables $Y_1 - Y_2$ and constraint forces $\Delta\lambda_{Yi}$ $(i = 1, 2)$. These os-

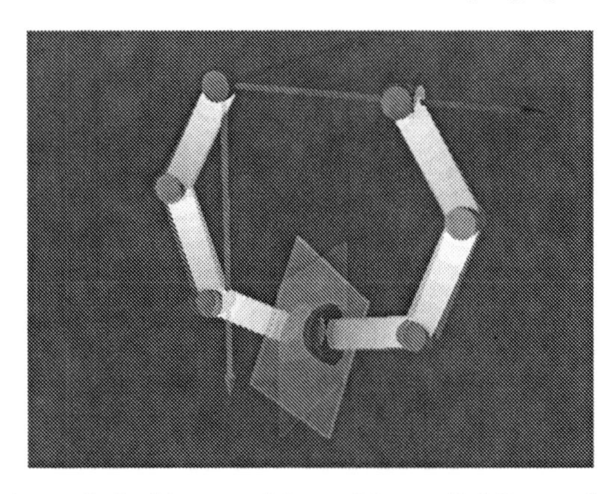

Fig. 6.22. Motions of pinching an object with parallel flat surfaces in 3-D space

cillations can be suppressed by letting γ_i ($i = 1, 2$) and γ_{02} become smaller or neglecting the effect of gravity. Spinning around the opposition axis is noteworthy, as seen in Figure 6.22 and the graphs of ω_x and φ of Figure 6.23, though the coefficient of viscous friction c_φ around the x-axis is taken to be considerably larger than the other viscous coefficients c_ψ and c_θ of the y- and z-axes. In this same simulation, Δx_1 and Δx_2 eventually converge to around 2.5 [mm] despite the light weight of the object and the small pressing force f_d ($= 0.02$ [N]), because the stiffness parameters k_i ($i = 1, 2$) of the soft fingertips are chosen small enough in comparison with the case of Figures 6.19, 6.20 and 6.21 together with Table 6.17. The transient behaviours of most physical variables in this simulation shown in Figure 6.23 resemble those of the corresponding variables in Figure 6.17 except for a noteworthy spinning motion around the x-axis in the former case. From the figure we can observe that $|Y_1 - Y_2|$ converges to around 0.5 [mm] and $|Z_1 - Z_2|$ converges to around 0.7 [mm] as $t \to \infty$. Both the maximum deformations Δx_1 and Δx_2 converge to around 2.5 [mm] as $t \to \infty$. All of these results suggest that for sufficiently long times the two contact areas of the fingertips interlace the thin and light flat object firmly.

Through computer simulations of such a pair of soft fingers, we observe that the choice of stiffness parameter k [Nm^{-2}] for the fingertip material is quite sensitive to the physical order of the object mass, and as well viscosities c_φ, c_ψ and c_θ. Looking at human fingers, these viscosities rely heavily on the physiological and structural characteristics of the finger skin and, in particular, the process of the finger print may increase these viscosities drastically. The speed of convergences toward force/torque balance depends crucially on the magnitudes of the viscosities c_φ, c_ψ and c_θ. Fingerprints play a crucial role in the regulation of the frictional characteristics of rolling contacts between fingertips and a grasped object.

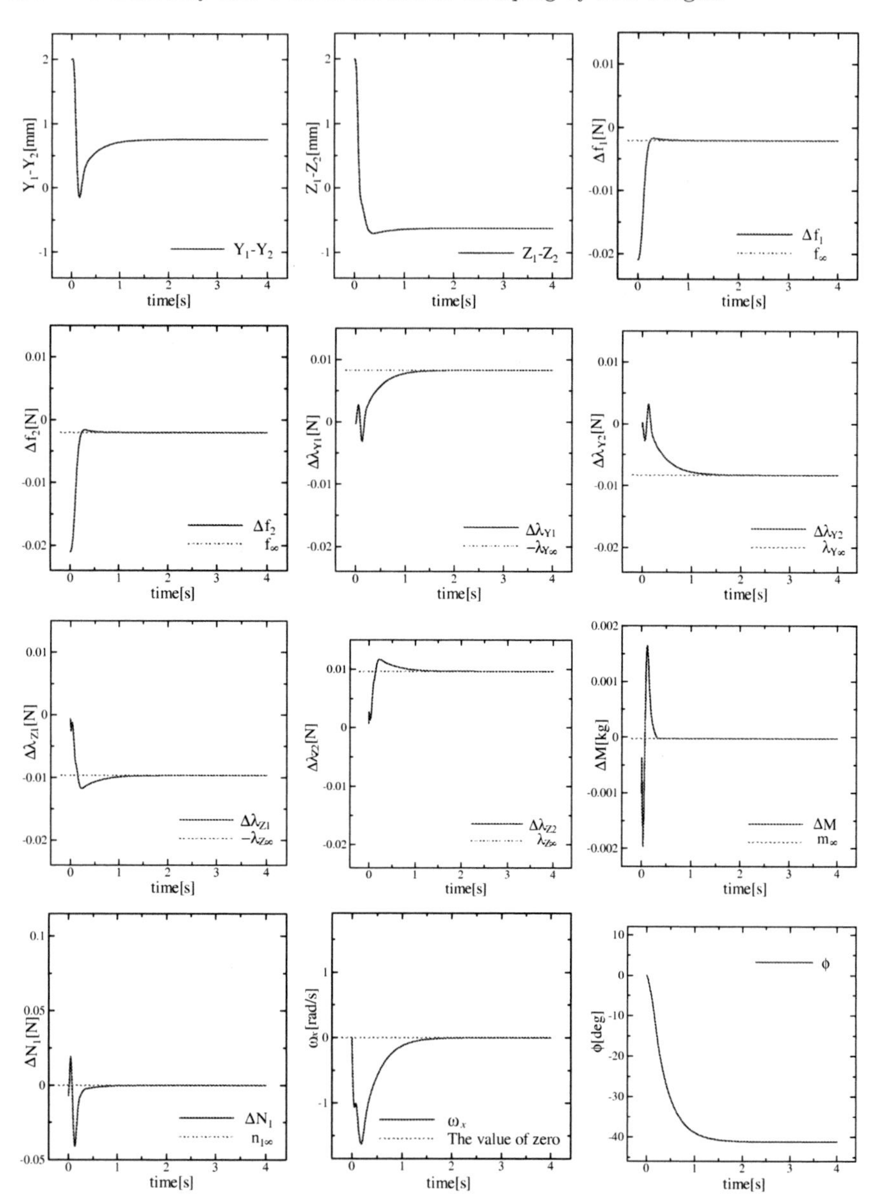

Fig. 6.23. The transient responses of the physical variables

A

Mathematical Supplements

Lemma 1

Two basic lemmata concerning the convergence of functions in t defined over $t \in [0, \infty)$ as $t \to \infty$ are presented, as referred to frequently in the text.

Lemma. If a differentiable scalar function $\omega(t)$ defined over $t \in [0, \infty)$ converges to 0 as $t \to \infty$ and its derivative $\dot{\omega}(t)$ $(= \mathrm{d}\omega(t)/\mathrm{d}t)$ is uniformly continuous in t, then $\dot{\omega}(t) \to 0$ as $t \to \infty$.

Proof. By denying the convergence of $\dot{\omega}(t)$ to zero as $t \to \infty$, it is possible to derive a contradiction. Since $\dot{\omega}(t)$ does not converge to zero as $t \to \infty$, there exist some positive constant $c > 0$ and a time sequence $\{t_k\}$ with $t_k \to \infty$ as $k \to \infty$ such that $|\dot{\omega}(t_k)| \geq c$ for any positive integer k. On the other hand, it follows from the uniform continuity of $\dot{\omega}(t)$ that for any $\varepsilon > 0$ there exists a number $\mu(\varepsilon) > 0$ such that $|t - s| < \mu(\varepsilon)$ implies $|\dot{\omega}(t) - \dot{\omega}(s)| \leq \varepsilon$. Since $|\dot{\omega}(t_k)| \geq c$, this shows that $|\dot{\omega}(t)| \geq c/2$ for any t satisfying $|t - t_k| \leq \mu(c/2)$, where we set $\varepsilon = c/2$. Next, we select a sub-sequence $\{t_{k(1)} = t_1, t_{k(2)}, t_{k(3)}, \cdots\}$ from the original sequence $\{t_k, i = 1, \cdots\}$ in such a way that

$$t_{k(l+1)} > t_{k(l)} + \mu(c/2), \quad l = 1, 2, \cdots, \tag{A.1}$$

where the number $k(l+1)$ is defined as the smallest integer $k(l) + r$ for which $t_{k(l)+r}$ becomes greater than $t_{k(l)}$. Then, if we set $\delta = \mu(c/2)$, it follows that

$$\left|\omega(t_{k(l)} + \delta) + \omega(t_{k(l)} - \delta)\right| = \left| \int_{t_{k(l)-\delta}}^{t_{k(l)+\delta}} \dot{\omega}(t)\mathrm{d}t \right| \geq c\delta. \tag{A.2}$$

Since $t_{k(l)} \to \infty$ as l increases and tends to infinity, Equation (A.2) contradicts the convergence of $\omega(t)$ to zero as $t \to \infty$.

Lemma 2

Lemma. If an n-dimensional vector-valued function $q(t)$ is bounded and uniformly continuous in t, and moreover $q(t) \in L^1(0,\infty)$ or $L^2(0,\infty)$, then $q(t) \to 0$ as $t \to \infty$.

Proof. Similarly to the proof of Lemma 1, we derive a contradiction by denying the convergence of $q(t)$ to 0 as $t \to \infty$. Then, there exist a constant $c > 0$ and an infinite sequence $\{t_k\}$ with $t_k \to \infty$ as $k \to \infty$ such that $\|q(t_k)\| > c$ for any k. On the other hand, from the uniform continuity of $q(t)$ it follows that for any $\varepsilon > 0$ there exists a number $\mu(\varepsilon) > 0$ such that $|t - s| < \mu(\varepsilon)$ implies

$$\|q(t) - q(s)\| \leq \varepsilon. \tag{A.3}$$

Since it is possible to take $\varepsilon = c/2$, it follows from Equation (A.3) and inequality $\|q(t_k)\| \geq c$ that, for any t satisfying $|t - t_k| \leq \mu(c/2)$,

$$\|q(t)\| \geq \frac{c}{2}. \tag{A.4}$$

Since it is possible to assume without loss of generality that $t_{k+1} > t_k + \mu(c/2)$ as remarked in the proof of Lemma A, it follows from Equation (A.4) that

$$\int_0^\infty \|q(t)\| \,\mathrm{d}t \geq \sum_{k=0}^{\infty} \frac{c}{2}\mu(c/2). \tag{A.5}$$

The right-hand side tends to infinity, which contradicts the assumption that $q(t) \in L^1(0,\infty)$. The proof for the case $q(t) \in L^2(0,\infty)$ is similar to the above argument.

B

A Bibliographic Note on the References

Chapter 1

The motif of the introductory chapter is actually indebted to the three books cited as [1-23], [1-1], and [1-15] in the references. The author first recognised the difficulty of "everyday physics" from the article authored by H.L. Dreyfus and S.E. Dreyfus entitled "Making a Mind versus Modeling the Brain: Artificial Intelligence Back at a Branchpoint", which is included in the book [1-23] as the second chapter, pp. 15–43. Later, this let the author claim in his own article [1-20] that robotics should be directed toward making everyday physics intelligible. Napier's book [1-1] attracted his attention to the human ability of precision prehension based on fingers–thumb opposability, which is one of the most crucial distinctions of humankind from primates. From the book [1-15] the author also noticed that modern psychologists have already noted the importance of dynamics of the physical interactions of human body movements with the environment in the development of infants. Regardless of this dynamics point of view, developmental psychology or anthropology has not explored every possibility for expressing the details of such dynamics in a mathematical form. Robotics, however, must design and make multi-fingered hands or multi-joint arms and implement programs in their central processing units to let them fulfill tasks. Around the beginning of the 21st century, this standpoint of robotics had reached a wall of silence in front of Bernstein's DOF problem. The author first came across the name of Bernstein, when he found it in the famous textbook of robotics [1-26] on page 303. At the same time, Chapter 6 of the book [1-26] was instructive for people working in research of multi-fingered hands. Difficulties in controlling such systems with many DOFs have been noted in the field of psychology, as in [1-26], as well as in the field of robotics as claimed in the elaborate survey papers by Shimoga [1-25] and Bicchi [1-24].

The last three sections of Chapter 1 summarise the fundamentals of robot dynamics by referring to the author's previous book [1-27].

Chapter 2

As for the geometry of immobilisation and force/torque closure (or form closure) the general results were given in the paper [2-1] for immobilisability and in [2-2] for form closure grasping. The research history of both problems in relation to multi-fingered hands is detailed in the survey papers [1-24] and [1-25] and the book [1-26] previously quoted. Computational problems for obtaining grasps with force/torque closure have been solved in [2-3] and [2-4]. The existence of force/torque balance in a dynamic sense for 2-D polygonal objects was first discussed in [2-5]. The testbed problem for immobilisation of a 2-D object pivoted at a point was first dealt with in the paper [2-6]. It was found first in the paper [2-7] in the case of 2-D grasping that the rolling contact constraint accompanied a tangential constraint force dynamically in the direction tangential to the object surface. Discussions on the stability of dynamic grasping with force/torque balance in Sections 2.5–2.8 were devised firstly in this book. Physical and mathematical essences of the problem of dynamic grasping is appparently captured by this simple setup of dual robot fingers with a single DOF. Stability of an equilibrium configuration of grasping satisfying force/torque balance can be regarded as an extension of the Lagrange–Dirichlet theorem [2-8], which says that an equilibrium configuration of a mechanical system is stable if it has a minimum potential relative to neighbouring positions. The original idea of the use of an artificial potential generating position feedback signals in stabilisation of point-to-point control of a robot arm was first presented in the paper [2-9].

Chapter 3

This chapter was prepared with the author's intention to help the reader gain a physical insight into rolling constraints that may play an important role in controlling physical interactions of a multi-DOF mechanical system with environments. Traditionally, rolling constraints have been considered to be a source of static friction because of the generic condition expressed as the zero relative velocity between contacting surfaces of two rigid bodies. Historically, control of physical interactions under rolling contact has drawn much attention from roboticists since the late 1990s as seen in the literature [3-1][3-2]. There are a great number of papers concerned with the kinematics of rolling contact as surveyed by Bicchi [1-24], including the papers treating a class of ball-plate control problems [3-3][3-4][3-5][3-6][3-7]. Notwithstanding the abundant literature on kinematic studies of rolling, there is a dearth of papers discussing the dynamic aspects of rolling constraints that may accompany constraint forces between contacting objects. Through the derivation of dynamics of the testbed problem for dynamic immobilisation, the importance of indirect control of shear forces arising tangentially to the object through rolling has been disclosed. Differently from the kinematic analysis of rolling, this approach can express the crucial role of internal forces/torques through the closed-loop dynamics of the system. The reason for the introduction of the Riemannian distance comes from the viewpoint that the configuration space under contact constraints can be treated locally as a Riemannian manifold and therefore the concept of neighbourhoods must be developed on the basis of the Riemannian metrics.

Chapter 4

Even in the case of planar grasp of 2-D objects by a pair of robot fingers, rolling constraints were analysed from the kinematics and motion planning as seen in the literature [4-1][4-2][4-3]. Stability of control of the overall fingers–object system toward an equilibrium configuration by using a coordinated control signal was first tackled in [4-4]. An expository paper about this planar grasp was presented in [4-5]. The effectiveness of a coordinated signal based on the opposable force was first shown in [4-6]. The concept of blind grasping was presented in [4-7] by applying the principle of superposition of control signals [4-8] to cope with a general robustness problem of grasping under the effect of gravity and the unknown geometry of a grasped object. A gain tuning method for choosing control gains is discussed in [4-9] by referring to Hill's model of force/velocity characteristics of muscles. Eponential convergence of the closed-loop dynamics to an equilibrium pose in the case of a rigid object with non-parallel surfaces was completed very recently [4-10].

Chapter 5

It has been pointed out by prominent roboticists [5-1][5-2][5-3][5-4] that one of the key difficulties in research of the grasp by a multifingered hand is the rolling contact between the fingertip and the object. However, analysis of a grasp with rolling contact has been restricted to kinematics and motion planning. In fact, initiated by Brockett's paper [5-5], there is a vast literature on a variety of ball–plate problems as [3-3][3-4][3-5][3-6][3-7] and one more recent publication [5-6]. In particular, Montanna [3-3] derived a set of equations, called the contact equations, and applied them to derive the velocity relationship between the relative motion of two fingers grasping an object. Nevertheless, most of the previous investigations for rolling contact did not step further toward the dynamics viewpoint beyond kinematics and statics, though it is well known that the velocity relationship on rolling contact can be expressed as a set of Pfaffian constraints, some of which accompany actual constraint forces [2-8]. The set of velocity relations between spherical fingertips and an object with parallel flat surfaces derived in this chapter was first reported in [5-7][5-9], which is a corrected version of the previous paper [5-8] with a naive treatment of 3-D grasping by means of a blind grasp. The dynamics of a rigid object with five variables treated in this chapter is related to the so-called Suslov problem that is concerned with the motion of a generalised rigid body with some of its body angular velocity components set equal to zero. The full dynamics of a rigid object with six variables contacted with a pair of spherical fingertips has been derived in a recent paper [5-10].

Chapter 6

Modelling of the stiffness characteristics of the area contact of the spherical visco-elastic fingertip based upon lumped parametrisation was first presented in [2-7]. The relation $f(\Delta x) = k\Delta x^2$, where f denotes the reproducing force of deformation, Δx the maximum deformation, and k the stiffness per square meter is coincident with experimental observations on the dynamic behaviours of a soft material reported by Shimoga [6-1]. In the case of a spherical fingertip made of elastic but hard and metallic material, it is well known [6-2][6-3][6-4] that $f(\Delta x) = k_0\Delta x^{3/2}$. An early experimental result on grasp of a rigid object by using a pair of single-DOF robot fingers was reported in [6-5]. Theoretical treatments of the problem were presented in [2-7] and [6-6]. Mathematically rigorous treatments of 2-D grasping by dual soft fingers were presented in [6-7][6-8]. The analysis and simulation on 2-D or 3-D grasping by means of soft fingers presented here will be published in [6-9] and elsewhere.

References

[1-1] Napier J (1993) Hands (Revised by Tuttle RH). Princeton Univ. Press, Princeton, New Jersey, USA

[1-2] Mackenzie CL, Iberall T (1994) The Grasping Hands. North-Holland, Amsterdam, The Netherlands

[1-3] IEEE Spectrum (2005), October issue

[1-4] Bernstein N (1967) Coordination and Regulation of Movements. Pergamon, New York, USA

[1-5] Latash ML, Turvey MT (eds.) (1996) Dexterity and Its Development. Lawrence Erlbaum, Mahmash, New Jersey, USA

[1-6] Feldman AG (1966) Functional tuning of the nervous system with control of movement or maintenance of steady posture. III. Mechanographic analysis of the execution by man of the simplest motor tasks. Biofizika 11:766–775

[1-7] Feldman AG (1986) Once more on the equilibrium-point hypothesis (λ-model) for motor control. J. Motor Behav. 18:17–54

[1-8] Bizzi E, Polit A, Morasso P (1976) Mechanisms underlying achievement of final head position. J. Neurophysiol. 39:435–444

[1-9] Hogan N (1984) An organizing principle for a class of voluntary movements. J. Neurosci. 4:2745–2754

[1-10] Morasso P (1981) Spatial control of arm movements. Exp. Brain Res. 42:223–227

[1-11] Ito M (1970) Neurophysiological aspects of the cerebellar motor control system. Int. J. Neurol. 7:162–176

[1-12] Ito M (1972) Neural design of the cerebellar motor control system. Brain Res. 40:81–84

[1-13] Bizzi E *et al.* (1992) Does the nervous system use equilibrium-point control to guide single and multiple joint movements? Behav. Brain Sci. 15:603–613

[1-14] Kawato M, Gomi H (1992) A computational model of four regions of the cerebellum based on feedback-error-learning. Biol. Cybernetics 68:95–103

[1-15] Thelen E, Smith LB (1995) A Dynamic Systems Approach to the Development of Cognition and Action. MIT Press, Cambridge, Massachusetts, USA

[1-16] Thelen E *et al.* (1993) The transition to reaching: mapping intention and intrinsic dynamics: Developmental biodynamics: brain, body behavior connections. Child Devel. 64:1058–1098

[1-17] Arimoto S, Sekimoto M, Hashiguchi H, Ozawa R (2005) Natural resolution of ill-posedness of inverse kinematics for redundant robots: A challenge to Bernstein's degrees-of-freedom problem. Adv. Robot. 19:401–434

[1-18] Arimoto S, Hashiguchi H, Sekimoto M, Ozawa R (2005) Generation of natural motions for redundant multi-joint systems: A differential-geometric approach based upon the principle of least actions. J. Robot. Syst. 22:583–605

[1-19] Arimoto S, Sekimoto M (2006) Human-like movements of robotic arms with redundant DOFs: Virtual spring/damper hypothesis to tackle the Bernstein problem. Proc. of the 2006 Int. Conf. on Robotics and Automation, May 15–19, Orlando, Florida, USA, 1860–1866

[1-20] Arimoto S (1999) Robotics research toward explication of everyday physics. Int. J. Robot. Res. 18-11:1056–1063

[1-21] Simon H (1965) The Shape of Automation for Men and Management. Harper and Row, New York, USA

[1-22] Minsky M (1977) Computation: Finite and Infinite Machines. Prentice-Hall, New York, USA

[1-23] Graubard SR (ed.) (1989) The Artificial Intelligence Debate. The MIT Press, Boston, USA

[1-24] Bicchi A (2000) Hands for dexterous manipulation and robust grasping: A difficult road towards simplicity. IEEE Trans. Robot. and Autom. 16–6:652–662

[1-25] Shimoga KB (1996) Robot grasp synthesis algorithms: A survey. Int. J. Robot. Res. 15–2:230–266

[1-26] Murray RM, Li Z, Sastry SS (1994) A Mathematical Introduction to Robotic Manipulation. CRC Press, Boca Raton, Florida, USA

[1-27] Arimoto S (1996) Control Theory of Nonlinear Mechanical Systems: A Passivity-based and Circuit-theoretical Approach. Oxford Univ. Press, Oxford, UK

[2-1] Czyzowics J, Stojmenovic I, Urrutia J (1991) Immobilizing a polytope. Lecture Notes Comput. Sci., Springer 519:214–227

[2-2] Mishra B, Schwartz JT, Sharin M (1987) On the existence and synthesis of multifinger positive grips. Algorithmica 2:541–548

[2-3] Liu YH (1999) Qualitative test and force optimization of 3-D frictional form-closure grasps using linear programming. IEEE Trans. Robot. and Autom. 15–1:163–173

[2-4] Liu YH (2000) Computing n-finger form-closure grasps of polygonal objects. Int. J. Robot. Res. 18–2:149–158

[2-5] Arimoto S, Yoshida M, Bae JH, Tahara K (2003) Dynamic force/torque balance of 2D polygonal objects by a pair of rolling contacts and sensor-motor coordination. J. Robot. Syst. 20–9:517–537

[2-6] Arimoto S, Bae JH, Hashiguchi H, Ozawa R (2004) Natural resolution of ill-posedness of inverse kinematics for redundant robots under constraints. Commun. Inform. and Syst. 4–1:1-28

[2-7] Arimoto S, Nguyen PTA, Han HY, Doulgeri Z (2000) Dynamics and control of a set of dual fingers with soft tips. Robotica 18–1:71-80

[2-8] Meirovitch A (1970) Methods of Analytical Dynamics. McGraw-Hill, New York, USA

[2-9] Takegaki M, Arimoto S (1981) A new feedback method for dynamic control of manipulators. Trans. ASME J. Dyn. Syst. Meas. Control 103–2:119–125

[3-1] Cole ABA, Hauser JE, Sastry SS (1989) Kinematics and control of multi-fingered hands with rolling contact. IEEE Trans. Autom. Control 34–4:398–404

[3-2] Paljug E, Yun X, Kumar V (1994) Control of rolling contacts in multi-arm manipulation. IEEE Trans. Robot. Autom. 10:441–452

[3-3] Montanna DJ (1988) The kinematics of contact and grasp. Int. J. Robot. Res. 7–3:17–32

[3-4] Li Z, Canny J (1990) Motion of two rigid bodies with rolling constraint. IEEE Trans. Robot. Autom. 6–1:62–72

[3-5] Bicchi A, Sorrentino R (1995) Dexterous manipulation through rolling. Proc. of IEEE Int. Conf. on Robotics and Automation, Nagoya, Japan, 452–457

[3-6] Bicchi A, Marigo A, Prattichizzo D (1999) Dexterity through rolling: manipulation of unknown objects. Proc. of the 1999 IEEE Int. Conf. on Robotics and Automation, Detroit, Michigan, USA, 1583–1588

[3-7] Marigo A, Bicchi A (2000) Rolling bodies with regular surface: Controllability theory and applications. IEEE Trans. Autom. Control 45–9: 1586–1599

[4-1] Fearing RS (1986) Simplified grasping and manipulation with dexterous robot hands. IEEE J. Robot. Autom. 2–4:188–195

[4-2] Montanna DJ (1992) Contact stability for two-fingered grasps. IEEE Trans. Robot. Autom. 8–4:421–430

[4-3] Howard WS, Kumar V (1994) Stability of planar grasps. Proc. of IEEE Int. Conf. on Robotics and Automation, San Diego, USA, 2822–2927

[4-4] Arimoto S, Tahara K, Bae JH, Yoshida M (2003) A stability theory on a manifold: Concurrent realization of grasp and orientation control of an object by a pair of robot fingers. Robotica 21–2:163–178

[4-5] Arimoto S (2004) Intelligent control of multi-fingered hands. Annu. Rev. Control 28–1:75–85

[4-6] Ozawa R, Arimoto S, Nakamura S, Bae JH (2005) Control of an object with parallel surfaces by a pair of finger robots without object sensing. IEEE Trans. Robot. 21–5:965–976

[4-7] Arimoto S, Ozawa R, Yoshida M (2005) Two-dimensional stable blind grasping under the gravity effect. Proc. of the 2005 IEEE Int. Conf. on Robotics and Automation, April 18–22, Barcelona, Spain, 1208–1214

[4-8] Arimoto S, Tahara K, Yamaguchi M, Nguyen PTA, Han HY (2001) Principle of superposition for controlling pinch motions by means of robot fingers with soft tips. Robotica 19–1:21–28

[4-9] Bae JH, Arimoto S (2004) Important role of force/velocity characteristics in sensory-motor coordination for control design of object manipulation by a multi-fingered robot hand. Robotica 22:479–491

[4-10] Arimoto S, Yoshida M, Bae JH (2007) Stability of two-dimensional blind grasping under the gravity effect and rolling constraints. To be published in Robotica 25

[5-1] Roth B, Kerr J (1986) Analysis of multifingered hands. Int. J. Robot. Res. 4–4:3–17

[5-2] Maekawa H, Tanie K, Komoriya K (1997) Kinematics, statics and stiffness effect of 3D grasp by multifingered hand with rolling contact at the fingertip. Proc. of the 1997 IEEE Int. Conf. on Robotics and Automation, Albuquerque, New Mexico, 78–85

[5-3] Han L, Guan YS, Li ZX, Shi Q, Trinkle JC (1997) Dexterous manipulation with rolling contacts. Proc. of the 1997 IEEE Int. Conf. on Robotics and Automation, Albuquerque, New Mexico, 992–997

[5-4] Cherif M, Gupta K (1999) Planning quasi-static fingertip manipulations for reconfiguring objects. IEEE Trans. Robot. Autom. 15–5:837–848

[5-5] Brockett R, Dai L (1992) Non-holonomic kinematics and the role of elliptic functions in constructive controllability. In: Li Z & Canny J (eds.) Nonholonomic Motion Planning. Kluwer Academic, Boston, 1–21

[5-6] Oriolo G, Vendittelli M (2005) A framework for the stabilization of general nonholonomic systems with an application to the plate-ball mechanism. IEEE Trans. Robot. 21–2:162–175

[5-7] Arimoto S, Yoshida M, Bae JH (2007) Modeling of 3-D object manipulation by multi-joint robot fingers under non-holonomic constraints and stable blind grasping. JSME (Japan Society of Mechanical Engineers) J. Syst. Des. Dyn., A Special Issue on Asian Conf. on Multi-body Dynamics 2006, 1–3: 434–446

[5-8] Arimoto S, Yoshida M, Bae JH (2006) Stable "Blind grasping" of a 3-D object under nonholonomic constraints. Proc. of the 2006 IEEE Int. Conf. on Robotics and Automation, May 15–19, Orlando, Florida, USA, 2124–2130

[5-9] Arimoto S (2007) A differential-geometric approach for 2-D and 3-D object grasping and manipulation. to be published in Annu. Rev. Control, Autumn issue of 2007

[5-10] Arimoto S, Yoshida M (2008) Modeling and control of three-dimensional grasp by a pair of robot fingers. SICE J. Control Meas. Syst. Integration 1–1:

[6-1] Shimoga KB, Goldenberg AA (1996) Soft robotic fingertips Part II: Modeling and impedance regulation. Int. J. Robot. Res. 15–4:335–350

[6-2] Kao I, Yang F (2004) Stiffness and contact mechanics for soft fingers in grasping and manipulation. IEEE Trans. Robot. Autom. 20–1:132–135

[6-3] Hertz H (1882) On the contact of rigid elastic solids and on hardness. MacMillan, New York, USA

[6-4] Walton K (1976) The oblique compression of two elastic spheres. J. Mech. Phys. Solids 26:139–150

[6-5] Han HY, Arimoto S, Tahara K, Yamaguchi M, Nguyen PTA (2001) Robotic pinching by means of a pair of soft fingers with sensory feedback. Proc. of the 2001 IEEE Int. Conf. on Robotics and Automation, May 21–26, Seoul, Korea, 97–102

[6-6] Doulgeri Z, Fasoulas J, Arimoto S (2002) Feedback control for object manipulation by a pair of soft tip fingers. Robotica 20–1:1-11

[6-7] Yoshida M, Arimoto S, Bae JH (2007) Stability analysis of 2-D object grasping by a pair of robot fingers with soft and hemispherical ends. in Proc. of 3rd IFAC Workshop on Lagrangian and Hamiltonian Methods for Nonlinear Control (LHMNLC'06), Springer

[6-8] Yoshida M, Arimoto S, Bae JH (2007) Blind grasp and manipulation of a rigid object by a pair of robot fingers with soft tips. Proc. of 2007 IEEE Int. Conf. on Robotics and Automation, April 10–14, Roma, Italy, 4707–4714

[6-9] Yoshida M, Arimoto S, Bae JH, Luo ZW (2007) Stable grasp of a 2D rigid object through rolling with soft fingers. to be published in Proc. Int. Symp. Robotics and Biomimetics 2007

Index

Printed in the United States
112970LV00001B/24/A

9 781848 000629